ENGINEERING BULLET

Simplified Design

Reinforced Concrete Buildings of Moderate Size and Height

Third Edition

Iyad M. Alsamsam
Mahmoud E. Kamara

An organization of cement companies to improve and extend the uses of portland cement and concrete through market development, engineering, research, education, and public affairs work.

5420 Old Orchard Road, Skokie, Illinois 60077-1083

www.cement.org

Third edition

Library of Congress Catalog Card Number 93-30929

ISBN 0-89312-129-0

This publication was prepared by the Portland Cement Association for the purpose of suggesting possible ways of reducing design time in applying the provisions contained in the ACI318-02 *Building Code Requirements for Structural Concrete.*

Simplified design procedures stated and illustrated throughout this publication are subject to limitations of applicability. While such limitations of applicability are, to a significant extent, set forth in the text of this publiction, no attempt has been made to state each and every possible limitation of applicability. Therefore, this publication is intended for use by professional personnel who are competent to evaluate the information presented herein and who are willing to accept responsibility for its proper application.

Portland Cement Association ("PCA") is a not-for-profit organization and provides this publication solely for the continuing education of qualified professionals. THIS PUBLICATION SHOULD ONLY BE USED BY QUALIFIED PROFESSIONALS who possess all required license(s), who are competent to evaluate the significance and limitations of the information provided herein, and who accept total responsibility for the application of this information. OTHER READERS SHOULD OBTAIN ASSISTANCE FROM A QUALIFIED PROFESSIONAL BEFORE PROCEEDING.

Foreword

The Building Code Requirements for Structural Concrete (ACI 318) is an authoritative document often adopted and referenced as a design and construction standard in state and municipal building codes around the country as well as in the specifications of several federal agencies, its provisions thus becoming law. Whether ACI 318 is enforced as part of building regulations or is otherwise utilized as a voluntary consensus standard, design professionals use this standard almost exclusively as the basis for the proper design and construction of buildings of reinforced concrete.

The ACI 318 standard applies to all types of building uses; structures of all heights ranging from the very tall high-rise down to single-story buildings; facilities with large areas as well as those of nominal size; buildings having complex shapes and those primarily designed as uncomplicated boxes; and buildings requiring structurally intricate or innovative framing systems in contrast to those of more conventional or traditional systems of construction. The general provisions developed to encompass all these extremes of building design and construction tend to make the application of ACI 318 complex and time consuming. However, this need not necessarily be the case, particularly in the design of reinforced concrete buildings of moderate size and height, as is demonstrated in the publication.

This book has been written as a timesaving aid for use by experienced professionals who consistently seek ways to simplify design procedures.

This third edition of the book is based on the ACI 318-02. Significant changes and additions have been introduced to the ACI 318 Code since the second edition was issued. Several new and sweeping changes were introduced: new load and resistance factors, and unified design provisions for reinforced and prestressed concrete.

New to this edition is Chapter 11 "Design Considerations for Earthquake Forces". The basic seismic design provisions of the International Building Code (2003 IBC) are introduced in this chapter. The wind load calculations in Chapter 2 were updated to comply with ASCE 7-02. Throughout the eleven chapters, equations, design aids, graphs, and code requirements have been updated to ACI 318-02. New design aids have been included.

In some of the example problems, the results obtained from the simplified design methods are compared to those obtained from PCA computer programs. These comparisons readily show that the simplified methods yield satisfactory results within the stated limitations.

Design professionals reading and working with the material presented in this book are encouraged to send in their comments to PCA, together with any suggestions for further design simplifications. PCA would also be grateful to any reader who would bring any errors or inconsistencies to our attention. Any suggestion for improvement is always genuinely welcome. Any errata to this book or other PCA publications may be found by checking http://www.cement.org/bookstore/errata.asp

Acknowledgments

The authors acknowledge their indebtedness to the authors and editors of the first and second editions of this book. Deep appreciation is due to many colleagues who reviewed and provided invaluable suggestions. In particular the authors would like to express their gratitude to PCA's regional structural engineers Attila B. Beres, Michael C. Mota, and Amy M. Trygestad for their review, valuable input and suggestions. Sincere appreciation and deep gratitude to Basile G. Rabbat for his valuable advices and assistance throughout the course of developing this book. Appreciation is also due to David N. Bilow for his encouragement and support during this project.

We sincerely appreciate the efforts of two consultants, Robert F. Mast for reviewing the technical contents of this book and David A. Fanella who reviewed and provided valuable comments to Chapter 11.

Thanks to Dale H. McFarlane who managed the production of this complex book including its many tables and figures and Diane F. Vanderlinde who was responsible for the word-processing of the entire manuscript. Their work and patience during this effort is greatly appreciated.

Table of Contents

Chapter 9—Design Considerations for Economical Formwork 9-1

Chapter 1

A Simplified Design Approach

1.1 THE BUILDING UNIVERSE

There is a little doubt that the construction of a very tall building, a large domed arena, or any other prominent mega structure attracts the interest of a great number of structural engineers. The construction of such structures usually represents the highest level of sophistication in structural design and often introduces daring new concepts and structural innovations as well as improvements in construction techniques.

Many structural engineers have the desire to become professionally involved in the design of such distinctive buildings during their careers. However, very few projects of this prestigious caliber are built in any given year. Truly, the building universe consists largely of low-rise and small-area buildings. Figure 1-1 shows the percentage of building floor area constructed in 2002 in terms of different building height categories. The figure shows that the vast majority of the physical volume of construction is in the 1- to 3-story height range.

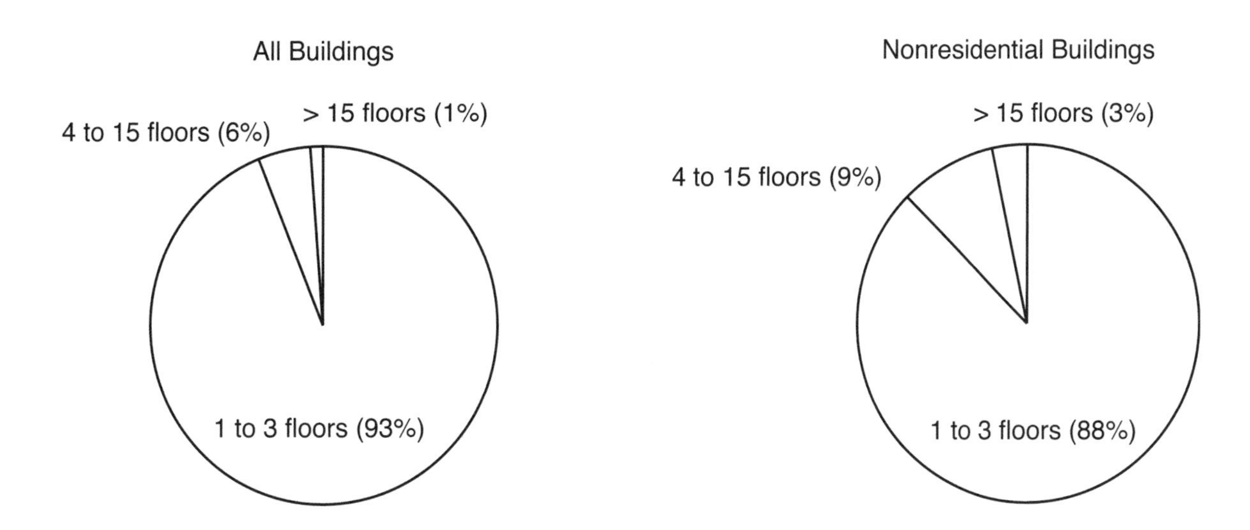

Figure 1-1 Floor Area of Construction, 2002

In the same way, Figure 1-2 shows the percentage of all building projects constructed in various size categories. Building projects less than 15,000 sq ft dominate the building market.

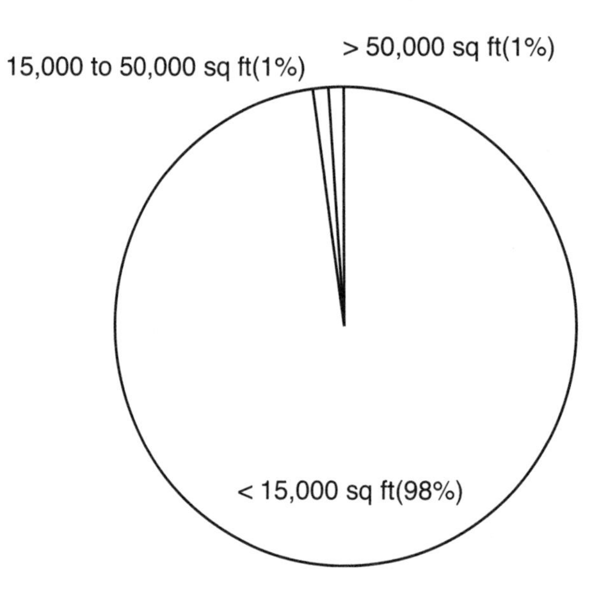

Figure 1-2 All Building Project Size. 2002

When all these statistics are considered, it becomes quickly apparent that while most engineers would like to work on prestigious and challenging high-rise buildings or other distinctive structures, it is far more likely that they will be called upon to design smaller and shorter buildings.

1.2 COST EFFICIENCIES

The benefit of efficient materials use is not sought nor realized in a low-rise building to the same degree as in a high-rise building. For instance, reducing a floor system thickness by an inch may save three feet of building height in a 36-story building and only 3 in. in a three-story building. The added design costs needed to make thorough studies in order to save the inch of floor depth may be justified by construction savings in the case of the 36-story building, but is not likely to be warranted in the design of the shorter building. As a matter of fact, the use of more materials in the case of the low-rise building may sometimes enable the engineer to simplify construction features and thereby effectively reduce the overall cost and time required to complete the building.

In reviewing cost studies of several nonresidential buildings, it was also noted that the cost of a building's frame and envelope represent a smaller percentage of the total building cost in low-rise buildings than in high-rise buildings.

In low-rise concrete construction, designs that seek to simplify concrete formwork will probably result in more economical construction than those that seek to optimize the use of reinforcing steel and concrete, since forming represents a significant part of the total frame costs. There is less opportunity to benefit from form repetition in a low-rise building than in a high-rise building.

Considering the responsibility of the engineer to provide a safe and cost-effective solution to the needs of the building occupant and owner, it becomes clear that, for the vast majority of buildings designed each year, there should be an extra effort made to provide for expediency of construction rather than efficiency of structural design. Often, the extra time needed to prepare the most efficient designs with respect to structural materials is not justified by building cost or performance improvements for low-rise buildings.

1.3 THE COMPLEX CODE

In 1956 the ACI 318 Code was printed in 73 small-size pages; by 2002, ACI 318 and 318R contained more than 440 large-size pages of Code and Commentary—a very substantial increase in the amount of printed material with which an engineer has to become familiar in order to design a concrete building. Furthermore, the code revision in 2002 has seen the most significant technical revisions since 1963. Several new and sweeping changes were introduced in the load and resistance factors, unified design provisions for reinforced and prestressed concrete design were introduced. For the first time, a new appendix on anchorage to concrete is provided along with another appendix on the strut-and-tie modeling and design techniques.

To find the reasons for the proliferation in code design requirements since 1956, it is useful to examine the extensive changes in the makeup of some of the buildings that required and prompted more complex code provisions.

1.3.1 Complex Structures Require Complex Designs

Advances in the technology of structural materials, new building systems, and new engineering procedures have resulted in the use of concrete in a new generation of more flexible structures, dramatically different from those for which the old codes were developed.

In the fifties, 3000 psi concrete was the standard in the construction industry. Today, concrete with 12,000 psi to 16,000 psi and higher strength is used for lower story columns and walls of very tall high-rise buildings. Grade 40 reinforcing steel has almost entirely been replaced by Grade 60 reinforcement. Today the elastic modulus of concrete plays an equally important role to compressive strength in building deflection calculations and serviceability checks.

Gradual switching in the 1963 and 1971 Codes from the Working Stress Design Method to the Strength Design Method permitted more efficient designs of the structural components of buildings. The size of structural sections (columns, beams and slabs) became substantially smaller and utilized less reinforcement, resulting in a 20 to 25% reduction in structural frame costs. In 2002 the working stress design method, long in Appendix A as an alternate design method was deleted.

While we have seen dramatic increases in strength of materials and greater cost efficiencies and design innovations made possible by the use of strength design method, we have, as a consequence, also created new and more complex problems. The downsizing of structural components has reduced overall building stiffness. A further reduction has resulted from the replacement of heavy exterior cladding and interior partitions with lightweight substitutes, which generally do not contribute significantly to building mass or stiffness. In particular, the drastic increase of stresses in the reinforcement at service loads from less than 20 ksi to nearly 40 ksi has caused a significantly wider spread of flexural cracking at service loads in slabs and beams, with consequent increases in their deflections.

When structures were designed by the classical working stress approach, both strength and serviceability of the structure were ensured by limiting the stresses in the concrete and the reinforcement, in addition to imposing limits on slenderness ratios of the members. The introduction of strength design with the resulting increase in member slenderness significantly lengthened the design process; in addition to designing for strength, a separate consideration of serviceability (deflections and cracking) became necessary.

We are now frequently dealing with bolder, larger, taller structures, which are not only more complex, but also more flexible. They are frequently for mixed use and as a result comprise more than one building and floor system. Their structural behavior is characterized by larger deformations relative to member dimensions than we had experienced in the past. As a consequence, a number of effects which heretofore were considered secondary and could be neglected now become primary considerations during the design process. In this category are changes in geometry of structures due to gravity and lateral loadings. The effects of shrinkage, creep, and temperature are also becoming significant and can no longer be neglected in tall or long structures because of their cumulative effects. Seismic codes continue to evolve and consolidate demanding consistent risk assessment and demanding more aggressive design and detailing requirements.

Building and material codes are consensus documents written and edited by committees which can lead to complications. Such committee is often hampered by the legal language these codes need in order to be adopted as a law. This format restricts what can be said and how to say it, which results in a complicated document that is not intended for easy reading and understanding.

1.4 A SIMPLE CODE

The more complex buildings undoubtedly require more complex design procedures to produce safe and economical structures. However, when we look at the reality of the construction industry as discussed at the beginning of this chapter, it makes little sense to impose on structures of moderate size and height intricate design approaches that were developed to assure safety in high complex structures. While the advances of the past decades have made it possible to build economical concrete structures soaring well over 1000 feet in height, the makeup of low-rise buildings has not changed all that significantly over the years.

It is possible to write a simplified code to be applicable to both moderately sized structures and large complex structures. However, this would require a technical conservatism in proportioning of members. While the cost of moderate structures would not be substantially affected by such an approach, the competitiveness of large complex structures could be severely impaired. To avoid such unnecessary penalties, and at the same time to stay within required safety limits, it is possible to extract from the complex code a simplified design approach that can be applied to specifically defined moderately sized structures. Such structures are characterized as having configurations and rigidity that eliminate sensitivity to secondary stresses and as having members proportioned with sufficient conservatism to be able to simplify complex code provisions.

1.5 PURPOSE OF SIMPLIFIED DESIGN

A complex code is unavoidable since it is necessary to address simple and complex structures in the same document. The purpose of this book is to give practicing engineers some way of reducing the design time required for smaller projects, while still complying with the letter and intent of the ACI Standard 318, *Building Code Requirements for Structural Concrete*.[1.1] The simplification of design with its attendant savings in design time result from avoiding building member proportioning details and material property selections which make it necessary to consider certain complicated provisions of the ACI Standard. These situations can often be avoided by making minor changes in the design approach. In the various chapters of this book, specific recommendations are made to accomplish this goal.

The simplified design procedures presented in this manual are an attempt to satisfy the various design considerations that need to be addressed in the structural design and detailing of primary framing members of a reinforced concrete building—by the simplest and quickest procedures possible. The simplified design material is intended for use by experienced engineers well versed in the design principles of reinforced concrete and completely familiar with the design provisions of ACI 318. It aims to arrange the information in the code in an organized step-by-step procedure for the building and member design. The formulae and language avoid complicated legal terminology without changing the intent or the objective of the code. As noted above, this book is written solely as a time saving design aid; that is, to simplify design procedures using the provisions of ACI 318 for reinforced concrete buildings of moderate size and height.

1.6 SCOPE OF SIMPLIFIED DESIGN

The simplified design approach presented in this book should be used within the general guidelines and limitations given in this section. In addition, appropriate guidelines and limitations are given within each chapter for proper application of specific simplifying design procedures.

- Type of Construction: Conventionally reinforced cast-in-place construction. Prestressed and precast construction are not addressed.

- Building Size: Buildings of moderate size and height with usual spans and story heights. Maximum building plan dimension should not exceed 250 ft to reduce effects of shrinkage and temperature to manageable levels.[1.2] Maximum building height should not exceed 7 stories to justify the economics of simplified design.

- Building Geometry: Buildings with standard shapes and geometry suitable for residential and commercial occupancies. Buildings should be free of plan and vertical irregularities as defined in most building codes (see chapter 11). Such irregularities demand more rigorous building analysis and detailed member design.

- Materials: Normal weight concrete $f'_c = 4000$ psi
  Deformed reinforcing bars $f_y = 60{,}000$ psi

Both material strengths are readily available in the market place and will result in members that are durable* and perform structurally well. Traditionally cost analyses have shown that for gravity loads, conventionally reinforced concrete floor systems with $f'_c = 4000$ psi are more economical than ones with higher concrete strengths. [1.3] One set of material parameters greatly simplifies the presentation of design aids. The 4000/60,000 strength combination is used in all simplified design expressions and design aids presented in this book with the following exceptions: the simplified thickness design for footings and the tables for development lengths consider concrete compressive strength of $f'_c = 3000$ psi and $f'_c = 4000$ psi. The use of $f'_c = 4000$ psi corresponds to a standard $\beta_1 = 0.85$. Further simplification is achieved in the minimum reinforcement equations by excluding the term $\sqrt{f'_c}$.

In most cases, the designer can easily modify the simplified design expressions for other material strengths. Also, welded wire fabric and lightweight concrete may be used with the simplified design procedures, with appropriate modification as required by ACI 318.

- Loading: Design dead load, live load, seismic and wind forces are in accordance with *American Society of Civil Engineers Minimum Design Loads for Buildings and Other Structures* (ASCE 7-02)[1.5], with reductions

  in live loads as permitted in ASCE 7-02. The building code having jurisdiction in the locality of construction should be consulted for any possible differences in design loads from those given in ASCE 7-02.

  If resistance to earth or liquid pressure, impact effects, or structural effects of differential settlement, shrinkage, or temperature change need to be included in design, such effects are to be included separately, in addition to the effects of dead load, live load, and lateral forces (see ACI 9.2). Also, effects of forces due to snow loads, rain loads (ponding), and fixed service equipment (concentrated loads) are to be considered separately where applicable (ACI 8.2). Exposed exterior columns or open structures may require consideration of temperature change effects, which are beyond the scope of this manual. Additionally, the durability requirements given in ACI Chapter 4 must be considered in all cases (see Section 1.7 of this book).

- Design Method: All simplified design procedures comply with provisions of Building Code Requirements for Structural Concrete (ACI 318-02), using appropriate load factors and strength reduction factors as specified in ACI 9.2 and 9.3. References to specific ACI Code provisions are noted (e.g., ACI 9.2 refers to ACI 318-02, Section 9.2).

* *This applies to members which are not exposed to 1) freezing and thawing in a moist condition, 2) deicing chemicals and 3) severe levels of sulfates (see ACI Chapter 4).*

1.7 BUILDING EXAMPLES

To illustrate application of the simplified design approach presented in this book, two simple and regular shaped building examples are included. Example No. 1 is a 3-story building with one-way joist slab and column framing. Two alternate joist floor systems are considered: (1) standard pan joist and (2) wide-module pan joist. The beam column frame is used for lateral force resistance. Example No. 2 is a 5-story building with two-way flat plate and column framing. Two alternate lateral-force resisting systems are considered: (1) slab and column framing with spandrel beams and (2) structural shearwalls. In all cases, it is assumed that the members will not be exposed to freezing and thawing, deicing chemicals and severe levels of sulfates since this is the case in most enclosed multistory buildings. Therefore, a concrete compressive strength of f'_c = 4000 psi can be used for all members. ACI Chapter 4 should be consulted if one or more of these aspects must be considered. In some cases, f'_c must be larger than 4000 psi to achieve the required durability.

To illustrate simplified design, typical structural members of the two buildings (beams, slabs, columns, walls, and footings) are designed by the simplified procedures presented in the various chapters of this book. Guidelines for determining preliminary member sizes and required fire resistance are given in Section 1.8.

1.7.1 Building No. 1—3-Story Pan Joist Construction

(1) Floor system: one-way joist slab
Alternate (1)—standard pan joists
Alternate (2)—wide-module joists

(2) Lateral-force resisting system: beam and column framing

(3) Load data: roof LL = 12 psf
DL = 105 psf (assume 95 psf joists and beams + 10 psf roofing and misc.)

floors LL = 60 psf
DL = 130 psf (assume 100 psf joists and beams + 20 psf partitions + 10 psf ceiling and misc.)

(4) Preliminary sizing:

Columns interior = 18 × 18 in.
exterior = 16 × 16 in.

Width of spandrel beams = 20 in.
Width of interior beams = 36 in.

(5) Fire resistance requirements:

floors: Alternate (1)—1 hour
Alternate (2)—2 hours*

roof: 1 hour

columns: 1 hour**

Figure 1-3 shows the plan and elevation of Building #1.

* *In some cases, floors may be serving as an "occupancy separation" and may require a higher rating based on building type of construction. For example, there may be a mercantile or parking garage on the lowest floor.*

** *Columns and walls supporting two hour rated floor, as in Alternate (2), are required to have a two hour rating.*

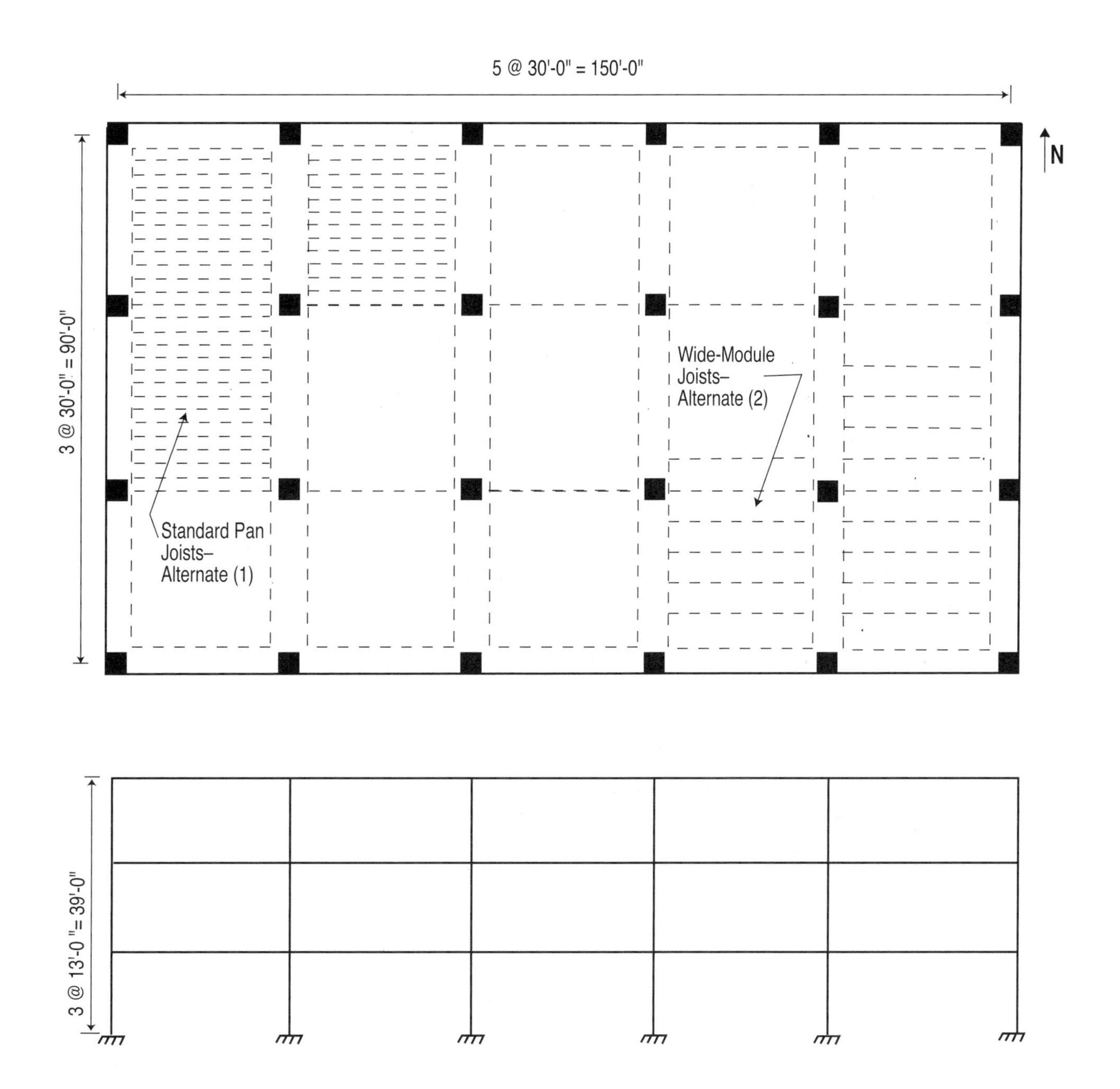

Figure 1-3 Plan and Elevation of Building #1

1.7.2 Building No. 2—5-Story Flat Plate Construction

(1) Floor system: two-way flat plate – with spandrel beams for Alternate (1)

(2) Lateral-force resisting system

Alternate (1)—slab and column framing-with spandrel beam
Alternate (2)—structural shearwalls

(3) Load data: roof LL = 20 psf
DL = 122 psf

floors LL = 50 psf
DL* = 142 psf (9 in. slab)
136 psf (8.5 in. slab)

(4) Preliminary sizing:

Slab (with spandrels) = 8.5 in.
Slab (without spandrels) = 9 in.

Columns interior = 16 × 16 in.
exterior = 12 × 12 in.
Spandrels = 12 × 20 in.

(5) Fire resistance requirements:

floors:	2 hours
roof:	1 hour
columns:	2 hours
shearwalls:**	2 hours

Figure 1-4 shows the plan and elevation of Building #2.

* *Assume 20 psf partitions + 10 psf ceiling and misc.*
** *Assume interior portions of walls enclose exit stairs.*

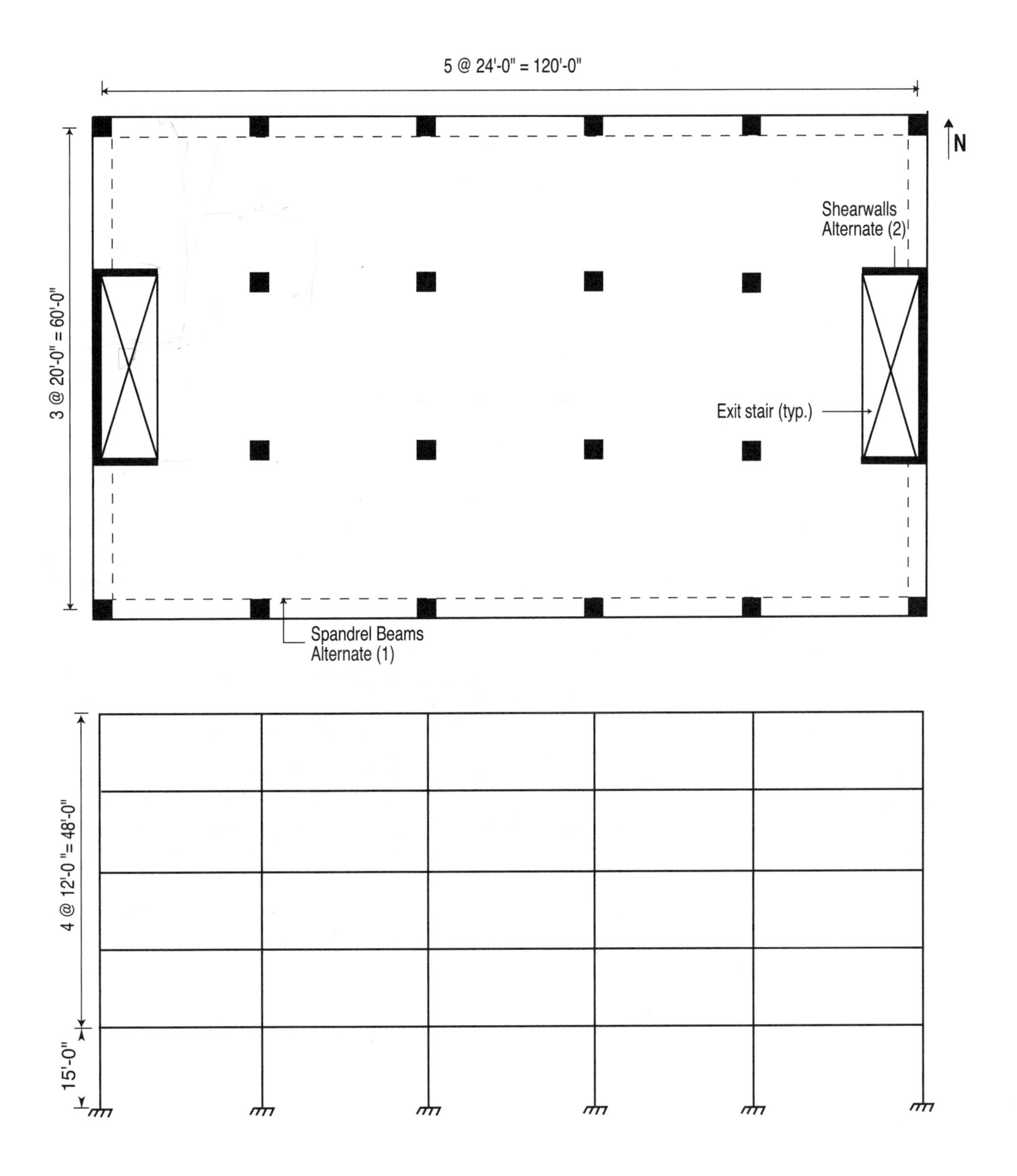

Figure 1-4 Plan and Elevation of Building #2

1.8 PRELIMINARY DESIGN

Preliminary member sizes are usually required to perform the initial frame analysis and/or to obtain initial quantities of concrete and reinforcing steel for cost estimating. Practical initial member sizes are necessary even when a computer analysis is used to determine the load effects on a structure. The guidelines for preliminary design given in the following sections are applicable to regular buildings of moderate size and height. These guidelines were used to obtain the preliminary sizes listed in Sections 1.7.1 and 1.7.2 for the two example buildings. Chapters 8 and 9 list additional guidelines to achieve overall economy.

1.8.1 Floor Systems

Various factors must be considered when choosing a floor system. The magnitude of the superimposed loads and the bay size (largest span length) are usually the most important variables to consider in the selection process. Fire resistance is also very important (see Section 1.8.5). Before specifying the final choice for the floor system, it is important to ensure that it has at least the minimum fire resistance rating prescribed in the governing building code.

In general, different floor systems have different economical span length ranges for a given total factored load. Also, each system has inherent advantages and disadvantages, which must be considered for a particular project. Since the floor system (including its forming) accounts for a major portion of the overall cost of a structure, the type of system to be utilized must be judiciously chosen in every situation.

Figures 1-5 through 1-7 can be used as a guide in selecting a preliminary floor system with $f'_c = 4000$ psi.[1.3] A relative cost index and an economical square bay size range are presented for each of the floor systems listed. The one-way joist system for Building #1 and the flat plate system for Building #2 are selected. In general, an exact cost comparison should be performed to determine the most economical system for a given building.

Once a particular floor system has been chosen, preliminary sizes must be determined for the members in the system. For one-way joists and beams, deflection will usually govern. Therefore, ACI Table 9.5(a) should be used to obtain the preliminary depth of members that are not supporting or attached to partitions and other construction likely to be damaged by deflection. The width of the member can then be determined by the appropriate simplified equation given in Chapter 3. Whenever possible, available standard sizes should be specified; this size should be repeated throughout the entire structure as often as possible. For overall economy in a standard joist system, the joists and the supporting beams must have the same depth. This also provides an optimum ceiling cavity to a uniform bottom of floor elevation with maximum clearance for building mechanical/electrical/plumbing (M/E/P) systems.

For flat plate floor systems, the thickness of the slab will almost always be governed by two-way (punching) shear. Figure 1-8 can be used to obtain a preliminary slab thickness based on two-way shear at an interior square column and $f'_c = 4000$ psi. For a total factored load w_u (psf) and the ratio of the floor tributary area, A, to the column area c_1^2, a value of d/c_1, can be obtained from the figure. Note that d is the distance from the compression face of the slab to the centroid of the reinforcing steel. The preliminary thickness of the slab h can be determined by adding 1.25 in. to the value of d (see Chapter 4).

It is important to note that the magnitude of the unbalanced moment at an interior column is usually small. However, at an edge column, the shear stress produced by the unbalanced moment can be as large as or larger than the shear stress produced by the direct shear forces. Consequently, in most cases, the preliminary slab thick-

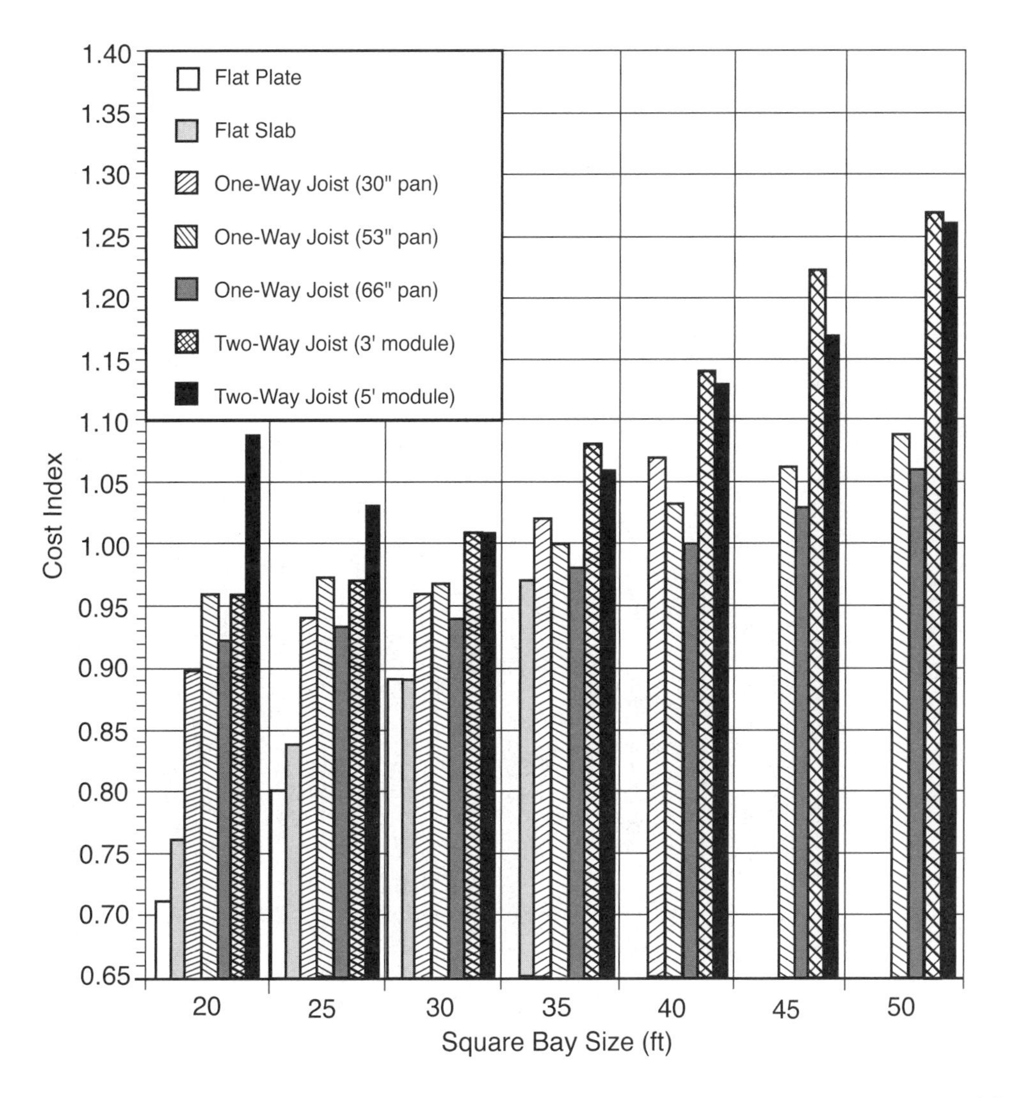

Figure 1-5 Relative Costs of Reinforced Concrete Floor systems, Live Load = 50 psf[1.3]

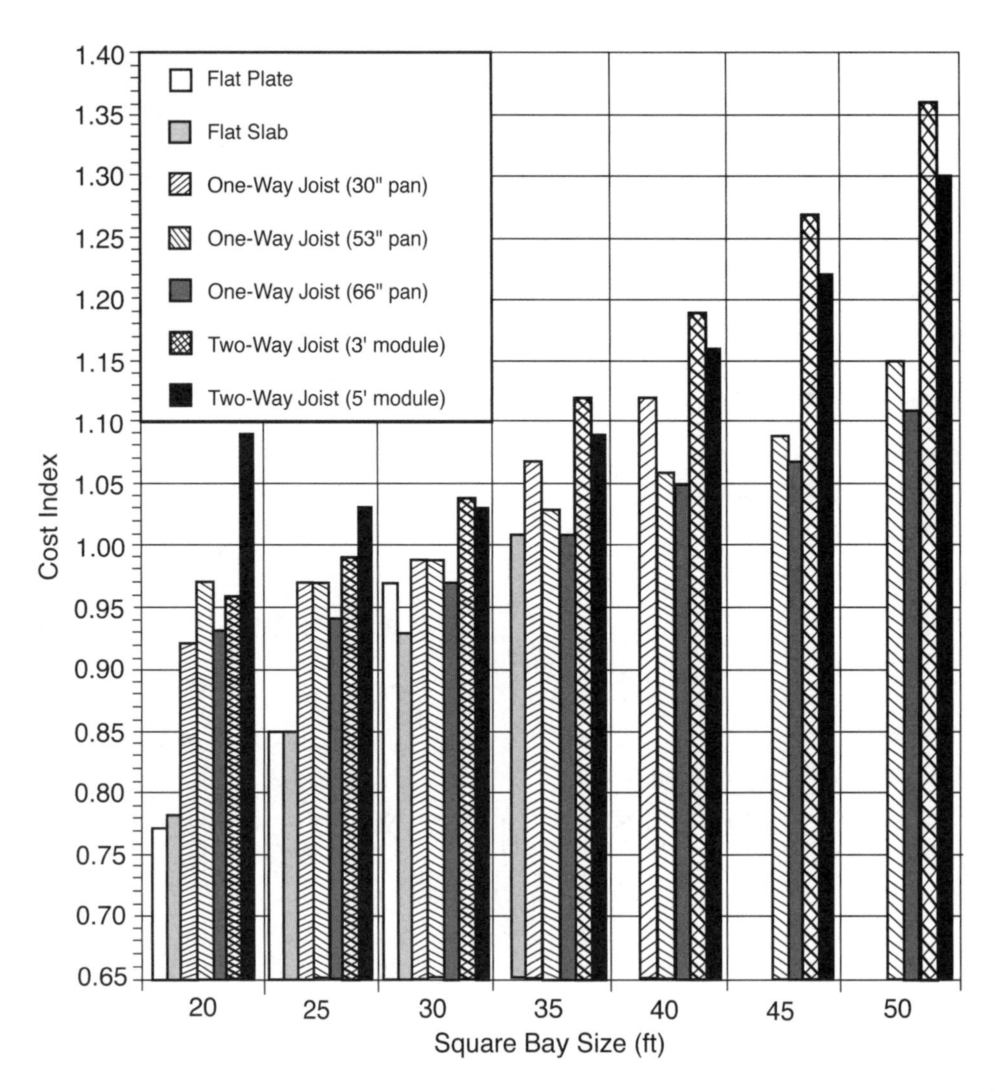

Figure 1-6 Relative Costs of Reinforced Concrete Floor systems, Live Load = 100 psf[1.3]

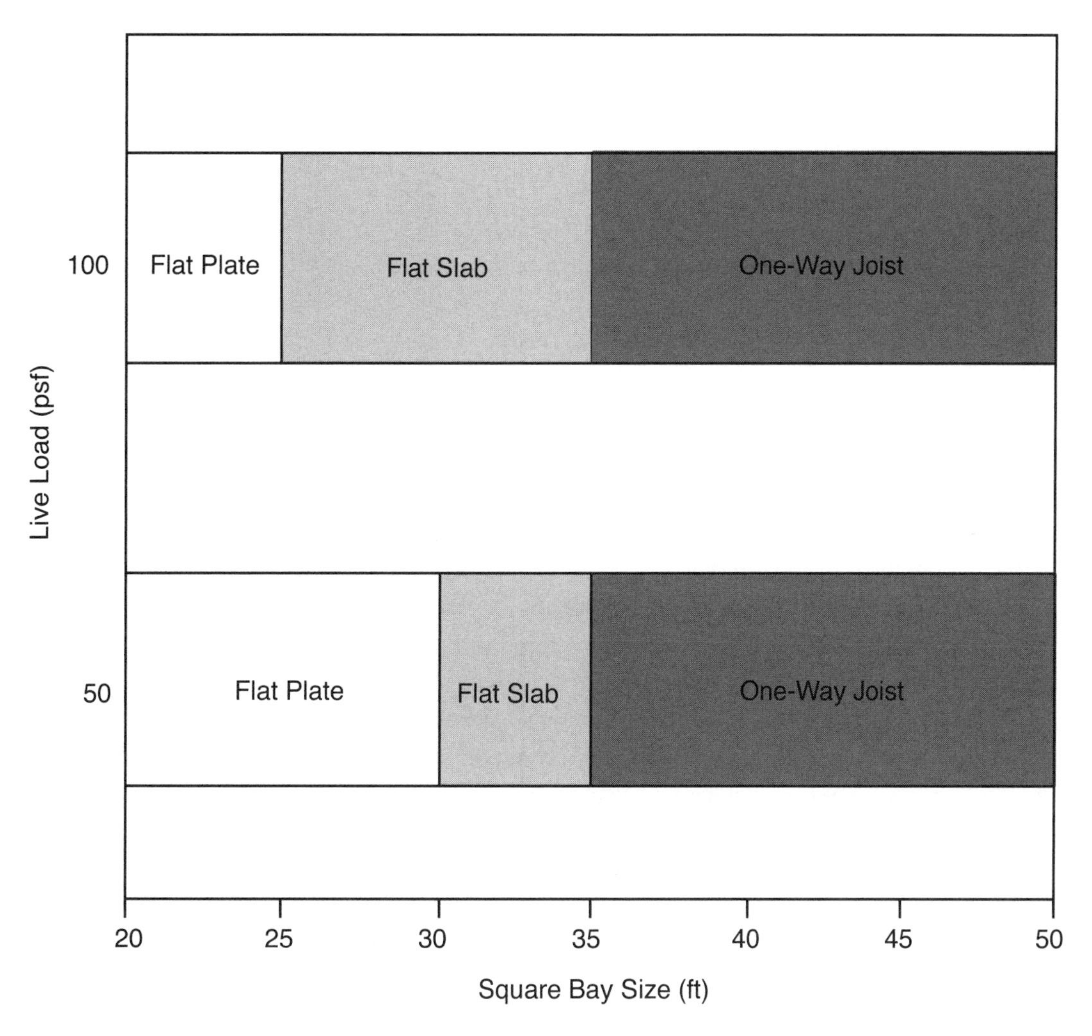

Insert Figure 1-7 Cost-Effective Concrete Floor Systems as a Function of span Length and Load[1.3]

ness determined from Fig. 1-8 will have to be increased in order to accommodate the additional shear stress at the edge columns. Exactly how much of an increase is required depends on numerous factors. In general, the slab thickness needs to be increased by about 15-20%; the shear stress can be checked at the edge columns after the nominal moment resistance of the column strip is determined (ACI 13.6.3.6).

When increasing the overall slab thickness is not possible or feasible, drop panels can be provided at the column locations where two-way shear is critical. Chapter 4 gives ACI 318 provisions for minimum drop panel dimensions.

In all cases, the slab thickness must be larger than the applicable minimum thickness given in ACI 9.5.3 for deflection control. Figure 4-3 may be used to determine the minimum thickness as a function of the clear span ℓ_n for the various two-way systems shown.

1.8.2 Columns

For overall economy, the dimensions of a column should be determined for the load effects in the lowest story of the structure and should remain constant for the entire height of the building; only the amounts of reinforcement should vary with respect to height.* The most economical columns usually have reinforcement ratios in the range of 1-2%. In general, it is more efficient to increase the column size than to increase the amount of reinforcement. This approach is taken to eliminate congestion of column reinforcement, which has to be critically evaluated along the lap splice length and to accommodate horizontal beam bars at the beam column intersections.

Columns in a frame that is braced by shearwalls (non-sway frame) are subjected to gravity loads only. Initial column sizes may be obtained from design aids such as the one given in Fig. 5-1: assuming a reinforcement ratio in the range of 1-2%, a square column size can be determined for the total factored axial load P_u in the lowest story. Once an initial size is obtained, it should be determined if the effects of slenderness need to be considered. If feasible, the size of the column should be increased so as to be able to neglect slenderness effects.
When a frame is not braced by shearwalls (sway frame), the columns must be designed for the combined effects of gravity and wind loads. In this case, a preliminary size can be obtained for a column in the lowest level from Fig. 5-1 assuming that the column carries gravity loads only. The size can be chosen based on 1% reinforcement in the column; in this way, when wind loads are considered, the area of steel can usually be increased without having to change the column size. The design charts given in Figs. 5-16 through 5-23 may also be used to determine the required column size and reinforcement of a given combination of factored axial loads and moments. Note that slenderness effects can have a significant influence on the amount of reinforcement that is required for a column in a sway frame; for this reason, the overall column size should be increased (if possible) to minimize the effects of slenderness.

1.8.3 Shearwalls

For buildings of moderate size and height, a practical range for the thickness of shearwalls is 8 to 10 in. The required thickness will depend on the length of the wall, the height of the building, and the tributary wind area of the wall. In most cases, minimum amounts of vertical and horizontal reinforcement are sufficient for both shear and moment resistance.

In the preliminary design stage, the shearwalls should be symmetrically located in the plan (if possible) so that torsional effects on the structure due to lateral loads are minimized.

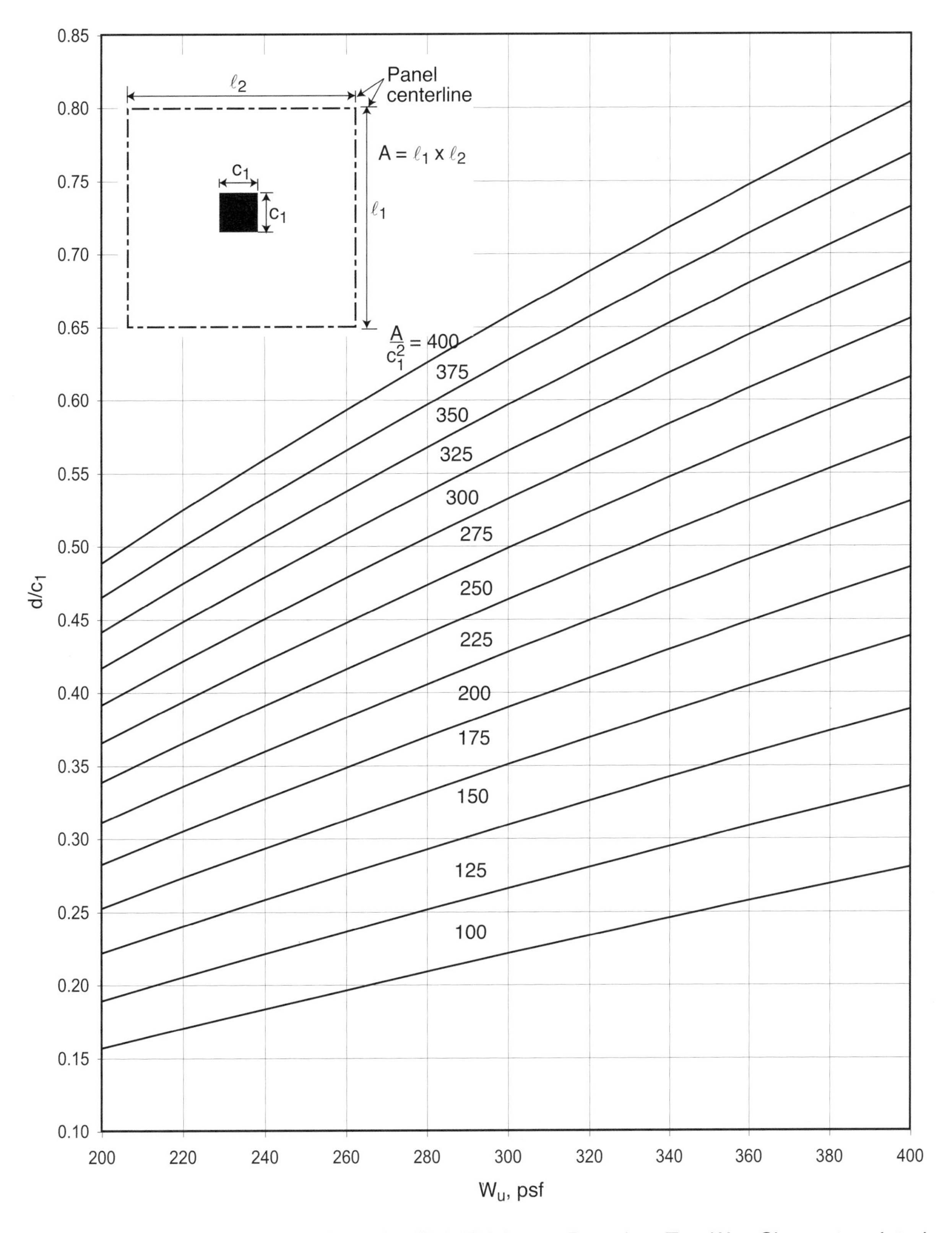

Figure 1-8 Preliminary Design Chart for Slab Thickness Based on Two-Way Shear at an Interior Square Column (f'_c = 4000 psi)

1.8.4 Footings

The required footing sizes can usually be obtained in a straightforward manner. In general, the base area is determined by dividing the total service (unfactored) loads from the column by the allowable (safe) soil pressure. In buildings without shearwalls, the maximum pressure due to the combination of gravity and wind loads must also be checked. The required thickness may be obtained for either a reinforced or a plain footing by using the appropriate simplified equation given in Chapter 7.

1.8.5 Fire Resistance

To insure adequate resistance to fire, minimum thickness and cover requirements are specified in building codes as a function of the required fire resistance rating. Two hours is a typical rating for most members; however, the local building code should be consulted for the ratings which apply to a specific project or special portions of projects.
Member sizes that are necessary for structural requirements will usually satisfy the minimum requirements for fire resistance as well (see Tables 10-1 and 10-2). Also, the minimum cover requirements specified in ACI 7.7 will provide at least a three-hour fire resistance rating for restrained floor members and columns (see Tables 10-3, 10-4, and 10-6).

It is important to check the fire resistance of a member immediately after a preliminary size has been obtained based on structural requirements. Checking the fire resistance during the preliminary design stage eliminates the possibility of having to completely redesign the member (or members) later on.

In the examples that appear in the subsequent chapters, the applicable fire resistance requirements tabulated in Chapter 10 are checked for all members immediately after the preliminary sizes are obtained. The required fire resistance ratings for both example buildings are listed in Section 1.7.

1.9 FROM ACI 318-02 TO ACI 318-05

As of this printing, several improvements and clarifications have been proposed for the 2005 edition of ACI 318, "Building Code Requirements for Structural Concrete and Commentary." These revisions can be categorized as technical changes and notation changes intended to make the code easier to use. For the 2005 revision, ACI Committee 318 decided to unify the notation to achieve the following main goals: provide a unique definition for each notation; consolidate similar notations as appropriate and eliminate unnecessary ones; move the list of notation from old Appendix E to Chapter 2; and remove notation list from beginning of each chapter.

In contrast with unprecedented significant technical changes incorporated in the 2002 revision of the building code, the technical changes in ACI 318-05 are limited. To be consistent with current practice, various editorial changes were made in the code including clarifications and updates to reference standards. Also included are changes concerning the spacing limits for crack control, an alternative design method for torsion, and construction joint location for post-tensioned slabs.

The design procedures and design aids provided in this publication are suitable for use with both versions of the building code: ACI 318-02 and ACI 318-05. When using the ACI 318-05, the designer may have to make minor adjustments in the design documentation where the notation has been changed.

References

1.1 *Building Code Requirements for Structural Concrete ACI 318-02 and Commentary—ACI 318R-02*, American Concrete Institute, Detroit, Michigan, 2002.

1.2 *Building Movements and Joints*, EB086, Portland Cement Association, Skokie, Illinois 1982, 64 pp.

1.3 *Concrete Floor Systems–Guide to Estimating and Economizing*, SP041, Portland Cement Association, Skokie, Illinois, 2000, 41 pp.

1.4 *Long-Span Concrete Floor Systems*, SP339, Portland Cement Association, Skokie, Illinois, 2000, 97 pp.

1.5 *American Society of Civil Engineers Minimum Design Loads for Buildings and Other Structures*, ASCE 7-02, American Society of Civil Engineers, New York, N.Y., 2003.

Chapter 2

Simplified Frame Analysis

2.1 INTRODUCTION

The final design of the structural components in a building frame is based on maximum moment, shear, axial load, torsion and/or other load effects, as generally determined by an elastic frame analysis (ACI 8.3). For building frames of moderate size and height, preliminary and final designs will often be combined. Preliminary sizing of members, prior to analysis, may be based on designer experience, design aids, or simplified sizing expressions suggested in this book.

Analysis of a structural frame or other continuous construction is usually the most time consuming part of the total design. For gravity load analysis of continuous one-way systems (beams and slabs), the approximate moments and shears given by ACI 8.3.3 are satisfactory within the span and loading limitations stated. For cases when ACI 8.3.3 is not applicable, a two-cycle moment distribution method is accurate enough. The speed and accuracy of the method can greatly simplify the gravity load analysis of building frames with usual types of construction, spans, and story heights. The method isolates one floor at a time and assumes that the far ends of the upper and lower columns are fixed. This simplifying assumption is permitted by ACI 8.8.3.

For lateral load analysis of a sway frame, the Portal Method may be used. It offers a direct solution for the moments and shears in the beams (or slabs) and columns, without having to know the member sizes or stiffnesses.

The simplified methods presented in this chapter for gravity load analysis and lateral load analysis are considered to provide sufficiently accurate results for buildings of moderate size and height. However, determinations of load effects by computer analysis or other design aids are equally applicable for use with the simplified design procedures presented in subsequent chapters of this book.

2.2 LOADING

2.2.1 Service Loads

The first step in the frame analysis is the determination of design (service) loads and lateral forces (wind and seismic) as called for in the general building code under which the project is to be designed and constructed. For the purposes of this book, design live loads (and permissible reductions in live loads) and wind loads are based on *Minimum Design Loads for Buildings and Other Structures*, ASCE7-02.[2.1] References to specific ASCE Standard requirements are noted (ASCE 4.2 refers to ASCE 7-02, Section 4.2). For a specific project, however, the governing general building code should be consulted for any variances from ASCE 7-02.

Design dead loads include member self-weight, weight of fixed service equipment (plumbing, electrical, etc.) and, where applicable, weight of built-in partitions. The latter may be accounted for by an equivalent uniform load of not less than 20 psf, although this is not specifically defined in the ASCE Standard (see ASCE Commentary Section 3.2).

Design live loads will depend on the intended use and occupancy of the portion or portions of the building being designed. Live loads include loads due to movable objects and movable partitions temporarily supported by the building during maintenance. In ASCE Table 4-1, uniformly distributed live loads range from 40 psf for residential use to 250 psf for heavy manufacturing and warehouse storage. Portions of buildings, such as library floors and file rooms, require substantially heavier live loads. Live loads on a roof include maintenance equipment, workers, and materials. Also, snow loads, ponding of water, and special features, such as landscaping, must be included where applicable.

Occasionally, concentrated live loads must be included; however, they are more likely to affect individual supporting members and usually will not be included in the frame analysis (see ASCE 4.3).

Design wind loads are usually given in the general building code having jurisdiction. For both example buildings here, the calculation of wind loads is based on the procedure presented in ASCE 6.0. Design for seismic loads is discussed in Chapter 11.

2.2.1.1 Example: Calculation of Wind Loads – Building #2

For illustration of the ASCE procedure, wind load calculations for the main wind-force resisting system of building #2 (5-story flat plate) are summarized below.

Wind-force resisting system:

Alternate (1) - Slab and column framing with spandrel beams
Alternate (2) - Structural walls

(1) Wind load data

Assuming the building is classified as closed (ASCE 6.5.9) and located in Midwest in flat open terrain.

Basic wind speed V = 90 mph	ASCE Figure 6-1
Occupancy category II	ASCE Table 1-1
Exposure category C	ASCE 6.5.6.3
Wind directionality factor $K_d = 0.85$	ASCE Table 6-4
Importance factor I = 1	ASCE Table 6-1
Topographic factor $K_{zt} = 1$	ASCE 6.5.7
Gust effect factor G = 0.85	ASCE 6.5.8
External pressure coefficient C_p	ASCE Figure 6-6

Windward - both directions	$C_p = 0.8$
Leeward - E-W direction (L/B = 120/60 = 2)	$C_p = -0.3$
N-S direction (L/B = 60/120 = 0.5)	$C_p = -0.5$

Velocity pressure exposure coefficients K_z (ASCE 6.5.6.6) at various story heights are summarized in Table 2-1

Velocity pressure $q_z = 0.00256K_zK_{zt}K_dV^2I$	ASCE 6.5.10
Design wind pressure $p_z = q_zGC_p$	ASCE 6.5.12.2

(2) Design wind pressure in the N-S Direction

Table 2-1 contains a summary of the design pressures calculations for wind in N-S direction.

Table 2-1 Design Pressures for N-S Direction

Level	Height above ground level, z (ft)	K_z	q_z (psf)	Windward design pressure q_zGC_p (psf)	Leeward design pressure q_hGC_p (psf)	Total design pressure (psf)
Roof	63	1.148	20.2	13.8	-8.6	22.4
4	51	1.098	19.4	13.2	-8.6	21.8
3	39	1.038	18.3	12.4	-8.6	21.0
2	27	0.961	16.9	11.5	-8.6	20.1
1	15	0.849	15.0	10.2	-8.6	18.8

(3) Wind loads in the N-S direction

The equivalent wind loads at each floor level are calculated as follows:

Alternate (1) Slab and column framing
Interior frame (24 ft bay width)

Roof = $22.4 \times 6.0 \times 24/1000$ = 3.2 kips
4th = $21.8 \times 12 \times 24/1000$ = 6.3 kips
3rd = $21.0 \times 12 \times 24/1000$ = 6.1 kips
2nd = $20.1 \times 12 \times 24/1000$ = 5.8 kips
1st = $18.8 \times 13.5 \times 24/1000$ = 6.1 kips

Alternate (2) Structural walls
Total for entire building (121 ft width)

Roof = $22.4 \times 6.0 \times 121/1000$ = 16.2 kips
4th = $21.8 \times 12 \times 121/1000$ = 31.6 kips
3rd = $21.0 \times 12 \times 121/1000$ = 30.6 kips
2nd = $20.1 \times 12 \times 121/1000$ = 29.2 kips
1st = $18.8 \times 13.5 \times 121/1000$ = 30.7 kips

(4) Wind loads in the E -W direction

Using the same procedure as for the N-S direction, the following wind loads are obtained for the E-W direction:

Alternate (1) Slab and column framing
Interior frame (24 ft bay width)

Roof = 2.3 kips
4th = 4.4 kips
3rd = 4.2 kips
2nd = 4.0 kips
1st = 4.1 kips

Alternate (2) Structural walls
Total for entire building (121 ft width)

Roof = 6.9 kips
4th = 13.4 kips
3rd = 12.9 kips
2nd = 12.2 kips
1st = 12.6 kips

2.2.1.2 Example: Calculation of Wind Loads – Building #1

Wind load calculations for the main wind-force resisting system of Building #1 (3-story pan joist framing) are summarized below.

Wind-force resisting system: Beam and column framing:

(1) Wind load data

Assuming the building is classified as closed (ASCE 6.5.9) and located in hurricane prone region.

Basic wind speed V = 145 mph	ASCE Figure 6-1
Occupancy category II	ASCE Table 1-1
Exposure category D	ASCE 6.5.6.3
Wind directionality factor K_d = 0.85	ASCE Table 6-4
Importance factor I = 1	ASCE Table 6-1
Topographic factor K_{zt} = 1	ASCE 6.5.7
Gust effect factor G = 0.85	ASCE 6.5.8
External pressure coefficient C_p	ASCE Figure 6-6

Windward - both directions $C_p = 0.8$
Leeward - E-W direction (L/B = 150/90 = 1.67) $C_p = -0.37$
N-S direction (L/B = 90/150 = 0.6) $C_p = -0.5$

Design pressures calculations for wind in N-S direction are summarized in Table 2-2

Table 2-2 Design Pressures for N-S Direction

Level	Height above ground level, z (ft)	K_z	q_z (psf)	Windward design pressure q_zGC_p (psf)	Leeward design pressure q_hGC_p (psf)	Total design pressure (psf)
Roof	39	1.216	55.7	37.8	-23.7	61.5
2	27	1.141	51.9	35.3	-23.7	59.0
1	15	1.030	47.1	32.1	-23.7	55.8

(2) Summary of wind loads

N-S & E-W directions (conservatively use N-S wind loads in both directions):
Interior frame (30 ft bay width)

Roof = 12 kips
2nd = 23.1 kips
1st = 21.7 kips

Note: The above loads were determined using design wind pressures computed at each floor level.

2.2.2 Live Load Reduction for Columns, Beams, and Slabs

Most general building codes permit a reduction in live load for design of columns, beams and slabs to account for the probability that the total floor area "influencing" the load on a member may not be fully loaded simultaneously. Traditionally, the reduced amount of live load for which a member must be designed has been based on a tributary floor area supported by that member. According to ASCE 7-02, the magnitude of live load reduction is based on an influence area rather than a tributary area. The influence area is a function of the tributary area for the structural member. The influence area for different structural members is calculated by multiplying the tributary area for the member A_T, by the coefficients K_{LL} given in Table 2-3, see ASCE 4.8.

The reduced live load L per square foot of floor area supported by columns, beams, and two-way slabs having an influence area ($K_{LL}A_T$) of more than 400 sq ft is:

$$L = L_0\left(0.25 + \frac{15}{\sqrt{K_{LL}A_T}}\right) \qquad \text{ASCE (Eq. 4-1)}$$

where L_o is the unreduced design live load per square foot. The reduced live load cannot be taken less than 50% for members supporting one floor, or less than 40% of the unit live load L_o otherwise. For other limitations on live load reduction, see ASCE 4.8.

Using the above expression for reduced live load, values of the reduction multiplier as a function of influence area are given in Table 2-4.

Table 2-3 Live Load Element Factor K_{LL}

Element	K_{LL}
Interior columns	4
Exterior column without cantilever slabs	4
Edge column with cantilever slabs	3
Corner columns with cantilever slabs	2
Edge beams without cantilever slabs	2
Interior beams	2
All other members not identified above including: Edge beams with cantilever slabs Cantilever beams Two-way slabs	1

The above limitations on permissible reduction of live loads are based on ASCE 4.8. The governing general building code should be consulted for any difference in amount of reduction and type of members that may be designed for a reduced live load.

Table 2-4 Reduction Multiplier (RM) for Live Load = $\left(0.25 + \frac{15}{\sqrt{K_{LL}A_T}}\right)$

Influence Area $K_{LL}A_T$	RM	Influence Area $K_{LL}A_T$	RM
400[a]	1.000	5600	0.450
800	0.780	6000	0.444
1200	0.683	6400	0.438
1600	0.625	6800	0.432
2000	0.585	7200	0.427
2400	0.556	7600	0.422
2800	0.533	8000	0.418
3200	0.515	8400	0.414
3600	0.500[b]	8800	0.410
4000	0.487	9200	0.406
4800	0.467	10000	0.400[c]
5200	0.458		

[a]No live load reduction is permitted for influence area less than 400 sq ft.
[b]Maximum reduction permitted for members supporting one floor only.
[c]Maximum absolute reduction.

2.2.2.1 Example: Live Load Reductions for Building #2

For illustration, typical influence areas for the columns and the end shear walls of Building #2 (5-story flat plate) are shown in Fig. 2-1. Corresponding live load reduction multipliers are listed in Table 2-5.

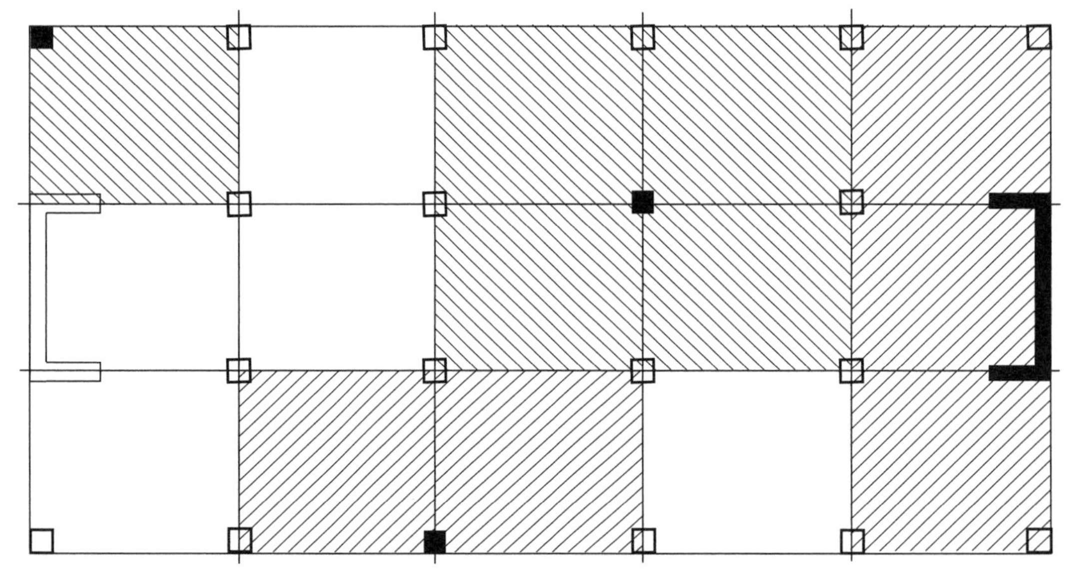

Figure 2-1 Typical Influence Areas

Table 2-5 Reduction Multiplier (RM) for Live Loads, Building #2

	Interior Columns		Edge Columns		Corner Columns		End Shear Wall Units	
Story	$K_{LL}A_T$ (ft^2)	RM	$K_{LL}A_T$ (ft^2)	RM	$K_{LL}A_T$ (ft^2)	RM	$K_{LL}A_T$ (ft^2)	RM
5 (roof)		*		*		*		*
4	1920	0.59	960	0.73	480	0.94	1440	0.65
3	3840	0.49	1920	0.59	960	0.73	2880	0.53
2	5760	0.45	2880	0.53	1440	0.65	4320	0.48
1	7680	0.42	3840	0.49	1920	0.59	5760	0.48

*No reduction permitted for roof live load (ASCE 4.8.2); the roof should not be included in the influence areas of the floors below.

For the interior columns, the reduced live load is $L = 0.42L_o$ at the first story (A_T = 4 bay areas × 4 stories = 20 × 24 × 4 × 4 = 7680 sq ft). The two-way slab may be designed with an RM = 0.94 (A_T = 480 sq ft for one bay area). Shear strength around the interior columns is designed for an RM = 0.59 (A_T = 1920 sq ft for 4 bay areas), and around an edge column for an RM = 0.73 (A_T = 960 sq ft for 2 bay areas). Spandrel beams could be designed for an RM = 0.94 (one bay area). If the floor system were a two-way slab with beams between columns, the interior beams would qualify for an RM = 0.73 (2 bay areas).

2.2.3 Factored Loads

The strength design method using factored loads to proportion members is used exclusively in this book. The design (service) loads must be increased by specified load factors (ACI 9.2), and factored loads must be combined in load combinations depending on the types of loads being considered. The method requires that the design strength of a member at any section should equal or exceed the required strength calculated by the code-specified factored load combinations. In general,

Design Strength ≥ Required Strength

or

Strength Reduction Factor (ϕ) × Nominal Strength ≥ Load Factor × Service Load Effects

All structural members and sections must be proportioned to meet the above criterion under the most critical load combination for all possible load effects (flexure, axial load, shear, etc.).

For structures subjected to dead, live, wind, and earthquake loads only, the ACI applicable load combinations are summarized in Table 2-6:

The strength reduction factors are listed below:

Tension-controlled sections	0.90
Compression-controlled sections	
Members with spiral reinforcement conforming to 10.9.3	0.70
Other reinforced members	0.65
Shear and torsion	0.75
Bearing on concrete	0.65

Table 2-6 ACI Load Combinations for Building Subjected for Dead, Live Wind, and Earthquake Loads

ACI Equations	Garages, places of public assembly and all areas where L is greater than 100 lb/ft^2	Other buildings
Eq. (9-1)	$U = 1.4D$	$U = 1.4D$
Eq. (9-2)	$U = 1.2D + 1.6L + 0.5L_r$	$U = 1.2D + 1.6L + 0.5L_r$
Eq. (9-3)	$U = 1.2D + 1.6L_r + 1.0L$	$U = 1.2D + 1.6L_r + 0.5L$
	$U = 1.2D + 1.6L_r \pm 0.8W$	$U = 1.2D + 1.6L_r \pm 0.8W$
Eq. (9-4)	$U = 1.2D \pm 1.6W + 1.0L + 0.5L_r$	$U = 1.2D \pm 1.6W + 0.5L + 0.5L_r$
Eq. (9-5)	$U = 1.2D \pm 1.0E + 1.0L$	$U = 1.2D \pm 1.0E + 0.5L$
Eq. (9-6)	$U = 0.9D \pm 1.6W$	$U = 0.9D \pm 1.6W$
Eq. (9-7)	$U = 0.9D \pm 1.0E$	$U = 0.9D \pm 1.0E$

D = dead loads, or related internal moments and forces
L = live loads, or related internal moments and forces
L_r = roof live load, or related internal moments and forces
U = required strength to resist factored loads or related internal moments and forces
W = wind load, or related internal moments and forces
E = earthquake loads, or related internal moments and forces

For design of beams and slabs, the factored load combinations used most often are:

$$U = 1.4D \qquad \text{ACI Eq. (9-1)}$$
$$U = 1.2D + 1.6L \qquad \text{ACI Eq. (9-2)}$$

ACI Eq. (9-1) seldom controls—only when the live load to dead load ratio (L/D) is less than 0.125.

For a frame analysis with live load applied only to a portion of the structure, i.e., alternate spans (ACI 8.9), the factored loads to be applied would be computed separately using the appropriate load factor for each load.

The designer has the choice of multiplying the service loads by the load factors before computing the factored load effects (moments, shears, etc.), or computing the effects from the service loads and then multiplying the effects by the load factors. For example, in the computation of bending moment for dead and live loads $U = 1.2D + 1.6L$, the designer may (1) determine $w_u = 1.2\ w_d + 1.6\ w_\ell$ and then compute the factored moments using w_u; or (2) compute the dead and live load moments using service loads and then determine the factored moments as $M_u = 1.2\ M_d + 1.6\ M_\ell$. Both analysis procedures yield the same answer. It is important to note that the second alternative is much more general than the first; thus, it is more suitable for computer analysis, especially when more than one load combination must be investigated.

2.3 FRAME ANALYSIS BY COEFFICIENTS

The ACI Code provides a simplified method of analysis for both one-way construction (ACI 8.3.3) and two-way construction (ACI 13.6). Both simplified methods yield moments and shears based on coefficients. Each method will give satisfactory results within the span and loading limitations stated in Chapter 1. The direct design method for two-way slabs is discussed in Chapter 4.

2.3.1 Continuous Beams and One-Way Slabs

When beams and one-way slabs are part of a frame or continuous construction, ACI 8.3.3 provides approximate moment and shear coefficients for gravity load analysis. The approximate coefficients may be used as long as all of the conditions illustrated in Fig. 2-2 are satisfied: (1) There must be two or more spans, approximately equal in length, with the longer of two adjacent spans not exceeding the shorter by more than 20 percent; (2) loads must be uniformly distributed, with the service live load not more than 3 times the dead load ($L/D \leq 3$); and (3) members must have uniform cross section throughout the span. Also, no redistribution of moments is permitted (ACI 8.4). The moment coefficients defined in ACI 8.3.3 are shown in Figs. 2-3 through 2-6. In all cases, the shear in end span members at the interior support is taken equal to $1.15w_u\ell_n/2$. The shear at all other supports is $w_u/2$ (see Fig. 2-7). $w_u\ell_n$ is the combined factored load for dead and live loads, $w_u = 1.2w_d + 1.6\ w_\ell$. For beams, w_u is the uniformly distributed load per unit length of beam (plf), and the coefficients yield total moments and shears on the beam. For one-way slabs, w_u is the uniformly distributed load per unit area of slab (psf), and the moments and shears are for slab strips one foot in width. The span length ℓ_n is defined as the clear span of the beam or slab. For negative moment at a support with unequal adjacent spans, ℓ_n is the average of the adjacent clear spans. Support moments and shears are at the faces of supports.

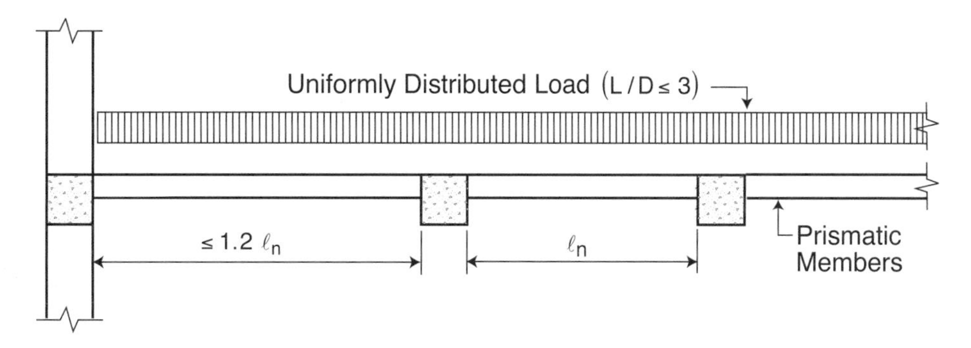

Figure 2-2 Conditions for Analysis by Coefficients (ACI 8.3.3)

2.3.2 Example: Frame Analysis by Coefficients

Determine factored moments and shears for the joists of the standard pan joist floor system of Building #1 (Alternate (1)) using the approximate moment and shear coefficients of ACI 8.3.3. Joists are spaced at 3 ft on center.

(1) Data: Width of spandrel beam = 20 in.
Width of interior beams = 36 in.
Floors: LL = 60 psf
DL = 130 psf
$w_u = 1.2\ (130) + 1.6\ (60) = 252$ psf $\times$ 3 ft = 756 plf

(2) Factored moments and shears using the coefficients from Figs. 2-3, 2-4, and 2-7 are summarized in Fig. 2-8.

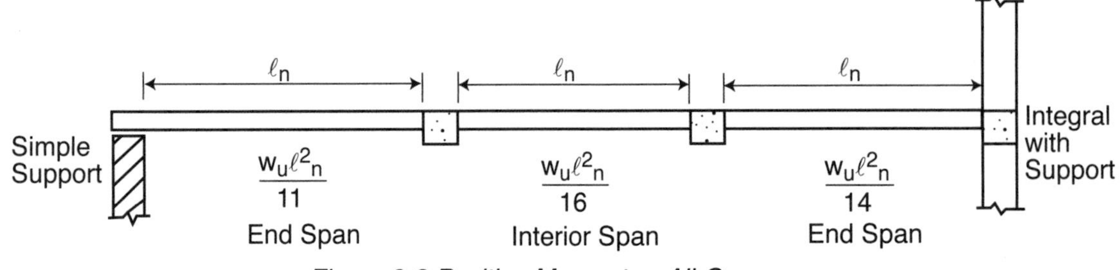

Figure 2-3 Positive Moments—All Cases

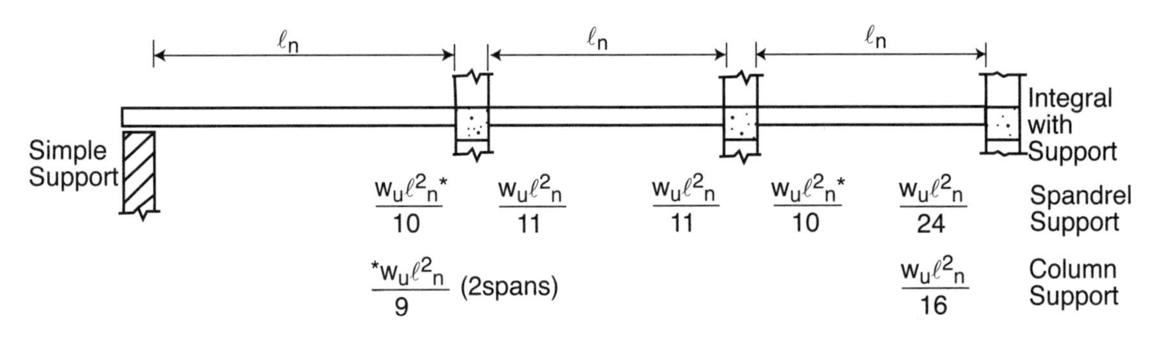

Figure 2-4 Negative Moments—Beams and Slabs

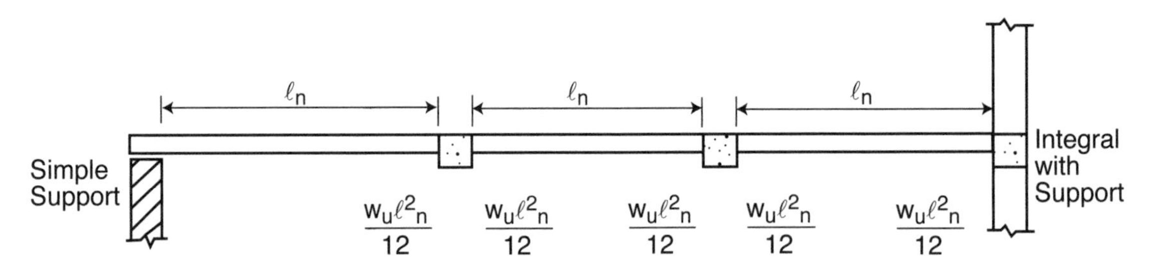

Figure 2-5 Negative Moments—Slabs with spans < 10 ft

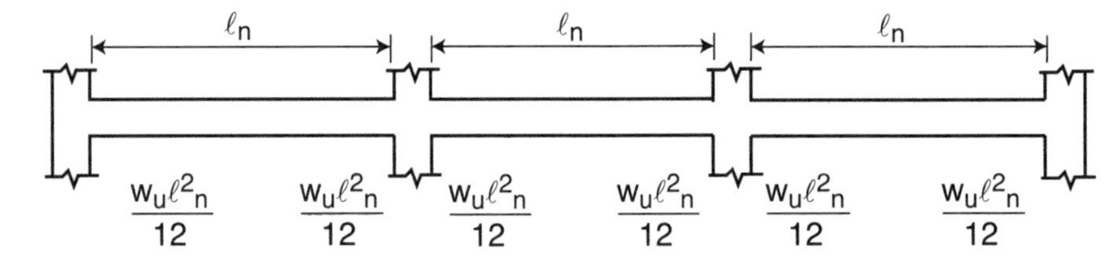

Figure 2-6 Negative Moments—Beams with Stiff Columns

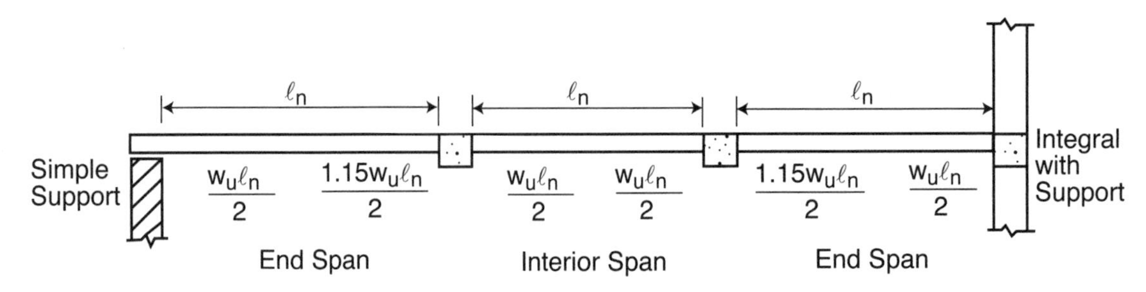

Figure 2-7 End Shears—All Cases

30'-0" | 30'-0" | 15'-0"

8" | 1'-0" | ℓ_n = 27.5' | 3'-0" | ℓ_n = 27.0' | 3'-0" | ℓ_n = 27.0'

Sym. about ℄

	Span 1		Span 2		Span 3
Total Load	$w_u\ell_n$ = 0.76 x 27.5 = 20.9 kips		0.76 x 27.0 = 20.5 kips		0.76 x 27.0 = 20.5 kips
Coefficient from Fig. 2-7	1/2	1.15/2	1/2	1/2	1/2
Shear V_u	10.5 kips	12 kips	10.3 kips	10.3 kips	10.3 kips
$w_u\ell_n^2$	574.8 ft-kips		554 ft-kips		554 ft-kips
Coefficient from Fig. 2-3	1/14		1/16		1/16
Pos. M_u	41.1 ft-kips		34.6 ft-kips		34.6 ft-kips
ℓ_n	27.5 ft	27.25 ft*	27.25 ft*	27.0 ft	27.0 ft
$w_u\ell_n^2$	574.8 ft-kips	564.3 ft-kips	564.3 ft-kips	554 ft-kips	554 ft-kips
Coefficient from Fig. 2-3	1/24	1/10	1/11	1/11	1/11
Neg. M_u	24 ft-kips	56.3 ft-kips	51.3 ft-kips	50.4 ft-kips	50.4 ft-kips

*Average of adjacent clear spans

Figure 2-8 Factored Moments and Shears for the Joist Floor System of Building #1 (Alternate 1)

2.4 FRAME ANALYSIS BY ANALYTICAL METHODS

For continuous beams and one-way slabs not meeting the limitations of ACI 8.3.3 for analysis by coefficients, an elastic frame analysis must be used. Approximate methods of frame analysis are permitted by ACI 8.3.2 for "usual" types of buildings. Simplifying assumptions on member stiffnesses, span lengths, and arrangement of live load are given in ACI 8.6 through 8.9.

2.4.1 Stiffness

The relative stiffnesses of frame members must be established regardless of the analytical method used. Any reasonable consistent procedure for determining stiffnesses of columns, walls, beams, and slabs is permitted by ACI 8.6.

The selection of stiffness factors will be considerably simplified by the use of Tables 2-7 and 2-8. The stiffness factors are based on gross section properties (neglecting any reinforcement) and should yield satisfactory results for buildings within the size and height range addressed in this book. In most cases where an analytical procedure is required, stiffness of T-beam sections will be needed. The relative stiffness values K given in Table 2-8 allow for the effect of the flange by doubling the moment of inertia of the web section (b_wh). For values of h_f/h between 0.2 and 0.4, the multiplier of 2 corresponds closely to a flange width equal to six times the web width. This is considered a reasonable allowance for most T-beams.[2.2] For rectangular beam sections; the tabulated values should be divided by 2. Table 2-8 gives relative stiffness values K for column sections. It should be noted that column stiffness is quite sensitive to changes in column size. The initial judicious selection of column size and uniformity from floor to floor is, therefore, critical in minimizing the need for successive analyses.

As is customary for ordinary building frames, torsional stiffness of transverse beams is not considered in the analysis. For those unusual cases where equilibrium torsion is involved, a more exact procedure may be necessary.

2.4.2 Arrangement of Live Load

According to ACI 8.9.1, it is permissible to assume that for gravity load analysis, the live load is applied only to the floor or roof under consideration, with the far ends of the columns assumed fixed. In the usual case where the exact loading pattern is not known, the most demanding sets of design forces must be investigated. Figure 2-9 illustrates the loading patterns that should be considered for a three-span frame.

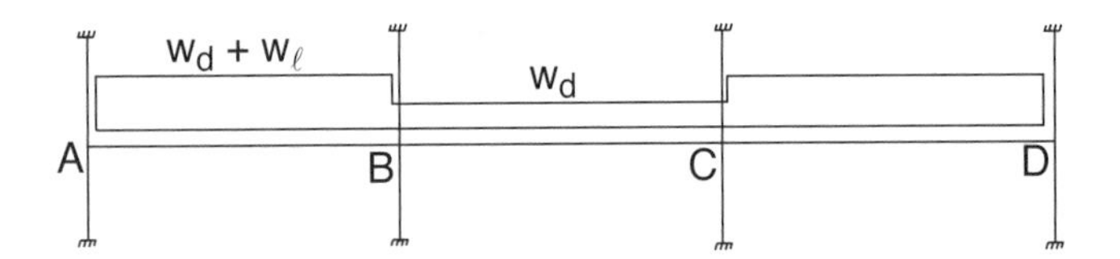

(1) Loading pattern for negative moment at support A and positive moment in span AB

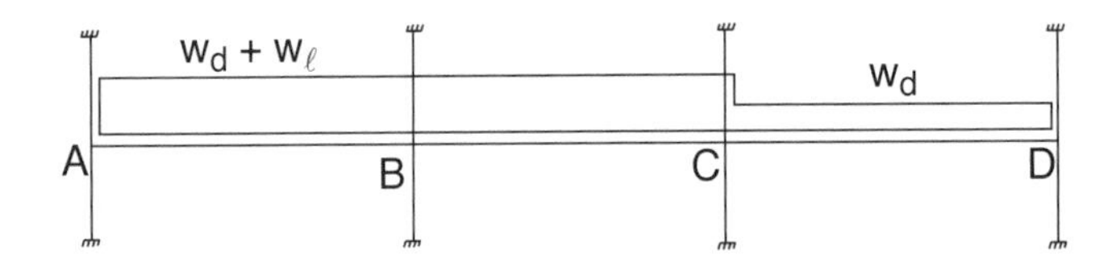

(2) Loading pattern for negative moment at support B

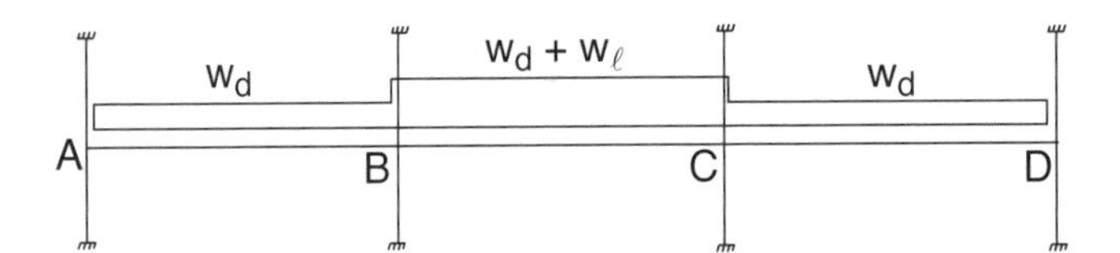

(3) Loading pattern for positive moment in span BC

Figure 2-9 Partial Frame Analysis for Gravity Loading

2.4.3 Design Moments

When determining moments in frames or continuous construction, the span length shall be taken as the distance center-to-center of supports (ACI 8.7.2). Moments at faces of supports may be used for member design purposes (ACI 8.7.3). Reference 2.2 provides a simple procedure for reducing the centerline moments to face of support moments, which includes a correction for the increased end moments in the beam due to the restraining effect of the column between face and centerline of support. Figure 2-10 illustrates this correction. For beams and slabs subjected to uniform loads, negative moments from the frame analysis can be reduced by $w_u\ell_a^2/6$. A companion reduction in the positive moment of $w_u\ell_a^2/12$ can also be made.

Table 2-7 Beam Stiffness Factors

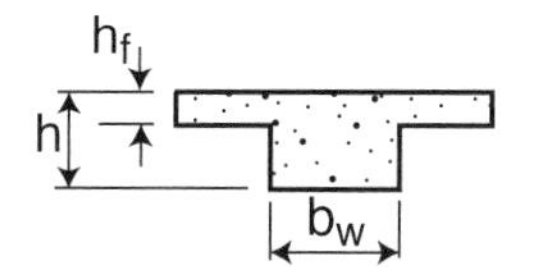

Moment of inertia, excluding overhanging flanges:

$$I = \frac{b_w h^3}{12} \qquad K^* = \frac{2I}{10\ell}$$

Moment of inertia of T-section ≅ 2I

Values of K for T-beams

h	b_w	I	Span of beam, ℓ (ft)** 8	10	12	14	16	20	24	30
	6	256	6	5	4	4	3	3	2	2
	8	341	9	7	6	5	4	3	3	2
	10	427	11	9	7	6	5	4	4	3
8	11_	491	12	10	8	7	6	5	4	3
	13	555	14	11	9	8	7	6	5	4
	15	640	16	13	11	9	8	6	5	4
	17	725	18	15	12	10	9	7	6	5
	19	811	20	16	14	12	10	8	7	5
	6	500	13	10	8	7	6	5	4	3
	8	667	17	13	11	10	8	7	6	4
	10	833	21	17	14	12	10	8	7	6
10	11_	958	24	19	16	14	12	10	8	6
	13	1083	27	22	18	15	14	11	9	7
	15	1250	31	25	21	18	16	13	10	8
	17	1417	35	28	24	20	18	14	12	9
	19	1583	40	32	26	23	20	16	13	11
	6	864	22	17	14	12	11	9	7	6
	8	1152	29	23	19	16	14	12	10	8
	10	1440	36	29	24	21	18	14	12	10
12	11_	1656	41	33	28	24	21	17	14	11
	13	1872	47	37	31	27	23	19	16	12
	15	2160	54	43	36	31	27	22	18	14
	17	2448	61	49	41	35	31	25	20	16
	19	2736	68	55	46	39	34	27	23	18
	6	1372	34	27	23	20	17	14	11	9
	8	1829	46	37	30	26	23	18	15	12
	10	2287	57	46	38	33	29	23	19	15
14	11_	2630	66	53	44	38	33	26	22	18
	13	2973	74	59	50	42	37	30	25	20
	15	3430	86	69	57	49	43	34	29	23
	17	3887	97	78	65	56	49	39	32	26
	19	4345	109	87	72	62	54	43	36	29
	6	2048	51	41	34	29	26	20	17	14
	8	2731	68	55	46	39	34	27	23	18
	10	3413	85	68	57	49	43	34	28	23
16	11_	3925	98	79	65	56	49	39	33	26
	13	4437	111	89	74	63	55	44	37	30
	15	5120	128	102	85	73	64	51	43	34
	17	5803	145	116	97	83	73	58	48	39
	19	6485	162	130	108	93	81	65	54	43
	6	2916	73	58	49	42	36	29	24	19
	8	3888	97	78	65	56	49	39	32	26
	10	4860	122	97	81	69	61	49	41	32
18	11_	5589	140	112	93	80	70	56	47	37
	13	6318	158	126	105	90	79	63	53	42
	15	7290	182	146	122	104	91	73	61	49
	17	8262	207	165	138	118	103	83	69	55
	19	9234	231	185	154	132	115	92	77	62
	6	4000	100	80	67	57	50	40	33	27
	8	5333	133	107	89	76	67	53	44	36
	10	6667	167	133	111	95	83	67	56	44
20	11_	7667	192	153	128	110	96	77	64	51
	13	8667	217	173	144	124	108	87	72	58
	15	10000	250	200	167	143	125	100	83	67
	17	11333	283	227	189	162	142	113	94	76
	19	12667	317	253	211	181	158	127	106	84
	6	5324	133	106	89	76	67	53	44	36
	8	7099	177	142	118	101	89	71	59	47
	10	8873	222	177	148	127	111	89	74	59
22	11_	10204	255	204	170	146	128	102	85	68
	13	11535	288	231	192	165	144	115	96	77
	15	13310	333	266	222	190	166	133	111	89
	17	15085	377	302	251	215	189	151	126	101
	19	16859	421	337	281	241	211	169	141	112

h	b_w	I	Span of beam, ℓ (ft)** 8	10	12	14	16	20	24	30
	8	9216	230	185	155	130	115	90	75	60
	10	11520	290	230	190	165	145	115	95	75
	11_	13248	330	265	220	190	165	130	110	90
24	13	14976	375	300	250	215	185	150	125	100
	15	17280	430	345	290	245	215	175	145	115
	17	19584	490	390	325	280	245	195	165	130
	19	21888	545	440	365	315	275	220	180	145
	21	24192	605	485	405	345	300	240	200	160
	8	11717	295	235	195	165	145	115	100	80
	10	14647	365	295	245	210	185	145	120	100
	11_	16844	420	335	280	240	210	170	140	110
26	13	19041	475	380	315	270	240	190	160	125
	15	21970	550	440	365	315	275	220	185	145
	17	24899	620	500	415	355	310	250	205	165
	19	27892	695	555	465	400	350	280	230	185
	21	30758	770	615	515	440	385	310	255	205
	8	14635	365	295	245	210	185	145	120	100
	10	18293	455	365	305	260	230	185	150	120
	11_	21037	525	420	350	300	265	210	175	140
28	13	23781	595	475	395	340	295	240	200	160
	15	27440	685	550	455	390	345	275	230	185
	17	31099	775	620	520	445	390	310	260	205
	19	34757	870	695	580	495	435	350	290	230
	21	38416	960	770	640	550	480	385	320	255
	8	18000	450	360	300	255	225	180	150	120
	10	22500	565	450	375	320	280	225	190	150
	11_	25875	645	520	430	370	325	260	215	175
30	13	29250	730	585	490	420	365	295	245	195
	15	33750	845	675	565	480	420	340	280	225
	17	38250	955	765	640	545	480	385	320	255
	19	42750	1070	855	715	610	535	430	355	285
	21	47250	1180	945	790	675	590	475	395	315
	8	31104	780	620	520	445	390	310	260	205
	10	38880	970	780	650	555	485	390	325	260
	11_	44712	1120	895	745	640	560	445	375	300
36	13	50544	1260	1010	840	720	630	505	420	335
	15	58320	1460	1170	970	835	730	585	485	390
	17	66096	1650	1320	1100	945	825	660	550	440
	19	73872	1850	1480	1230	1060	925	740	615	490
	21	81648	2040	1630	1360	1170	1020	815	680	545
	8	49392	1230	990	825	705	615	495	410	330
	10	61740	1540	1230	1030	880	770	615	515	410
	11_	71001	1780	1420	1180	1010	890	710	590	475
42	13	80262	2010	1610	1340	1150	1000	805	670	535
	15	92610	2320	1850	1540	1320	1160	925	770	615
	17	104958	2620	2100	1750	1500	1310	1050	875	700
	19	117306	2930	2350	1950	1680	1470	1170	975	780
	21	129654	3240	2590	2160	1850	1620	1300	1080	865
	8	73728	1840	1470	1230	1050	920	735	615	490
	10	92160	2300	1840	1540	1320	1150	920	770	615
	11_	105984	2650	2120	1770	1510	1320	1060	885	705
48	13	119808	3000	2400	2000	1710	1500	1200	1000	800
	15	138240	3460	2760	2300	1970	1730	1380	1150	920
	17	156672	3920	3130	2610	2240	1960	1570	1310	1040
	19	175104	4380	3500	2920	2500	2190	1750	1460	1170
	21	193536	4840	3870	3230	2760	2420	1940	1610	1290
	8	104976	2620	2100	1750	1500	1310	1050	875	700
	10	131220	3280	2620	2190	1880	1640	1310	1090	875
	11_	150903	3770	3020	2510	2160	1890	1510	1260	1010
54	13	170586	4260	3410	2840	2440	2130	1710	1420	1140
	15	196830	4920	3490	3280	2810	2460	1970	1640	1310
	17	223074	5580	4460	3720	3190	2790	2230	1860	1490
	19	249318	6230	4990	4160	3560	3120	2490	2080	1660
	21	275562	6890	5510	4590	3940	3440	2760	2300	1840

* Coefficient 10 introduced to reduce magnitude of relative stiffness values

** Center-to-center distance between supports

Table 2-8 Beam Stiffness Factors

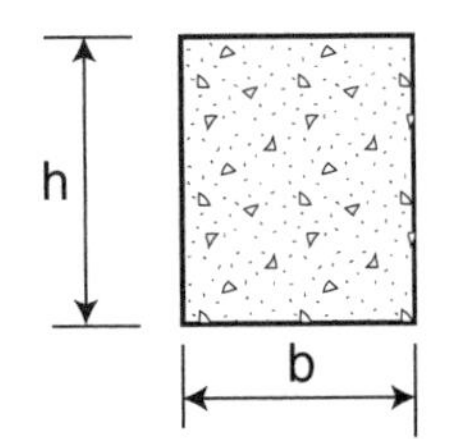

$$I = \frac{bh^3}{12} \qquad K^* = \frac{I}{10\ell_c}$$

Values of K for T-beams

h	b	I	Span of beam, ℓ (ft)** 8	9	10	11	12	14	16	20
	10	427	5	5	4	4	4	3	3	2
	12	512	6	6	5	5	4	4	3	3
	14	597	7	7	6	5	5	4	4	3
8	18	766	10	9	8	7	6	5	5	4
	22	939	12	10	9	9	8	7	6	5
	26	1109	14	12	11	10	9	8	7	6
	30	1280	16	14	13	12	11	9	8	6
	36	1536	19	17	15	14	13	11	10	8
	10	833	10	9	8	8	7	6	5	4
	12	1000	13	11	10	9	8	7	6	5
	14	1167	15	13	12	11	10	8	7	6
10	18	1500	19	17	15	14	13	11	9	8
	22	1833	23	20	18	17	15	13	11	9
	26	2167	27	24	22	20	18	16	14	11
	30	2500	31	28	25	23	21	18	16	13
	36	3000	38	33	30	27	25	21	19	15
	10	1440	18	16	14	13	12	10	9	7
	12	1728	22	19	17	16	14	12	11	9
	14	2016	25	22	20	18	17	14	13	10
12	18	2592	32	29	26	24	22	19	16	13
	22	3168	40	35	32	29	26	23	20	16
	26	3744	47	42	37	34	31	27	23	19
	30	4320	54	48	43	39	36	31	27	22
	36	5184	65	58	52	47	43	37	32	26
	10	2287	29	25	23	21	19	16	14	11
	12	2744	34	30	27	25	23	20	17	14
	14	3201	40	36	32	29	27	23	20	16
14	18	4116	51	46	41	37	34	29	26	21
	22	5031	63	56	50	46	42	36	31	25
	26	5945	74	66	59	54	50	42	37	30
	30	6860	86	76	69	62	57	49	43	34
	36	8232	103	91	82	75	69	59	51	41
	10	3413	43	38	34	31	28	24	21	17
	12	4096	51	46	41	37	34	29	26	20
	14	4779	60	53	48	43	40	34	30	24
16	18	6144	77	68	61	56	51	44	38	31
	22	7509	94	83	75	68	63	54	47	38
	26	8875	111	99	89	81	74	63	55	44
	30	10240	128	114	102	93	85	73	64	51
	36	12288	154	137	123	112	102	88	77	61
	10	4860	61	54	49	44	41	35	30	24
	12	5832	73	65	58	53	49	42	36	29
	14	6804	85	76	68	62	57	49	43	34
18	18	8748	109	97	87	80	73	62	55	44
	22	10692	134	119	107	97	89	76	67	53
	26	12636	158	140	126	115	105	90	79	63
	30	14580	182	162	146	133	122	104	91	73
	36	17496	219	194	175	159	146	125	109	87
	10	6667	83	74	67	61	56	48	42	33
	12	8000	100	89	80	73	67	57	50	40
	14	9333	117	104	93	85	78	67	58	47
20	18	12000	150	133	120	109	100	86	75	60
	22	14667	183	163	147	133	122	105	92	73
	26	17333	217	193	173	158	144	124	108	87
	30	20000	250	222	200	182	167	143	125	100
	36	24000	300	267	240	218	200	171	150	120
	10	8873	111	99	89	81	74	63	55	44
	12	10648	133	118	106	97	89	76	67	53
	14	12422	155	138	124	113	104	89	78	62
22	18	15972	200	177	160	145	133	114	100	80
	22	19521	244	217	195	177	163	139	122	98
	26	23071	288	256	231	210	192	165	144	115
	30	26620	333	296	266	242	222	190	166	133
	36	31944	399	355	319	290	266	228	200	160

h	b	I	Span of beam, ℓ (ft)** 8	9	10	11	12	14	16	20
	12	13824	175	155	140	125	115	100	85	70
	14	16128	200	180	160	145	135	115	100	80
	18	20738	260	230	205	190	175	150	130	105
24	22	25344	315	280	255	230	210	180	160	125
	26	29952	375	335	300	270	250	215	185	150
	30	34560	430	385	345	315	290	245	215	175
	36	41472	520	460	415	375	345	295	260	205
	42	48384	605	540	485	440	405	345	300	240
	12	17576	220	195	175	160	145	125	110	90
	14	20505	255	230	205	185	170	145	130	105
	18	26364	330	295	265	240	220	190	165	130
26	22	32223	405	360	320	295	270	230	200	160
	26	38081	475	425	380	345	315	270	240	190
	30	43940	550	490	440	400	365	315	275	220
	36	52728	660	585	525	480	440	375	330	265
	42	61516	770	685	615	560	515	440	385	310
	12	21952	275	245	220	200	185	155	135	110
	14	25611	320	285	255	235	215	185	160	130
	18	32928	410	365	330	300	275	235	205	165
28	22	40245	505	445	400	365	335	285	250	200
	26	47563	595	530	475	430	395	340	295	240
	30	54880	685	610	550	500	455	390	345	275
	36	65856	825	730	660	600	550	470	410	330
	42	76832	960	855	770	700	640	550	480	385
	12	27000	340	300	270	245	225	195	170	135
	14	31500	395	350	315	285	265	225	195	160
	18	40500	505	450	405	370	340	290	255	205
30	22	49500	620	550	495	450	415	355	310	250
	26	58500	730	650	585	530	490	420	365	295
	30	67500	845	750	675	615	565	480	420	340
	36	81000	1010	900	810	735	675	580	505	405
	42	94500	1180	1050	945	860	790	675	590	475
	12	32768	410	365	330	300	275	235	205	165
	14	38229	480	425	380	350	320	275	240	190
	18	49152	615	545	490	445	410	350	305	245
32	22	60075	750	670	600	545	500	430	375	600
	26	70997	885	790	710	645	590	505	445	355
	30	81920	1020	910	820	745	685	585	510	410
	36	98304	1230	1090	985	895	820	700	615	490
	42	114688	1430	1270	1150	1040	955	820	715	575
	12	39304	490	435	395	355	330	280	245	195
	14	45855	575	510	460	415	380	330	285	230
	18	58956	735	655	590	535	490	420	370	295
34	22	72057	900	800	720	655	600	515	450	360
	26	85159	1060	945	850	775	710	610	530	425
	30	98260	1230	1090	985	895	820	700	615	490
	36	117912	1470	1310	1180	1070	980	840	735	590
	42	137564	1720	1530	1380	1250	1150	985	860	690
	12	46656	585	520	465	425	390	335	290	235
	14	54432	680	605	545	495	455	390	340	270
	18	69984	875	780	700	635	585	500	435	350
36	22	85536	1070	950	855	780	715	610	535	430
	26	101088	1260	1120	1010	920	840	720	630	505
	30	116640	1460	1300	1170	1060	970	835	730	585
	36	139968	1750	1560	1400	1270	1170	1000	875	700
	42	163296	2040	1810	1630	1480	1360	1170	1020	815
	12	54872	685	610	550	500	460	390	345	275
	14	64017	800	710	640	580	535	455	400	320
	18	82308	1030	915	825	750	685	590	515	410
38	22	100599	1260	1120	1010	915	840	720	630	505
	26	118889	1490	1320	1190	1080	990	850	745	595
	30	137180	1710	1520	1370	1250	1140	980	855	685
	36	164616	2060	1830	1650	1500	1370	1180	1030	825
	42	192052	2400	2130	1920	1750	1600	1370	1200	960

* Coefficient 10 introduced to reduce magnitude of relative stiffness values

** Center-to-center distance between supports

Figure 2-10 Correction Factors for Span Moments [2.2]

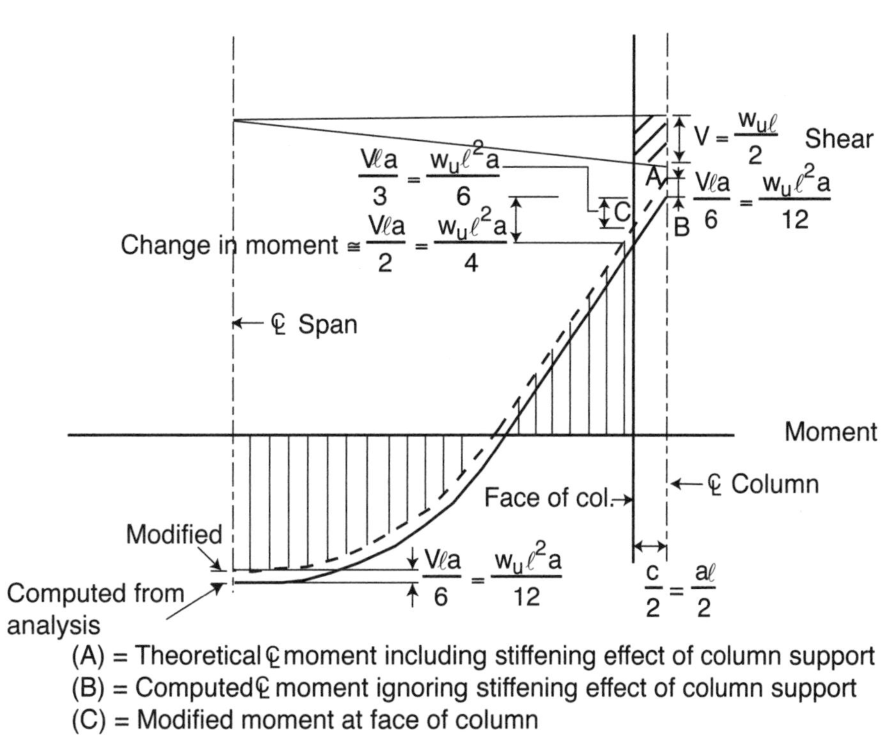

(A) = Theoretical ℄ moment including stiffening effect of column support
(B) = Computed ℄ moment ignoring stiffening effect of column support
(C) = Modified moment at face of column
w_u = uniformly distributed factored load (plf)
ℓ = span length center-to-center of supports
c = width of column support
a = c/ℓ

2.4.4 Two-Cycle Moment Distribution Analysis for Gravity Loading

Reference 2.2 presents a simplified two-cycle method of moment distribution for ordinary building frames. The method meets the requirements for an elastic analysis called for in ACI 8.3 with the simplifying assumptions of ACI 8.6 through 8.9.

The speed and accuracy of the two-cycle method will be of great assistance to designers. For an in-depth discussion of the principles involved, the reader is directed to Reference 2.2.

2.5 COLUMNS

In general, columns must be designed to resist the axial loads and maximum moments from the combination of gravity and lateral loading.

For interior columns supporting two-way construction, the maximum column moments due to gravity loading can be obtained by using ACI Eq. (13-4) (unless a general analysis is made to evaluate gravity load moments from alternate span loading). With the same dead load on adjacent spans, this equation can be written in the following form:

$$M_u = 0.07[w_d(\ell_n^2 - \ell_n'^2) + 0.5w_\ell \ell_n^2]\ell_2$$

where:

w_d = uniformly distributed factored dead load, psf
w_ℓ = uniformly distributed factored live load (including any live load reduction; see Section 2.2.2), psf
ℓ_n = clear span length of longer adjacent span, ft
ℓ'_n = clear span length of shorter adjacent span, ft
ℓ_2 = length of span transverse to ℓ_n and ℓ'_n, measured center-to-center of supports, ft

For equal adjacent spans, this equation further reduces to:

$$M_u = 0.07(0.5 w_\ell \ell_n^2)\ell_2 = 0.035 w_\ell \ell_n^2 \ell_2^2$$

The factored moment M_u can then be distributed to the columns above and below the floor in proportion to their stiffnesses. Since the columns will usually have the same cross-sectional area above and below the floor under consideration, the moment will be distributed according to the inverse of the column lengths.

2.6 LATERAL LOAD ANALYSIS

For frames without shear walls, the lateral load effects must be resisted by the "sway" frame. For low-to-moderate height buildings, lateral load analysis of a sway frame can be performed by either of two simplified methods: the Portal Method or the Joint Coefficient Method. Both methods can be considered to satisfy the elastic frame analysis requirements of the code (ACI 8.3). The two methods differ in overall approach. The Portal Method considers a vertical slice through the entire building along each row of column lines. The method is well suited to the range of building size and height considered in this book, particularly to buildings with a regular rectangular floor plan. The Joint Coefficient Method considers a horizontal slice through the entire building, one floor at a time. The method can accommodate irregular floor plans, and provision is made to adjust for a lateral loading that is eccentric to the centroid of all joint coefficients (centroid of resistance). The Joint Coefficient Method considers member stiffnesses, whereas the Portal Method does not.

The Portal Method is presented in this book because of its simplicity and its intended application to buildings of regular shape. If a building of irregular floor plan is encountered, the designer is directed to Reference 2.2 for details of the Joint Coefficient Method.

2.6.1 Portal Method

The Portal Method considers a two-dimensional frame consisting of a line of columns and their connecting horizontal members (slab-beams), with each frame extending the full height of the building. The frame is considered to be a series of portal units. Each portal unit consists of two story-high columns with connecting slab-beams. Points of contraflexure are assumed at mid-length of beams and mid-height of columns. Figure 2-11 illustrates the portal unit concept applied to the top story of a building frame, with each portal unit shown separated (but acting together).

The lateral load W is divided equally between the three portal units. The shear in the interior columns is twice that in the end columns. In general, the magnitude of shear in the end column is W/2n, and in an interior column it is W/n, where n is the number of bays. For the case shown with equal spans, axial load occurs only in the end columns since the combined tension and compression due to the portal effect results in zero axial loads in the interior columns. Under the assumptions of this method, however, a frame configuration with unequal spans will have axial load in those columns between the unequal spans, as well as in the end columns. The general term for axial load in the end columns in a frame of n bays with unequal spans is:

$$\frac{Wh}{2n\ell_1} \text{ and } \frac{Wh}{2n\ell_n},\ \ell_n = \text{length of bay n}$$

The axial load in the first interior column is:

$$\frac{Wh}{2n\ell_1} - \frac{Wh}{2n\ell_2}$$

and, in the second interior column:

$$\frac{Wh}{2n\ell_2} - \frac{Wh}{2n\ell_3}$$

Column moments are determined by multiplying the column shear with one-half the column height. Thus, for joint B in Fig. 2-11, the column moment is (W/3) (h/2) = Wh/6. The column moment Wh/6 must be balanced by equal moments in beams BA and BC, as shown in Fig. 2-12.

Note that the balancing moment is divided equally between the horizontal members without considering their relative stiffnesses. The shear in beam AB or BC is determined by dividing the beam end moment by one-half the beam length, $(Wh/12)(\ell/2) = Wh/6\ell$.

The process is continued throughout the frame taking into account the story shear at each floor level.

2.6.2 Examples: Wind Load Analyses for Buildings #1 and #2

For Building #1, determine the moments, shears, and axial forces using the Portal Method for an interior frame resulting from wind loads acting in the N-S direction. The wind loads are determined in Section 2.2.1.2.

Moments, shears, and axial forces are shown directly on the frame diagram in Fig. 2-13. The values can be easily determined by using the following procedure:

(1) Determine the shear forces in the columns:

For the end columns:

3rd story: V = 12.0 kips/6 = 2.0 kips
2nd story V = (12.0 kips + 23.1 kips)/6 = 5.85 kips
1st story: V = (12.0 kips + 23.1 kips + 21.7 kips)/6 = 9.50 kips

The shear forces in the interior columns are twice those in the end columns.

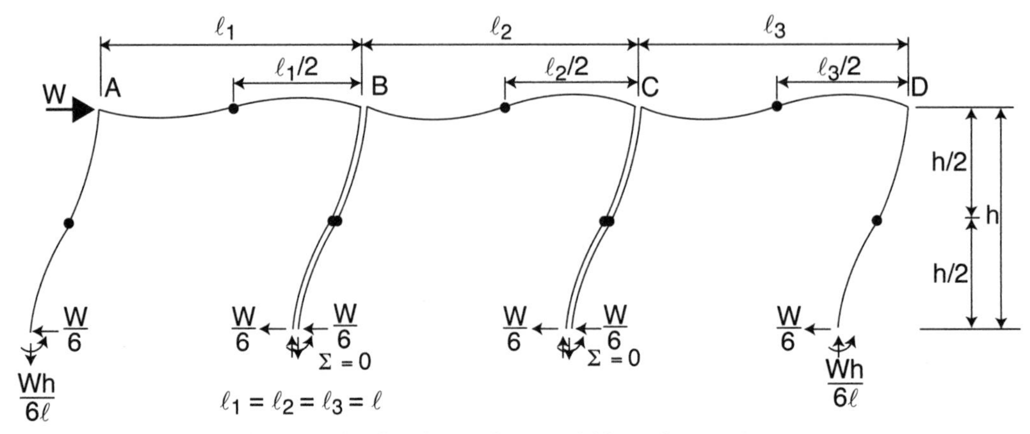

• Assumed inflection point at mid-length members

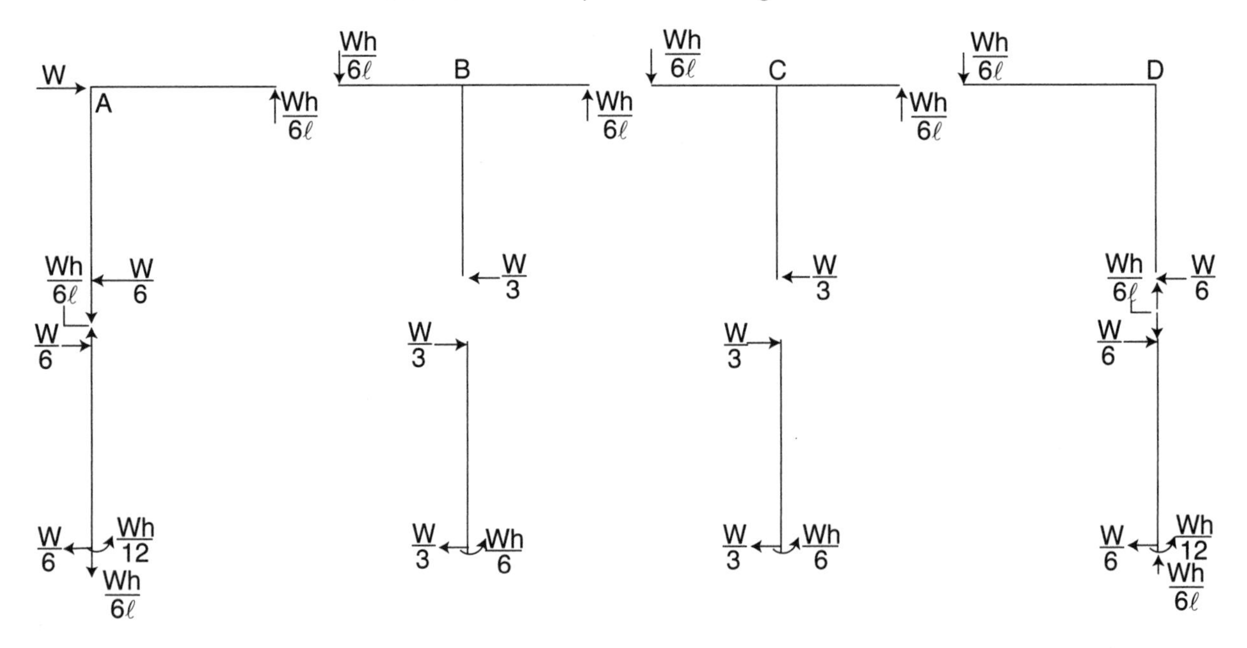

Figure 2-11 Portal Method

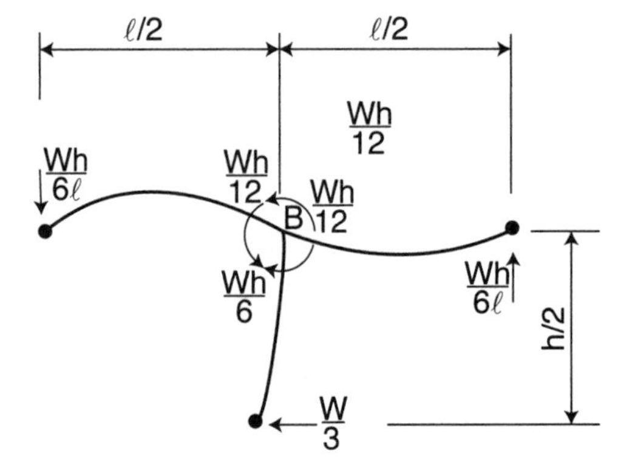

Figure 2-12 Joint Detail

(2) Determine the axial loads in the columns:

For the end columns, the axial loads can be obtained by summing moments about the column inflection points at each level. For example, for the 2nd story columns:

$$\Sigma M = 0 : 12(13 + 6.5) + 23.1\,(6.5) - P\,(90) = 0$$
$$P = 4.27 \text{ kips}$$

For this frame, the axial forces in the interior columns are zero.

(3) Determine the moments in the columns:

The moments can be obtained by multiplying the column shear force by one-half of the column length.

For example, for an exterior column in the 2nd story:

$$M = 5.85(13/2) = 38.03 \text{ ft-kips}$$

(4) Determine the shears and the moments in the beams:
These quantities can be obtained by satisfying equilibrium at each joint. Free-body diagrams for the 2nd story are shown in Fig. 2-14.

As a final check, sum moments about the base of the frame:

$$\Sigma M = 0\text{:} \quad 12.0(39) + 23.1(26) + 21.7(13) - 10.91(90) - 2(61.53 + 123.07) = 0 \qquad \text{(checks)}$$

In a similar manner, the wind load analyses for an interior frame of Building #2 (5-story flat plate), in both the N-S and E-W directions are shown in Figs. 2-15 and 2-16, respectively. The wind loads are determined in Section 2.2.1.1.

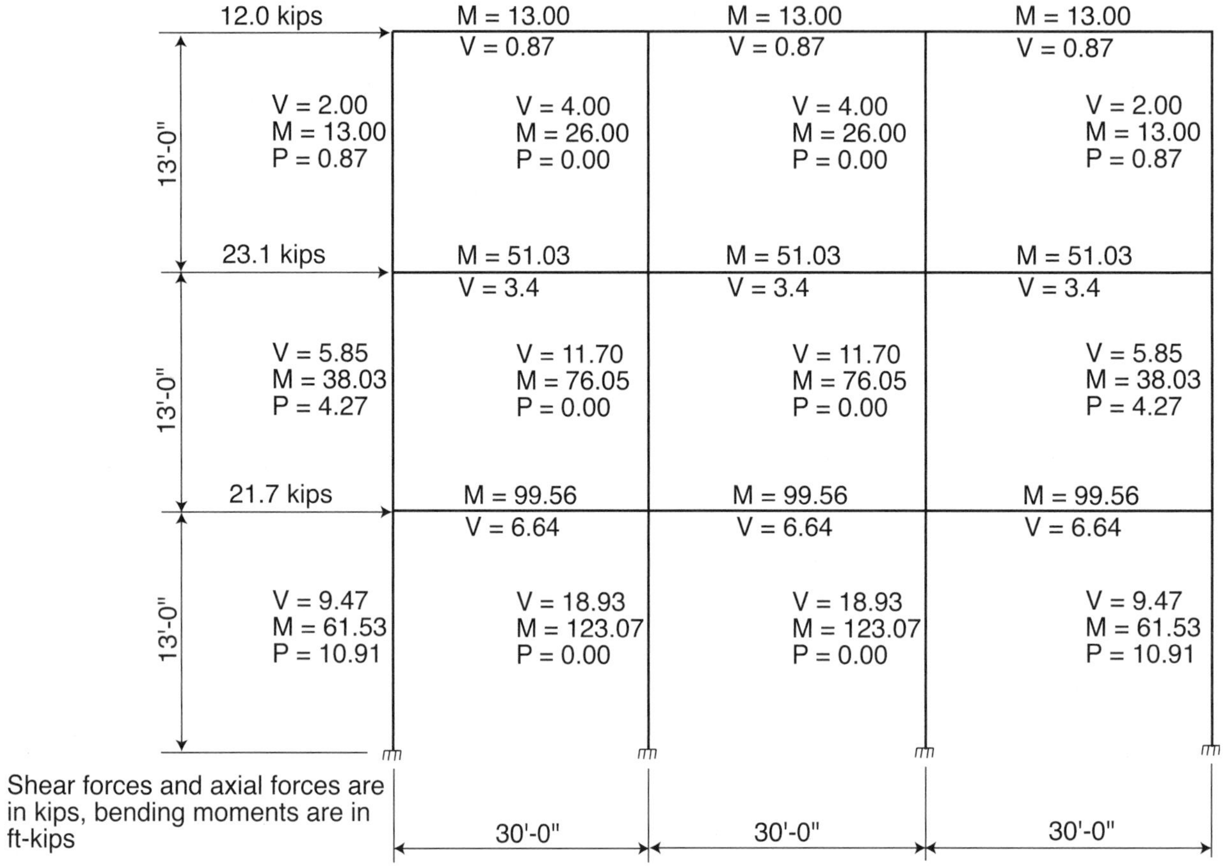

12.0 kips
M = 13.00 V = 0.87 | M = 13.00 V = 0.87 | M = 13.00 V = 0.87

13'-0"
V = 2.00 M = 13.00 P = 0.87 | V = 4.00 M = 26.00 P = 0.00 | V = 4.00 M = 26.00 P = 0.00 | V = 2.00 M = 13.00 P = 0.87

23.1 kips
M = 51.03 V = 3.4 | M = 51.03 V = 3.4 | M = 51.03 V = 3.4

13'-0"
V = 5.85 M = 38.03 P = 4.27 | V = 11.70 M = 76.05 P = 0.00 | V = 11.70 M = 76.05 P = 0.00 | V = 5.85 M = 38.03 P = 4.27

21.7 kips
M = 99.56 V = 6.64 | M = 99.56 V = 6.64 | M = 99.56 V = 6.64

13'-0"
V = 9.47 M = 61.53 P = 10.91 | V = 18.93 M = 123.07 P = 0.00 | V = 18.93 M = 123.07 P = 0.00 | V = 9.47 M = 61.53 P = 10.91

Shear forces and axial forces are in kips, bending moments are in ft-kips

30'-0" | 30'-0" | 30'-0"

Figure 2-13 Shear, Moments and Axial Forces Resulting from Wind Loads for an Interior Frame of Building #1 in the N-S Direction, using the Portal Method

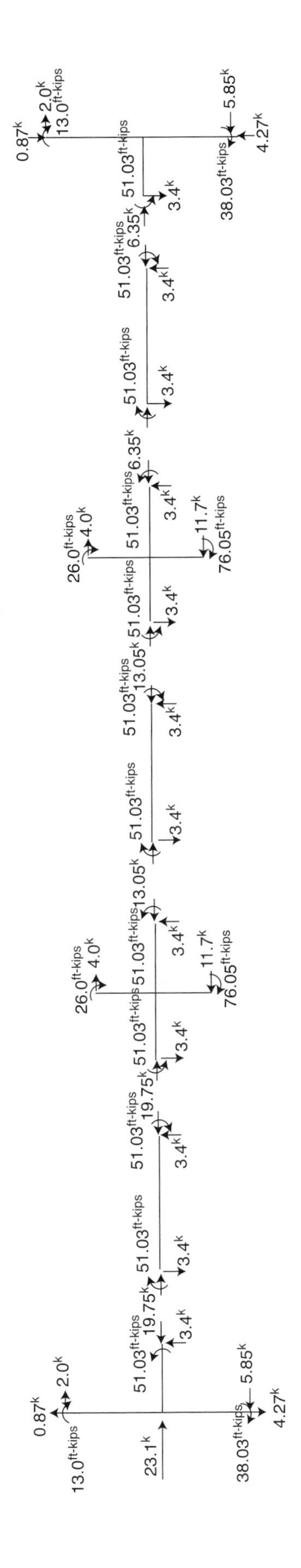

Figure 2-14 Shear Forces, Axial Forces, and Bending Moments at 2nd Story of Building #1

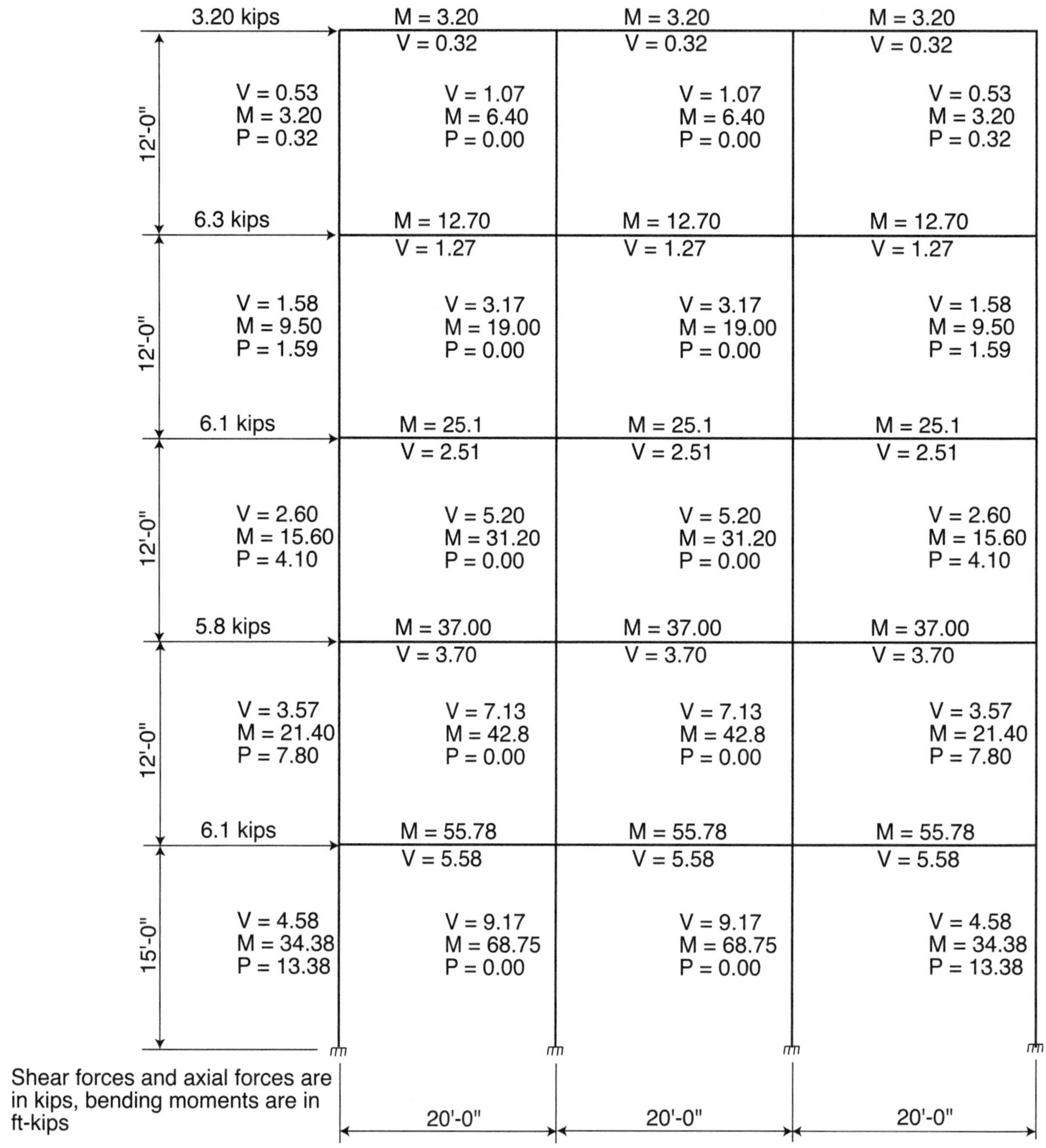

3.20 kips

M = 3.20 V = 0.32 | M = 3.20 V = 0.32 | M = 3.20 V = 0.32

12'-0"

V = 0.53 M = 3.20 P = 0.32 | V = 1.07 M = 6.40 P = 0.00 | V = 1.07 M = 6.40 P = 0.00 | V = 0.53 M = 3.20 P = 0.32

6.3 kips

M = 12.70 V = 1.27 | M = 12.70 V = 1.27 | M = 12.70 V = 1.27

12'-0"

V = 1.58 M = 9.50 P = 1.59 | V = 3.17 M = 19.00 P = 0.00 | V = 3.17 M = 19.00 P = 0.00 | V = 1.58 M = 9.50 P = 1.59

6.1 kips

M = 25.1 V = 2.51 | M = 25.1 V = 2.51 | M = 25.1 V = 2.51

12'-0"

V = 2.60 M = 15.60 P = 4.10 | V = 5.20 M = 31.20 P = 0.00 | V = 5.20 M = 31.20 P = 0.00 | V = 2.60 M = 15.60 P = 4.10

5.8 kips

M = 37.00 V = 3.70 | M = 37.00 V = 3.70 | M = 37.00 V = 3.70

12'-0"

V = 3.57 M = 21.40 P = 7.80 | V = 7.13 M = 42.8 P = 0.00 | V = 7.13 M = 42.8 P = 0.00 | V = 3.57 M = 21.40 P = 7.80

6.1 kips

M = 55.78 V = 5.58 | M = 55.78 V = 5.58 | M = 55.78 V = 5.58

15'-0"

V = 4.58 M = 34.38 P = 13.38 | V = 9.17 M = 68.75 P = 0.00 | V = 9.17 M = 68.75 P = 0.00 | V = 4.58 M = 34.38 P = 13.38

Shear forces and axial forces are in kips, bending moments are in ft-kips

20'-0" | 20'-0" | 20'-0"

Figure 2-15 Shears, Moments, and Axial Forces Resulting from Wind Loads for an Interior Frame of Building #2 in the N-S Direction, using the Portal Method

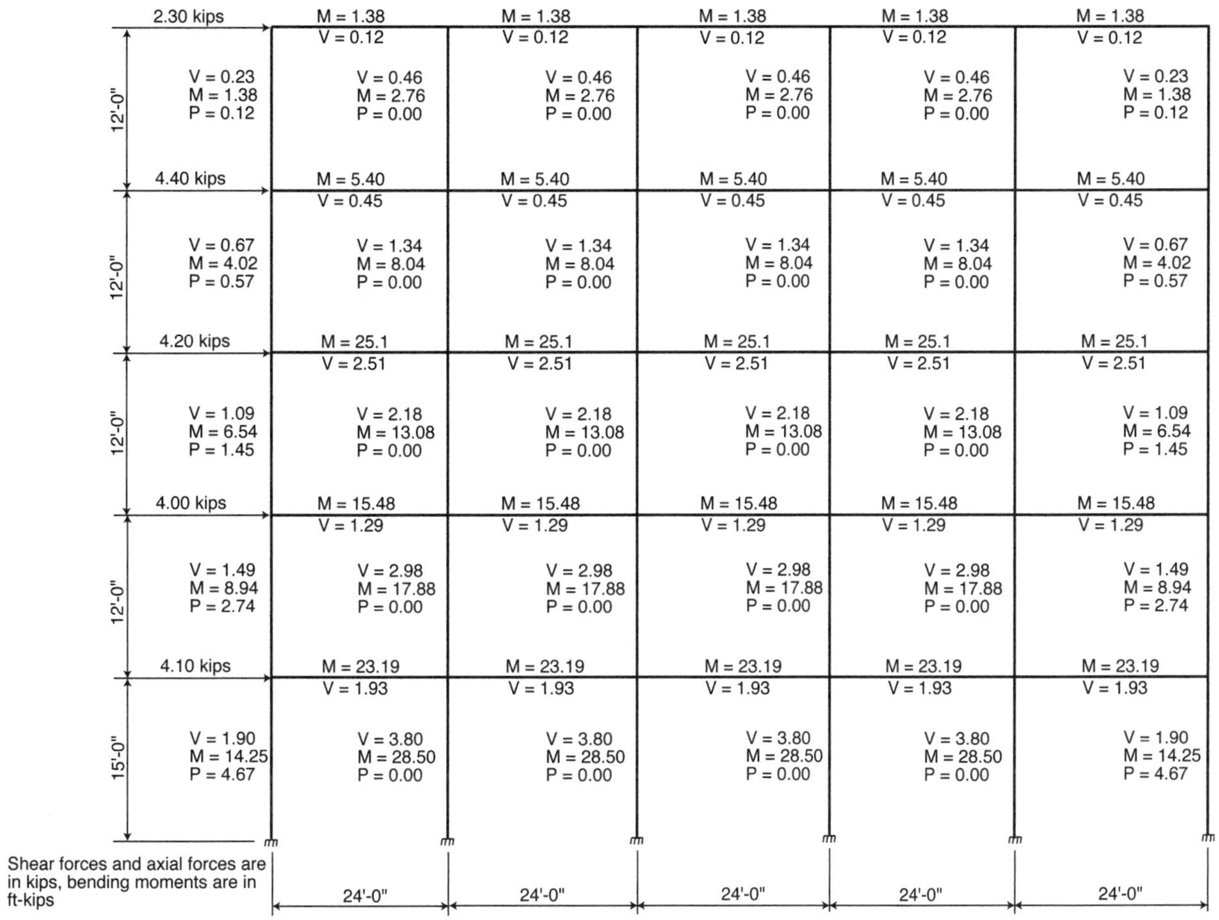

Figure 2-16 Shears, Moments, and Axial Forces Resulting from Wind Loads for an Interior Frame of Building #2 in the E-W Direction, using the Portal Method

References

2.1 *American Society of Civil Engineers Minimum Design Loads for Buildings and Other Structures*, ASCE 7-02, American Society of Civil Engineers, New York, N.Y., 2003.

2.2 *Continuity in Concrete Building Frames*, Portland Cement Association, Skokie, EB033, 1959, 56 pp.

Chapter 3

Simplified Design for Beams and Slabs

3.1 INTRODUCTION

The simplified design approach for proportioning beams and slabs (floor and roof members) is based in part on published articles,[3.1-3.6] and in part on simplified design aid material published by CRSI.[3.7,3.10] Additional data for design simplification are added where necessary to provide the designer with a total simplified design approach for beam and slab members. The design conditions that need to be considered for proportioning the beams and slabs are presented in the order generally used in the design process.

The simplified design procedures comply with the ACI 318-02 code requirements for both member strength and member serviceability. The simplified methods will produce slightly more conservative designs within the limitations noted. All coefficients are based on the Strength Design Method, using appropriate load factors and strength reduction factors specified in ACI 318. Where simplified design requires consideration of material strengths, 4000 psi concrete and Grade 60 reinforcement are used. The designer can easily modify the data for other material strengths.

The following data are valid for reinforced concrete flexural members with f'_c = 4000 psi and f_y = 60,000 psi:

modulus of elasticity for concrete	E_c = 3,600,000 psi	(ACI 8.5.1)
modulus of elasticity for rebars	E_s = 29,000,000 psi	(ACI 8.5.2)
minimum reinforcement ratio (beams, joists)	$\rho_b = 0.0033$	(ACI 10.5.1)
minimum reinforcement ratio (slabs)	$\rho_{min} = 0.0018$	(ACI 7.12.2) and (ACI 10.5.4)
maximum reinforcement ratio*	$\rho_{max} = 0.0206$	(ACI 10.3.5)
maximum useful reinforcement ratio**	$\rho_t = 0.0181$	(ACI 10.3.4)

3.2 DEPTH SELECTION FOR CONTROL OF DEFLECTIONS

Deflection of beams and one-way slabs need not be computed if the overall member thickness meets the minimum specified in ACI Table 9.5(a). Table 9.5(a) may be simplified to four values as shown in Table 3-1. The quantity is the clear span length for cast-in-place beam and slab construction. For design convenience, minimum thicknesses for the four conditions are plotted in Fig. 3-1.

Deflections are not likely to cause problems when overall member thickness meets or exceeds these values for uniform loads commonly used in the design of buildings. The values are based primarily on experience and are not intended to apply in special cases where beam or slab spans may be subject to heavily distributed loads or concen-

* *The ACI 318-02 does not provide direct maximum reinforcement ratio for beams and slabs. The maximum reinforcement ratio was derived from the strain profile as limited in ACI Section 10.3.5.*

** *For tension controlled sections with $\phi = 0.9$*

traded loads. Also, they are not intended to apply to members supporting or attached to nonstructural elements likely to be damaged by deflections (ACI 9.5.2.1).

Table 3-1 Minimum Thickness for Beams and One-Way Slabs

Beams and One-way Slabs	Minimum h
Simple Span Beams or Joists*	$\ell_n/16$
Continuous Beams or Joists	$\ell_n/18.5$
Simple Span Slabs*	$\ell_n/20$
Continuous Slabs	$\ell_n/24$

*Minimum thickness for cantilevers can be considered equal to twice that for a simple span.

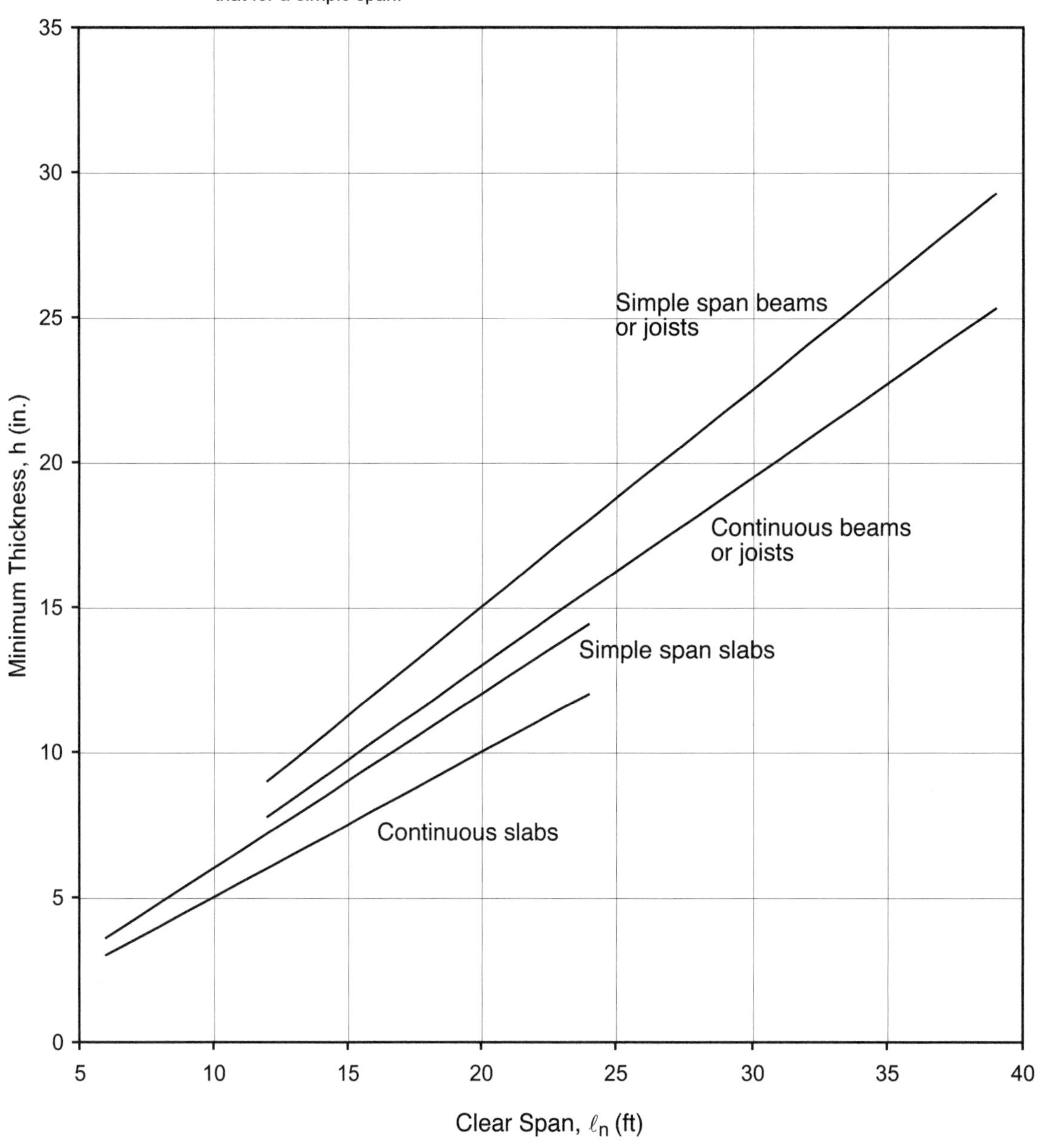

Figure 3-1 Minimum Thicknessees for Beams and One-Way Slabs

For roof beams and slabs, the values are intended for roofs subjected to normal snow or construction live loads only, and with minimal water ponding or drifting snow problems.

Prudent choice of steel percentage can also minimize deflection problems. Members will usually be of sufficient size, so that deflections will be within acceptable limits, when the tension reinforcement ratio ρ used in the positive moment regions does not exceed approximately one-half of the maximum value permitted. For f'_c = 4000 psi and f_y = 60,000 psi, one-half of ρ_{max} is approximately one percent (0.01).

Depth selection for control of deflections of two-way slabs is given in Chapter 4.

As a guide, the effective depth d can be calculated as follows:

For beams with one layer of bars	$d = h - (\cong 2.5 \text{ in.})$
For joists and slabs	$d = h - (\cong 1.25 \text{ in.})$

3.3 MEMBER SIZING FOR MOMENT STRENGTH

A simplified sizing equation can be derived using the strength design data developed in Chapter 6 of Reference 3.8. For our selected materials (f'_c = 4000 psi and f_y = 60,000 psi), the maximum reinforcement ratio $\rho_b = 0.0206$. As noted above, deflection problems are rarely encountered with beams having a reinforcement ratio ρ equal to about one-half of the maximum permitted.

Set $\rho = 0.5\rho_{max} = 0.0103$

$$Mu = \phi\, A_s f_y (d - a/2) = \phi \rho b d f_y (d - a/2)$$

$$a = A_s f_y / 0.85 f'_c b = \rho d f_y / 0.85 f'_c$$

$$M_u / \phi b d^2 = \rho f_y \left(1 - \frac{0.5 \rho f_y}{0.85 f'_c}\right) = R_n$$

$$R_n = \rho f_y \left(1 - \frac{0.5 \rho f_y}{0.85 f'_c}\right)$$

$$= 0.0103 \times 60{,}000 \left[\left(1 - \frac{0.5 \times 0.0103 \times 60{,}000}{0.85 \times 4000}\right)\right]$$

$$= 562 \text{ psi}$$

$$bd^2_{reqd} = \frac{M_u}{\varphi R_n} = \frac{M_u \times 12 \times 1000}{0.9 \times 562} = 23.7 M_u$$

For simplicity, set $bd^2_{reqd} = 20M_u$.

For f'_c = 4000 psi and f_y = 60,000 psi:

$$bd^2(\rho \cong 0.5\rho_{max}) = 20M_u$$

where M_u is in ft-kips and b and d are in inches

A similar sizing equation can be derived for other material strengths.

With factored moments M_u and effective depth d known, the required beam width b is easily determined using the sizing equation $bd^2 = 20M_u$. When frame moments vary, b is usually determined for the member which has the largest M_u; for economy, this width may be used for all similar members in the frame. Since slabs are designed by using a 1-ft strip (b = 12 in.), the sizing equation can be used to check the initial depth selected for slabs; it simplifies to $d = 1.3\sqrt{M_u}$.

If the depth determined for control of deflections is shallower than desired, a larger depth may be selected with a corresponding width b determined from the above sizing equation. Actually, any combination of b and d could be determined from the sizing equation with the only restriction being that the final depth selected must be greater than that required for deflection control (Table 3-1).

It is important to note that for minimum beam size with maximum reinforcement, the sizing equation becomes $bd^2_{min} = 14.6M_u$.

3.3.1 Notes on Member Sizing for Economy

- Use whole inches for overall beam dimensions; slabs may be specified in 1/2-in. increments.
- Use beam widths in multiples of 2 or 3 inches, such as 10, 12, 14, 16, 18, etc.
- Use constant beam size from span to span and vary reinforcement as required.
- Use wide flat beams (same depth as joist system) rather than narrow deep beams.
- Use beam width equal to or greater than the column width.
- Use uniform width and depth of beams throughout the building.

See also Chapter 9 for design considerations for economical formwork.

3.4 DESIGN FOR FLEXURAL REINFORCEMENT

A simplified equation for the area of tension steel As can be derived using the strength design data developed in Chapter 6 of Reference 3.8. An approximate linear relationship between R_n and ρ can be described by an equation in the form $M_n/bd^2 = \rho$ (constant), which readily converts to $A_s = M_u/\phi d$(constant). This linear equation for A_s is reasonably accurate up to about two-thirds of the maximum ρ. For $f'_c = 4000$ psi and $\rho = 2/3\ \rho_{rmax}$, the constant for the linear approximation is :

$$\frac{f_y}{12{,}000*}\left[1-\frac{0.5\rho f_y}{0.85f'_c}\right] = \frac{60{,}000}{12{,}000}\left[1-\frac{0.5(2/3\times 0.0206)(60)}{0.85\times 4}\right] = 4.39$$

*To convert M_u from ft-kips to in.-lbs

Therefore,

$$A_s = \frac{M_u}{\phi d(\text{constant})} = \frac{M_u}{0.9\times 4.39\times d} = \frac{M_u}{3.95d} \cong \frac{M_u}{4d}$$

For $f'_c = 4000$ psi and $f_y = 60{,}000$ psi:

$$A_s = \frac{M_u}{4d}**$$

**Note: this equation is in mixed units: M_u is in ft-kips, d is in in. and A_s is in sq in.

For all values of $\rho < 0.0125$, the simplified A_s equation is slightly conservative. The maximum deviation in A_s is less than + 10% at the minimum and maximum useful tension steel ratios.[3.9] For members with reinforcement ratios in the range of approximately 1% to 1.5%, the error is less than 3%.

The range of change in the constant in the denominator in the above equation is very narrow for different concrete strengths; this allows the use of the above equation for approximate reinforcement area estimination with other concrete strengths.

The simplified A_s equation is applicable for rectangular cross sections with tension reinforcement only. Members proportioned with reinforcement in the range of 1% to 1.5% will be well within the code limits for singly reinforced members. For positive moment reinforcement in flanged floor beams, A_s is usually computed for a rectangular compression zone; rarely will A_s be computed for a T-shaped compression zone.
The depth of the rectangular compression zone, a, is given by:

$$a = \frac{A_s f_y}{0.85 f'_c b_c}$$

where b_e = effective width of slab as a T-beam flange (ACI 8.10).

The flexural member is designed as a rectangular section whenever $h_f \geq a$ where h_f is the thickness of the slab (i.e., flange thickness).

3.5 REINFORCING BAR DETAILS

The minimum and maximum number of reinforcing bars permitted in a given cross section is a function of cover and spacing requirements given in ACI 7.6.1 and ACI 3.3.2 (minimum spacing for concrete placement), ACI 7.7.1 (minimum cover for protection of reinforcement), and ACI 10.6 (maximum spacing for control of flexural cracking). Tables 3-2 and 3-3 give the minimum and maximum number of bars in a single layer for beams of various widths; selection of bars within these limits will provide automatic code conformance with the cover and spacing requirements.

Table 3-2 Minimum Number of Bars in a Single Layer (ACI 10.6)

BAR	BEAM WIDTH (in.)												
SIZE	12	14	16	18	20	22	24	26	28	30	36	42	48
No. 4	2	2	3	3	3	3	3	4	4	4	5	5	6
No. 5	2	2	3	3	3	3	3	4	4	4	5	5	6
No. 6	2	2	3	3	3	3	3	4	4	4	5	5	6
No. 7	2	2	3	3	3	3	3	4	4	4	5	5	6
No. 8	2	2	3	3	3	3	3	4	4	4	5	5	6
No. 9	2	2	3	3	3	3	3	4	4	4	5	5	6
No. 10	2	2	3	3	3	3	3	4	4	4	5	5	6
No. 11	2	2	3	3	3	3	3	4	4	4	5	5	6

The values in Tables 3-2 are based on a cover of 2 in. to the main flexural reinforcement (i.e., 1.5 in. clear cover to the stirrups plus the diameter of a No. 4 stirrup). In general, the following equations can be used to determine the minimum number of bars n in a single layer for any situation (see Fig. 3-2):

$$n_{min} = \frac{b_w - 2(c_c + 0.5d_b)}{S} + 1$$

where

$$S = \frac{540}{f_s} - 2.5c_c \leq 12\left(\frac{36}{f_s}\right)$$

Table 3-3 Maximum Number of Bars in a Single Layer

BAR SIZE	BEAM WIDTH (in.)												
	12	14	16	18	20	22	24	26	28	30	36	42	48
No. 4	5	6	8	9	10	12	13	14	16	17	21	25	29
No. 5	5	6	7	8	10	11	12	13	15	16	19	23	27
No. 6	4	6	7	8	9	10	11	12	14	15	18	22	25
No. 7	4	5	6	7	8	9	10	11	12	13	17	20	23
No. 8	4	5	6	7	8	9	10	11	12	13	16	19	22
No. 9	3	4	5	6	7	8	8	9	10	11	14	17	19
No. 10	3	4	4	5	6	7	8	8	9	10	12	15	17
No. 11	3	3	4	5	5	6	7	8	8	9	11	13	15

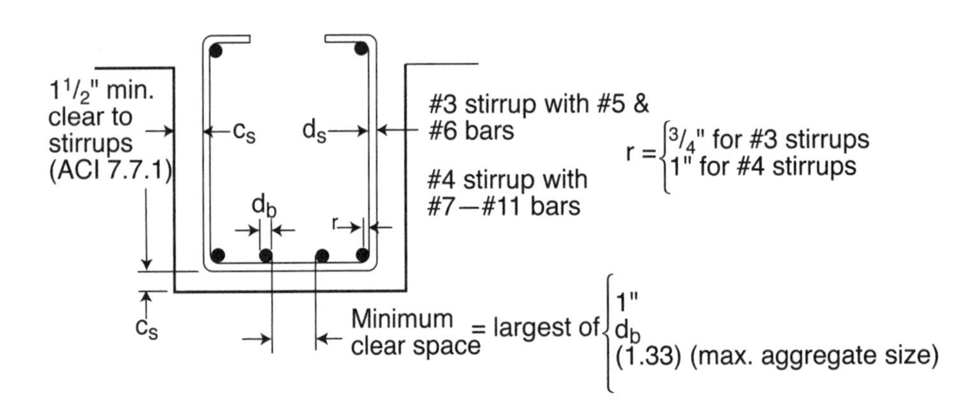

Figure 3-2 Cover and Spacing Requirements for Tables 3-2 and 3-3

where

b_w = beam width, in.

d_c = clear cover to tension reinforcement, in.

d_s = clear cover to stirrups, in.

d_b = diameter of main flexural bar, in.

The values obtained from the above equations should be rounded up to the next whole number.
The values in Table 3-3 can be determined from the following equation:

$$n = 1 + \frac{b_w - 2(c_s + d_s + r)}{(\text{minimum clear space}) + d_b}$$

where

$$r = \begin{cases} 3/4 \text{ in. for No. 3 stirrups} \\ 1 \text{ in. for No. 4 stirrups} \end{cases}$$

The minimum clear space between bars is defined in Fig. 3-2. The above equation can be used to determine the maximum number of bars in any general case; computed values should be rounded down to the next whole number.

Suggested temperature and shrinkage reinforcement for one-way floor and roof slabs is given in Table 3-4. The provided area of reinforcement (per foot width of slab) satisfies ACI 7.12.2. Bar spacing must not exceed 5h and 18 in. (where h = thickness of slab). The same area of reinforcement is also applied for minimum moment reinforcement in one-way slabs (ACI 10.5.4) at a maximum spacing of 3h, not to exceed 18 in. (ACI 7.6.5). As noted in Chapter 4, this same minimum area of steel applies for flexural reinforcement in each direction for two-way floor and roof slabs; in this case, the maximum spacing is 2h not to exceed 18 in. (ACI 13.3).

As an aid to designers, reinforcing bar data are presented in Tables 3-5 and 3-6.

See Chapter 8, Section 8.2, for notes on reinforcement selection and placement for economy.

Table 3-4 Temperature Reinforcement for One-Way Slabs

Slab Thickness h (in.)	A_s (req'd)* (in.^{2}/ft)	Suggested Reinforcement**
3-½	0.08	#3@16
4	0.09	#3@15
4-½	0.10	#3@13
5	0.11	#3@12
5-½	0.12	#4@18
6	0.13	#4@18
6-½	0.14	#4@17
7	0.15	#4@16
7-½	0.16	#4@15
8	0.17	#4@14
8-½	0.18	#4@13
9	0.19	#4@12
9-½	0.21	#5@18
10	0.22	#5@17

*A_s = 0.0018bh = 0.022h (ACI 7.12.2).
**For minimum moment reinforcement, bar spacing must not exceed 3h or 18 in. (ACI 7.6.5). For 3½ in. slab, use #3@10 in.; for 4 in. slab, use #3@12 in.; for 5½ in. slab, use #3@11 in. or #4 @16 in.

Table 3-5 Total Areas of Bars—A_s(in.^{2})

Slab Thickness h (in.)	A_s (req'd)* (in.^{2}/ft)	Suggested Reinforcement**
3-½	0.08	#3@16
4	0.09	#3@15
4-½	0.10	#3@13
5	0.11	#3@12
5-½	0.12	#4@18
6	0.13	#4@18
6-½	0.14	#4@17
7	0.15	#4@16
7-½	0.16	#4@15
8	0.17	#4@14
8-½	0.18	#4@13
9	0.19	#4@12
9-½	0.21	#5@18
10	0.22	#5@17

*A_s = 0.0018bh = 0.022h (ACI 7.12.2).

**For minimum moment reinforcement, bar spacing must not exceed 3h or 18 in. (ACI 7.6.5). For 3 1/2 in. slab, use #3@10 in.; for 4 in. slab, use #3@12 in.; for 5 1/2 in. slab, use #3@11 in. or #4@16 in.

Table 3-6 Areas of Bars per Foot Width of Slab—A_s(in.^{2}/ft)

Bar size	Bar diameter (in.)	Number of bars							
		1	2	3	4	5	6	7	8
#3	0.375	0.11	0.22	0.33	0.44	0.55	0.66	0.77	0.88
#4	0.500	0.20	0.40	0.60	0.80	1.00	1.20	1.40	1.60
#5	0.625	0.31	0.62	0.93	1.24	1.55	1.86	2.17	2.48
#6	0.750	0.44	0.88	1.32	1.76	2.20	2.64	3.08	3.52
#7	0.875	0.60	1.20	1.80	2.40	3.00	3.60	4.20	4.80
#8	1.000	0.79	1.58	2.37	3.16	3.95	4.74	5.53	6.32
#9	1.128	1.00	2.00	3.00	4.00	5.00	6.00	7.00	8.00
#10	1.270	1.27	2.54	3.81	5.08	6.35	7.62	8.89	10.16
#11	1.410	1.56	3.12	4.68	6.24	7.80	9.36	10.92	12.48

Bar size	Bar spacing (in.)												
	6	7	8	9	10	11	12	13	14	15	16	17	18
#3	0.22	0.19	0.17	0.15	0.13	0.12	0.11	0.10	0.09	0.09	0.08	0.08	0.07
#4	0.40	0.34	0.30	0.27	0.24	0.22	0.20	0.18	.017	0.16	0.15	0.14	0.13
#5	0.62	0.53	0.46	0.41	0.37	0.34	0.31	0.29	0.27	0.25	0.23	0.22	0.21
#6	0.88	0.75	0.66	0.59	0.53	0.48	0.44	0.41	0.38	0.35	0.33	0.31	0.29
#7	1.20	1.03	0.90	0.80	0.72	0.65	0.60	0.55	0.51	0.48	0.45	0.42	0.40
#8	1.58	1.35	1.18	1.05	0.95	0.86	0.79	0.73	0.68	0.63	0.59	0.56	0.53
#9	2.00	1.71	1.50	1.33	1.20	1.09	1.00	0.92	0.86	0.80	0.75	0.71	0.67
#10	2.54	2.18	1.91	1.69	1.52	1.39	1.27	1.17	1.09	1.02	0.95	0.90	0.85
#11	3.12	2.67	2.34	2.08	1.87	1.70	1.56	1.44	1.34	1.25	1.17	1.10	1.04

3.6 DESIGN FOR SHEAR REINFORCEMENT

In accordance with ACI Eq. (11-2), the total shear strength is the sum of two components: shear strength provided by concrete (ϕV_c) and shear strength provided by shear reinforcement (ϕV_s). Thus, at any section of the member, $V_u \leq \phi V_c + \phi V_s$. Using the simplest of the code equations for shear strength of concrete, V_c, specific values can be assigned to the two resisting components for a given set of material parameters and a specific cross section. Table 3-7 summarizes ACI 318 provisions for shear design.

Table 3-7 ACI Provisions for Shear Design (f'_c = 4000 psi)

		$V_u \leq \phi V_c / 2^*$	$\phi V_c \geq V_u > \phi V_c / 2$	$V_u > \phi V_c$
Required area of stirrups, A_v**		none	$\dfrac{50 b_w s}{f_y}$	$\dfrac{(V_u - \phi V_c)s}{\phi f_y d} \geq \dfrac{50 b_w s}{f_y}$
Stirrup spacing, s	Required	—	$\dfrac{A_v f_y}{50 b_w}$	$\dfrac{\phi A_v f_y d}{V_u - \phi V_c} \geq \dfrac{A_v f_y}{50 b_w}$
	Maximum***	—	$d/2 \leq 24$ in	$d/2 \leq 24$ in. for $(V_u - \phi V_c) \leq \phi 4\sqrt{f'_c} b_w d$
				$d/4 \leq 12$ in. for $(V_u - \phi V_c) > \phi 4\sqrt{f'_c} b_w d$

* Members subjected to shear and flexure only; $\phi V_c = \phi 2 \sqrt{f'_c} b_w d$, ϕ = 0.75 (ACI 11.3.1.1)
** $A_v = 2 \times A_b$ for U stirrups; $f_y \leq 60$ ksi (ACI 11.5.2)
*** Maximum spacing based on minimum shear reinforcement (= $A_v f_y / 50 b_w$) must also be considered (ACI 11.5.5.3).

The selection and spacing of stirrups can be simplified if the spacing is expressed as a function of the effective depth d (see Reference 3.3). According to ACI 11.5.4.1 and ACI 11.5.4.3, the practical limits of stirrup spacing vary from s = d/2 to s = d/4, since spacing closer than d/4 is not economical. With one intermediate spacing at d/3, the calculation and selection of stirrup spacing is greatly simplified. Using the three standard stirrup spacings noted above (d/2, d/3, and d/4), a specific value of ϕV_s can be derived for each stirrup size and spacing as follows:

For vertical stirrups:

$$\phi V_s = \frac{\phi A_v f_y d}{s} \qquad \text{ACI Eq. (11.15)}$$

By substituting d/n for s (where n = 2, 3, or 4), the above equation can be rewritten as:

$$\phi V_s = \phi A_v f_y n$$

Thus, for No.3 U-stirrups @ s = d/2 with f_y = 60,000 psi and ϕ = 0.75

$$\phi V_s = 0.75(0.22)60 \times 2 = 19.8 \text{ kips, say 20 kips}$$

The values ϕV_s given in Table 3-8 may be used to select shear reinforcement with Grade 60 rebars.

Table 3-8 Values of ϕV_s (f_y = 60 ksi) – reconstructed considering fi

s	#3 U-stirrups	#4 U-stirrups	#5 U-stirrups
d/2	20 kips	36 kips	56 kips
d/3	30 kips	54 kips	84 kips
d/4	40 kips	72 kips	112 kips

*Valid for stirrups with 2 legs (double the tabulated values for 4 legs, etc.)

It should be noted that these values of ϕV_s are not dependent on the member size nor on the concrete strength. The following design values are valid for f'_c = 4000 psi:

$$\text{Maximum } (\phi V_c + \phi V_s) = \phi 10\sqrt{f'_c}b_w d = 0.48\ b_w d \qquad \text{(ACI 11.5.6.8)}$$

$$\phi V_c = \phi 2\sqrt{f'_c}b_w d = 0.095\ b_w d \qquad \text{(ACI 11.3.1.1)}$$

$$\phi V_c / 2 = \phi\sqrt{f'_c}b_w d = 0.048\ b_w d \qquad \text{(ACI 11.5.5.1)}$$

Joists defined by ACI 8.11:

$$\phi V_c = \phi 2.2\sqrt{f'_c}b_w d = 0.104\ b_w d \qquad \text{(ACI 8.11.8)}$$

In the above equations, b_w and d are in inches and the resulting shears are in kips.

The design charts in Figs. 3-3 through 3-6 offer another simplified method for shear design. By entering the charts with values of d and $\phi V_s = V_u - \phi V_c$ for the member at the section under consideration, the required stirrup spacing can be obtained by locating the first line above the point of intersection of d and ϕV_s. Values for spacing not shown can be interpolated from the charts if desired. Also given in the charts the values for the minimum practical beam widths b_w that correspond to the maximum allowable $\phi V_s = \phi 8\sqrt{f'_c}b_w d$ for each given spacing s; any member which has at least this minimum b_w will be adequate to carry the maximum applied V_u. Fig. 3-6 can also be used to quickly determine if the dimensions of a given section are adequate: any member with an applied V_u which is less than the applicable $V_{u(max)}$ can carry this shear without having to increase the values of b_w and/or d. Once the adequacy of the cross-section has been verified, the stirrup spacing can be established by using Figs. 3-3 through 3-6. This spacing must then be checked for compliance with all maximum spacing criteria.

3.6.1 Example: Design for Shear Reinforcement

The example shown in Fig. 3-8 illustrates the simple procedure for selecting stirrups using design values for V_c and V_s.

(1) Design data: $f'_c = 4000$ psi, $f_y = 60{,}000$ psi, $w_u = 7$ kips/ft.

(2) Calculations:

V_u @ column centerline:	$w_u\ell/2 = 7 \times 24/2 = 84.0$ kips
V_u @ face of support:	$84 - 1.17(7) = 75.8$ kips
V_u @ d from support face (critical section):	$75.8 - 2(7) = 61.8$ kips
$(\phi V_c + \phi V_s)_{max}$:	$0.48\ b_wd = 0.48(12)(24) = 138.2$ kips
ϕV_c:	$0.095\ b_wd = 0.095(12)(24) = 27.4$ kips
$\phi V_c/2$:	$0.048\ b_wd = 0.048(12)(24) = 13.80$ kips

(3) Beam size is adequate for shear strength, since 138.2 kips > 61.8 kips (also see Fig. 3-6). ϕV_s (required) = 61.8 – 27.4 = 34.4 kips. From Table 3-8, No.4 @ d/2 = 12 in. is adequate for full length where stirrups are required since ϕV_s = 36 kips > 34.4 kips. Length over which stirrups are required is (75.8 – 13.8)/7 = 8.86 ft from support face.

Check maximum stirrup spacing:

$$\phi 4\sqrt{f'_c}\,b_wd = 0.19b_wd = 54.6 \text{ kips}$$

Since ϕV_s = 36 kips < 54.6 kips, the maximum spacing is the least of the following:

$$s_{max} = \begin{cases} d/2 = 12 \text{ in. (governs)} \\ 24 \text{ in.} \\ A_vf_y/50b_w = 40 \text{ in.} \end{cases}$$

Use 10-No.4 U-stirrups at 12 in. at each end of beam.

The problem may also be solved graphically as shown in Fig. 3-8. ϕV_s for No.3 stirrups at d/2, d/3, and d/4 are scaled vertically from ϕV_c. The horizontal intersection of the ϕV_s values (20 kips, 30 kips, and 40 kips) with the shear diagram automatically sets the distances where the No.3 stirrups should be spaced at d/2, d/3, and d/4. The exact numerical values for these horizontal distances are calculated as follows (although scaling from the sketch is close enough for practical design):

No.3	@ d/4 = 6 in.:	(75.8 – 57.4)/7 = 2.63 ft (31.5 in.)	use 6 @ 6 in.
	@ d/3 = 8 in.:	(57.4 – 47.4)/7 = 1.43 ft (17.0 in.)	use 2 @ 8 in.
	@ d/2 = 12 in.:	(47.4 – 13.8)/7 = 4.8 ft (57.6 in.)	use 5 @ 12 in.

A more practical solution may be to eliminate the 2 @ 8 in. and use 9 @ 6 in. and 5 @ 12 in.

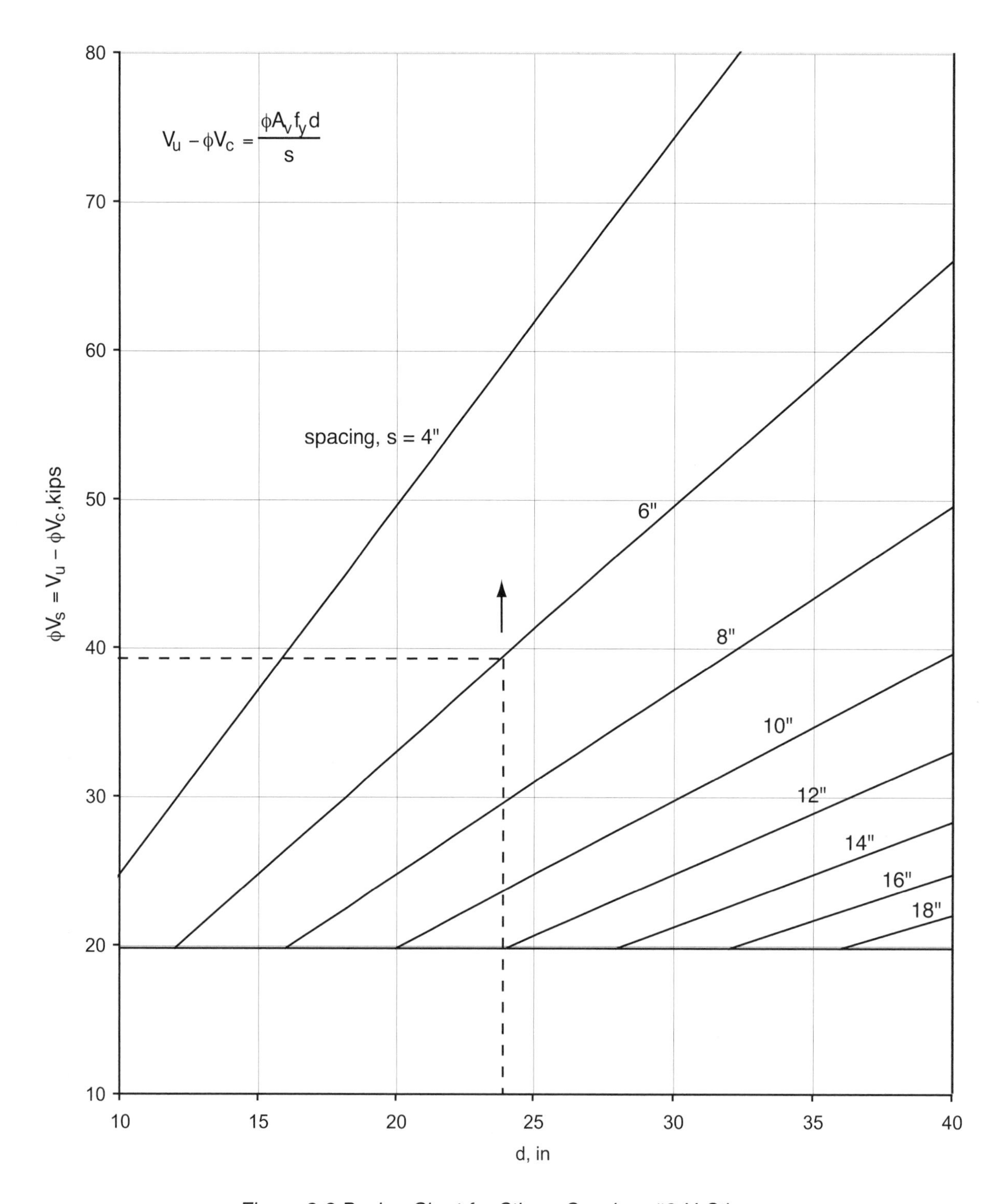

Figure 3-3 Design Chart for Stirrup Spacing, #3-U Stirrups

* *Horizontal line indicates ϕV_s for $s = d/2$.*

** *Minimum b_w corresponding to $\phi V_s = \phi 8\sqrt{f'_c}\, b_w d$ is less than 8 in. for all s*

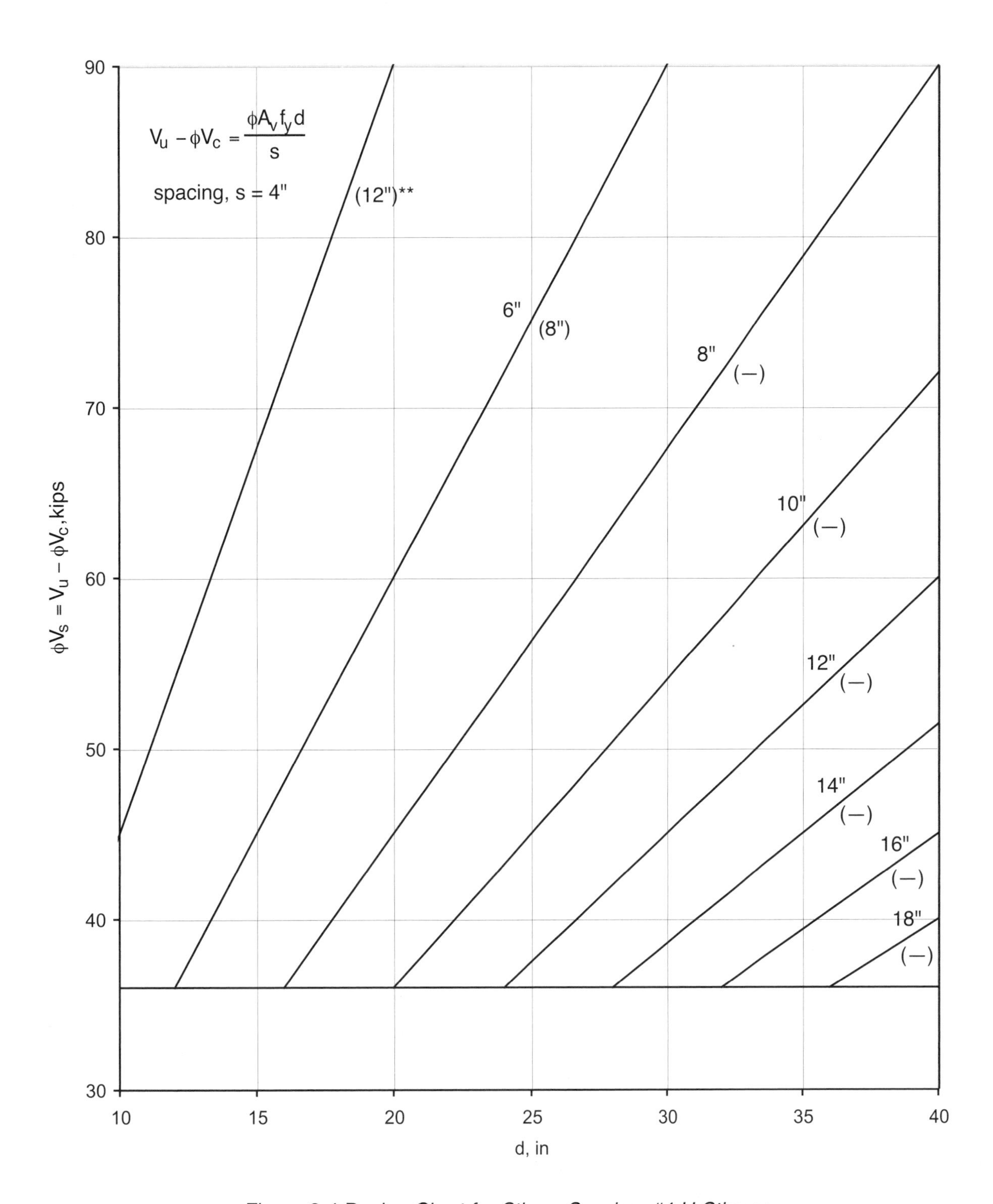

Figure 3-4 Design Chart for Stirrup Spacing, #4-U Stirrups

---

* *Horizontal line indicates ϕV_s for $s = d/2$.*

** *Values in ( ) indicate minimum practical b_w corresponding to $\phi V_s = \phi 8\sqrt{f_c'}\, b_w d$ for given s.*

(–) *Indicates minimum b_w corresponding to $\phi V_s = \phi 8\sqrt{f_c'}\, b_w d$ is less than 8 in. for given s.*

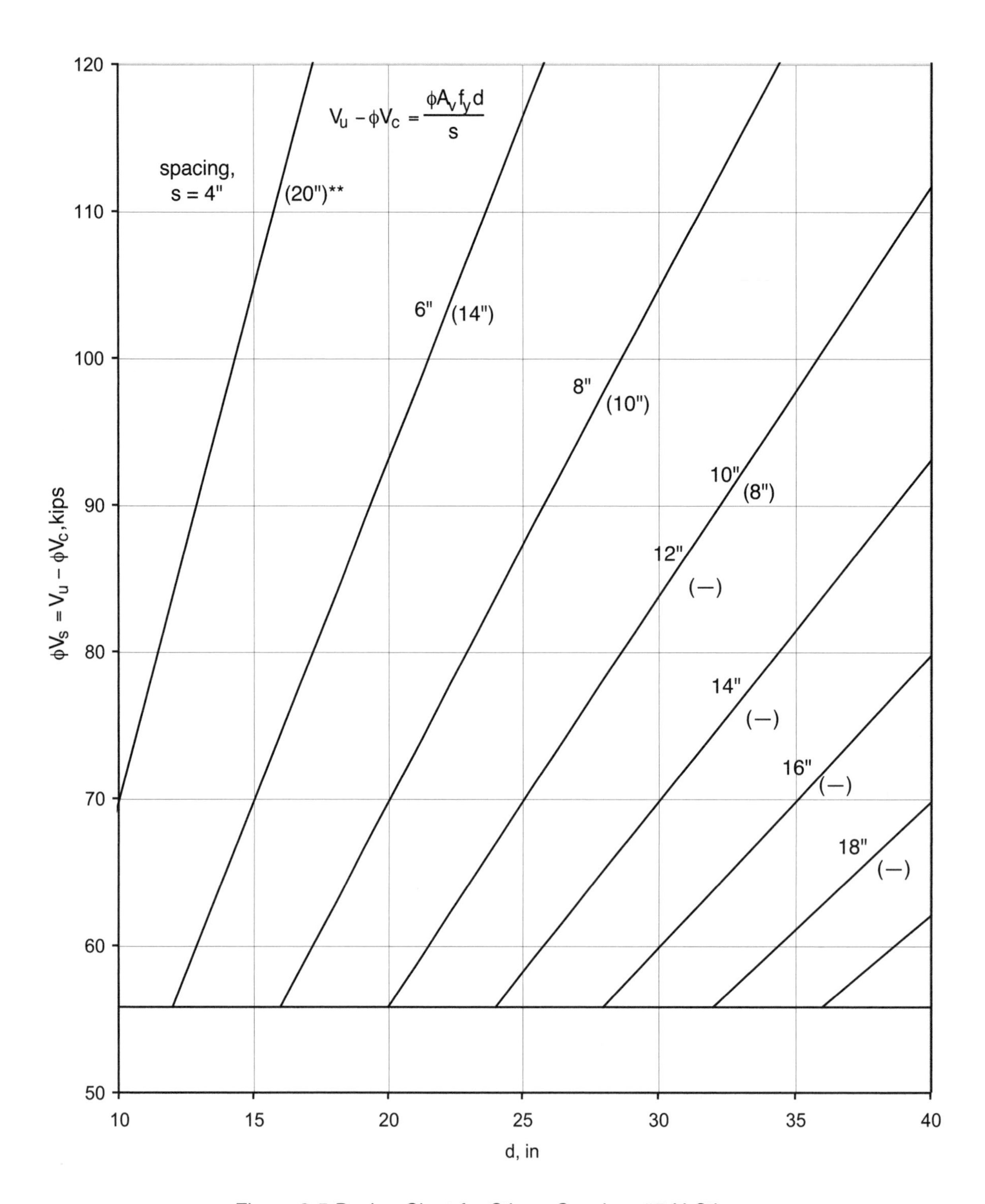

Figure 3-5 Design Chart for Stirrup Spacing, #5-U Stirrups

* *Horizontal line indicates ϕV_s for $s = d/2$.*

** *Values in ( ) indicate minimum practical b_w corresponding to $\phi V_s = \phi 8\sqrt{f'_c}\, b_w d$ for given s.*

(–) Indicates minimum b_w corresponding to $\phi V_s = \phi 8\sqrt{f'_c}\, b_w d$ is less than 8 in. for given s.

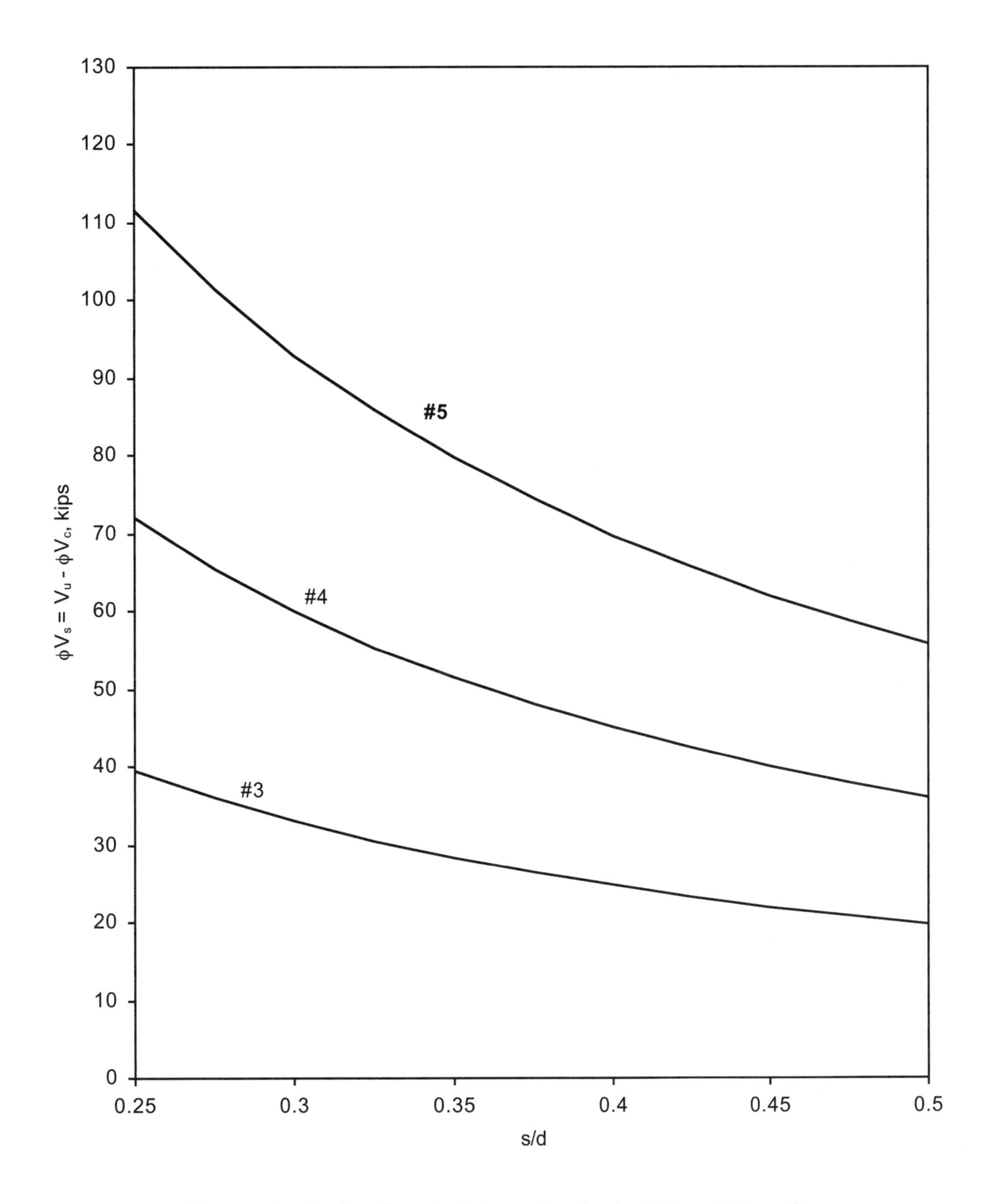

Figure 3-6 - Design Chart for Stirrups Spacing for Different Stirrup Sizes

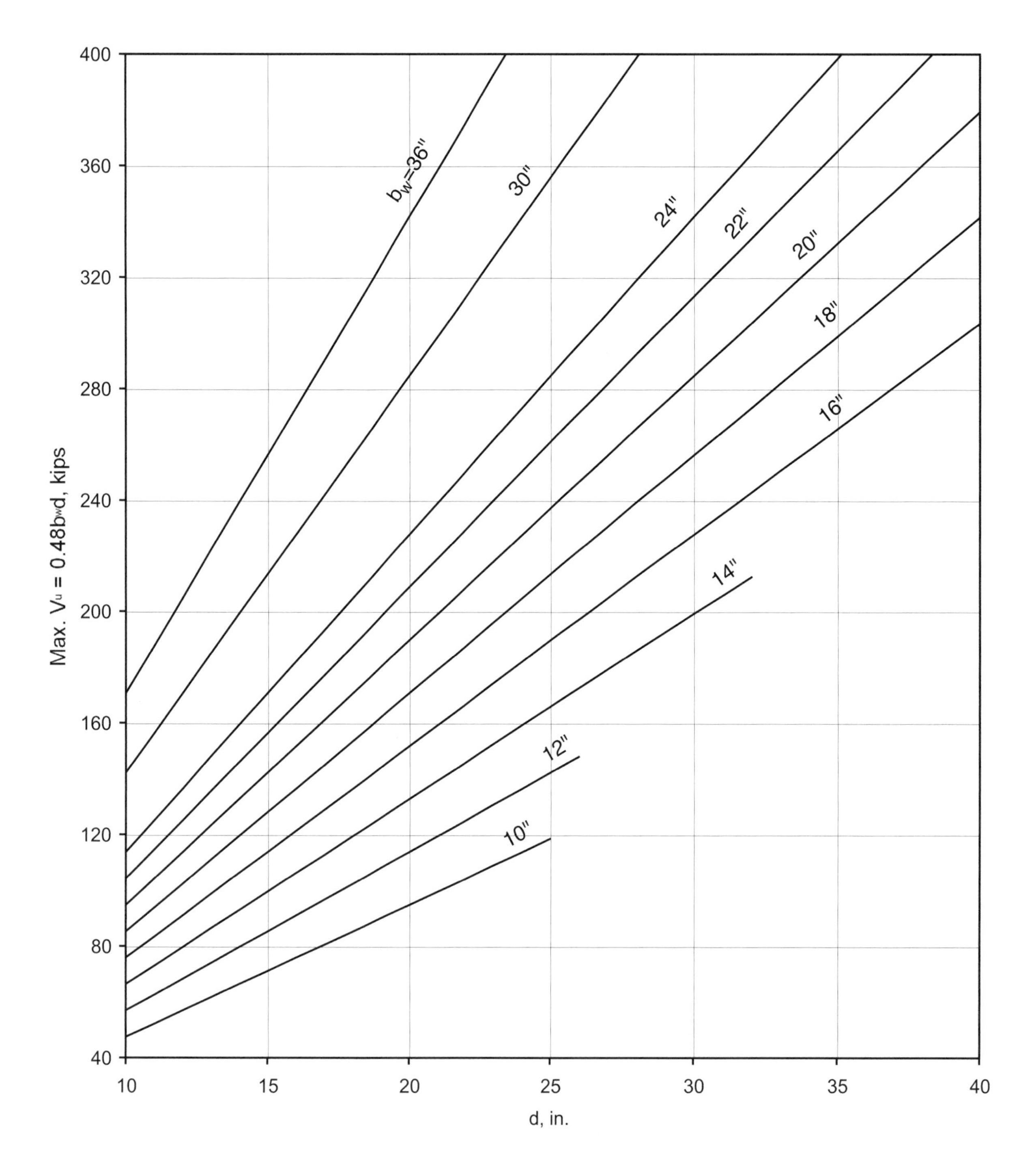

Figure 3-7 Design Chart for Maximum Allowable Shear Force

As an alternative, determine the required spacing of the No.3 U-stirrups at the critical section using Fig. 3-3. Enter the chart with d = 24 in. and ϕV_s = 38.9 kips. The point representing this combination is shown in the design chart. The line immediately above this point corresponds to a spacing of s = 6 in. which is exactly what was obtained using the previous simplified method.

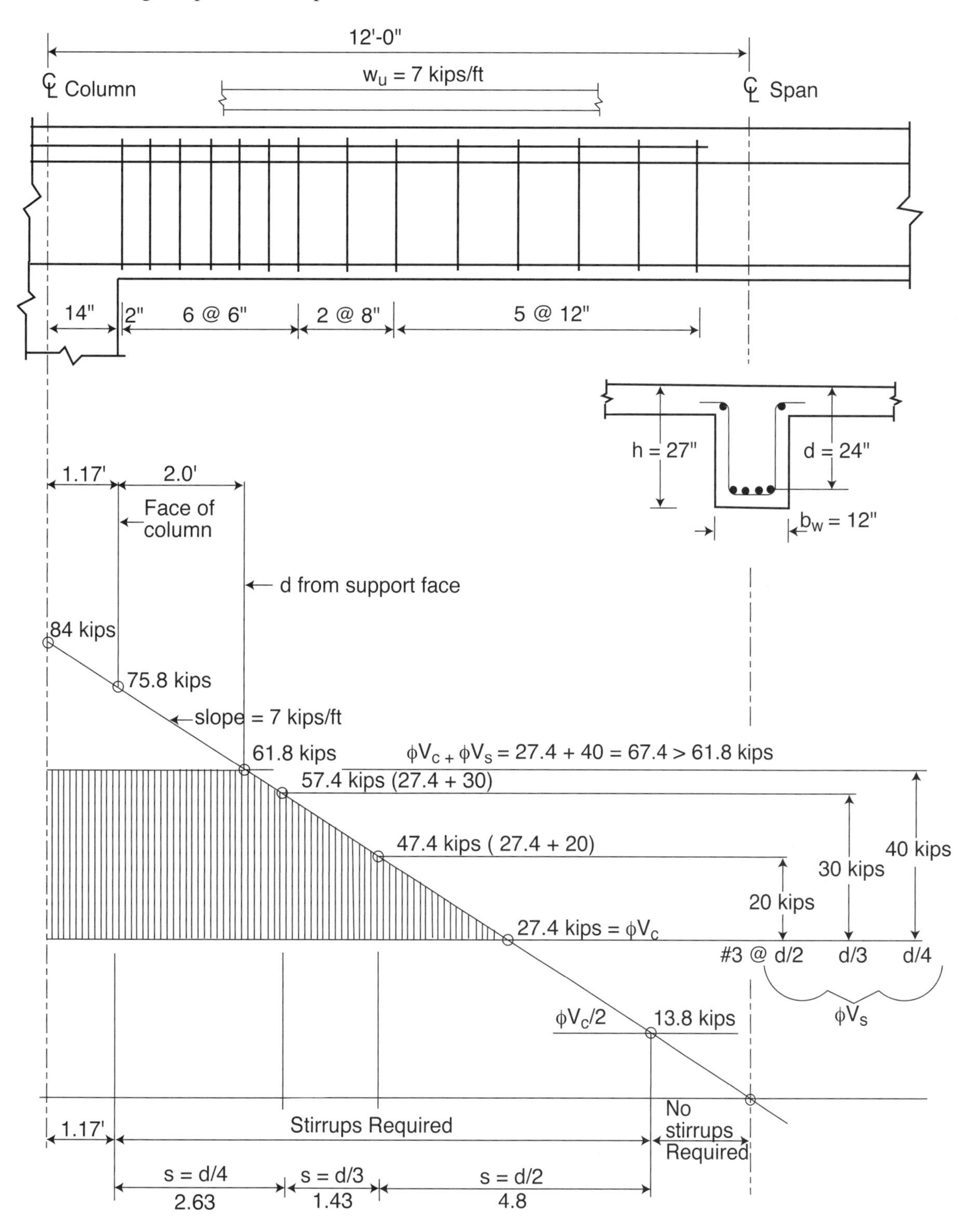

Figure 3-8 Simplified Method for Stirrup Spacing (Example 3.6.1)

3.6.2 Selection of Stirrups for Economy

Selection of stirrup size and spacing for overall cost savings requires consideration of both design time and fabrication and placing costs. An exact solution with an intricate stirrup layout closely following the variation in the shear diagram is not a cost-effective solution. Design time is more cost-effective when a quick, more conservative analysis is utilized. Small stirrup sizes at close spacings require disproportionately high costs in labor for fabrication and placement. Minimum cost solutions for simple placing should be limited to three spacings: the first stirrup located at 2 in. from the face of the support (as a minimum clearance), an intermediate spacing, and finally, a maximum spacing at the code limit of d/2. Larger size stirrups at wider spacings are more cost-effective (e.g., using No.4 for No.3 at double spacing, and No.5 and No.4 at 1.5 spacing) if it is possible to use them within the spacing limitations of d/2 and d/4.

In order to adequately develop the stirrups, the following requirements must all be satisfied (ACI 12.13): (1) stirrups shall be carried as close to the compression and tension surfaces of the member as cover requirements permit, (2) for No.5 stirrups and smaller, a standard stirrup hook (as defined in ACI 7.1.3) shall be provided around longitudinal reinforcement, and (3) each bend in the continuous portion of the stirrup must enclose a longitudinal bar. To allow for bend radii at corners of U stirrups, the minimum beam widths given in Table 3-9 should be provided.

Table 3-9 Minimum Beam Widths for Stirrups

Stirrup size	Minimum beam width (b_w)
#3	10 in.
#4	12 in.
#5	14 in.

Note that either the No.3 or the No.4 stirrup in the example of Fig. 3-7 can be placed in the 12 in. wide beam.

3.7 DESIGN FOR TORSION

For simplified torsion design for spandrel beams, where the torsional loading is from an integral slab, two options are possible:

(1) Size the spandrel beams so that torsion effects can be neglected (ACI 11.6.1), or

(2) Provide torsion reinforcement for a prescribed torque corresponding to the cracking torsional moment (ACI 11.6.2.2)

3.7.1 Beam Sizing to Neglect Torsion

Torsion can be neglected if the factored torque Tu is less than $\phi\sqrt{f'_c}\left(\frac{A_{cp}^2}{p_{cp}}\right)$ (ACI 11.6.1(a))

where:

A_{cp} = area enclosed by outside perimeter of concrete-cross section, in2.
p_{cp} = outside perimeter of concrete cross-section, in.
$\phi = 0.75$

For concrete with compressive strength = 4000 psi, torsion can be neglected if:

$$T_u < 0.004\left(\frac{A_{cp}^2}{p_{cp}}\right)$$

where A_{cp} and p_{cp} are in in.^{2} and in., respectively, and T_u in ft-kips.

For a rectangular sections having width and height equal to b and h, respectively, the torsion can be neglected if:

$$T_u < \phi\sqrt{f_c'}\left[\frac{h^2b^2}{2(h+b)}\right]$$

for concrete with f_c' = 4000 psi:

$$T_u < 0.002\left[\frac{h^2b^2}{h+b}\right]$$

where h and b are in inches (see Fig.3-11) and T_u ft-kip

For spandrel beam integral with a slab, torsion can be neglected if:

$$T_u < \frac{(bh+b_fh_f)^2}{2(h+b+b_f)}$$ (see Figure 3-10 for b_f and h_f)

A simplified sizing equation to neglect torsion effects can be derived based on the limiting factored torsional moment $T_u = \phi\sqrt{f_c'}\left(\frac{A_{cp}^2}{p_{cp}}\right)$ Total moment transferred from slab to spandrel = $0.3M_0$, (ACI 13.6.3.6) ACI 11.6.2.3 permits taking the torsional loading from a slab as uniformly distributed along the member. If the uniformly distributed torsional moments is t_u:

$$T_u = \frac{0.3M_o}{\ell}$$

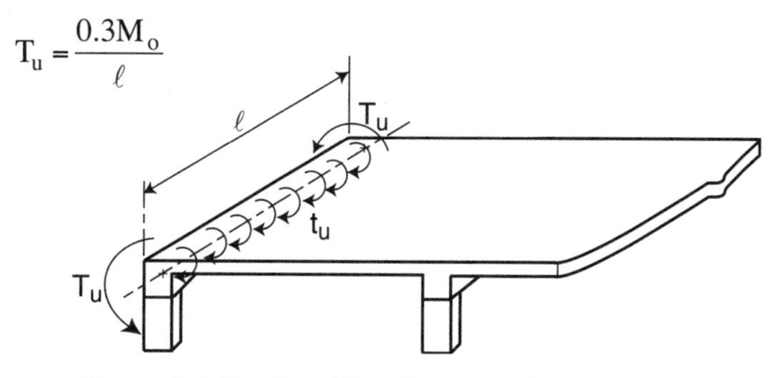

Figure 3-9 Torsional Loading on a Spandrel Beam

At the critical section, at distance d away from the support, the critical torsional moment is:

$$T_u = t_u\left(\frac{\ell}{2} - d\right) = \frac{0.3M_o}{\ell}\left(\frac{\ell}{2} - d\right) = 0.15M_o\left(1 - \frac{2d}{\ell}\right)$$

For preliminary sizing of spandrel to neglect torsion, assume d = 0.9h and $\frac{\ell}{h} = 20$ per Table 3-1

Therefore

$$\frac{d}{\ell} = \frac{0.9h}{20h} = 0.05$$

$$T_u = 0.15M_o(1 - 2 \times 0.05) = 0.135M_o$$

For preliminary sizing of spandrel beam to neglect torsion:

$$T_u \le \phi\sqrt{f_c'}\frac{A_{cp}^2}{p_{cp}}$$

$$0.135M_o \le 0.75\sqrt{4000}\frac{(bh + b_f h_f)^2}{2(h + b + b_f)}$$

$$\frac{(bh + b_f h_f)^2}{2(h + b + b_f)} > \frac{0.135(M_o \times 12000) \times 2}{0.75\sqrt{4000}} > 68.2M_o$$

for simplicity:

$$\frac{(bh + b_f h_f)^2}{2(h + b + b_f)} > 70M_o$$

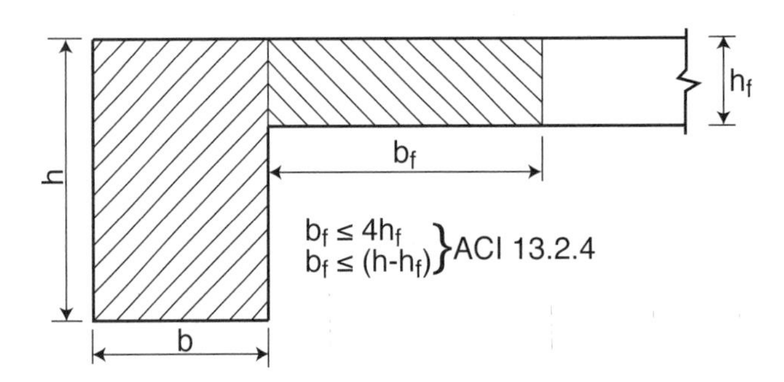

Figure 3-10 Torsional Section Properties

For one-way slab or one-way joists with spandrel beams, the exterior negative slab moment is equal to $w_u\ell_n^2/24$ (ACI 8.3.3). This negative moment can also be expressed as $0.33M_o$ where M_o = total static span moment for one-way slab = $w_u\ell_2\ell_n^2/8$. Thus for one-way system with spandrel beam, sizing to neglect torsion for the selected concrete (compressive strength = 4000 psi) reduces to:

$$\frac{(bh + b_f h_f)^2}{(h + b + b_f)} > 76M_o$$

The above sizing equations are mixed units: M_o ft-kips and section dimensions in inches.

Architectural or economic considerations may dictate a smaller spandrel size than that required to neglect torsion effects. For a specific floor framing system, both architectural and economic aspects of a larger beam size to neglect torsion versus a smaller beam size with torsion reinforcement (additional closed stirrups at close spacing combined with longitudinal bars) must be evaluated. If a smaller spandrel with torsion reinforcement is a more appropriate choice, Section 3.7.2 provides a simple method for the design of the torsion reinforcement.

3.7.1.1 Example: Beam Sizing to Neglect Torsion

Determine a spandrel beam size to neglect torsion effects for Building #2, Alternate (1) – slab and column framing with spandrel beams.

For N-S spandrels:
$\ell_2 = 20$ ft
$\ell_1 = 24$ ft
$\ell_n = 24 - (12 + 16)/(2 \times 12) = 22.83$ ft
$w_u = 1.2(136) + 1.6(50) = 243$ psf

$$M_o = w_u l_2 l_n^2/8 = 0.243 \times 20 \times 22.83^2/8 = 317 \text{ ft-kips}$$

For slab thickness $h_f = 8.5$ in., for monolithic construction (ACI 13.2.4) the portion of slab considered in beam design is the smaller of:

$4h_f$
$h - h_f$

For preliminary section calculations assume $b_f = 3h_f = 3(8.5) = 25.5$ in.

$$\frac{(bh + (25.5)(8.5))^2}{(h + b + 25.5)} = 70(317)$$

Some possible combinations of b and h that will satisfy the requirement for neglecting torsion are shown in Table 3-10.

Clearly large beam sizes are required to neglect torsion effects. It would be more economical to select a smaller beam and provide torsion reinforcement.

Table 3-10 Beam Dimensions for Neglecting Tension

b (in.)	h (required) in.	Possible selection b x h (in. x in.)
12	132.4	
14	98.7	
16	77.8	
18	64	
20	54.5	20 x 55
22	47.7	22 x 48
24	42.5	24 x 44
26	38.6	26 x 40
28	35.5	28 x 36
30	33	30 x 34
32	30.9	32 x 32
34	29.1	
36	27.6	

3.7.2 Simplified Design for Torsion Reinforcement

When required, torsion reinforcement must consist of a combination of closed stirrups and longitudinal reinforcement (ACI 11.6.4). For spandrel beams built integrally with a floor slab system (where reduction of torsional moment can occur due to redistribution of internal forces after cracking), a maximum torsional moment of $\phi 4\sqrt{f'_c}\left(\frac{A_{cp}^2}{p_{cp}}\right)$ may be assumed (ACI 11.6.2.2). The ACI provisions for members subjected to factored torsional moment T_u and factored shear force V_u can be summarized as follows:

(1) Check if the torsion can be neglected

$$T_u \leq \phi\sqrt{f'_c}\left(\frac{A_{cp}^2}{p_{cp}}\right) \qquad \text{(ACI 11.6.1)}$$

(2) If torsion cannot be neglected, for members in statically indeterminate structures where redistribution of forces can occur, calculate the maximum factored torsional moment at the critical section to be considered (ACI 11.6.2.2):

$$T_u = \phi 4\sqrt{f'_c}\left(\frac{A_{cp}^2}{p_{cp}}\right) \qquad \text{(ACI 11.6.2.2)}$$

(3) Design the torsional reinforcement for a torque equal to the smaller of the factored torsional moment from an elastic analysis or the value computed in Step 2. Consider only the spandrel beam reinforcement to resist torsion. Ignore portion of slab framing in spandrel. If portion of slab is included, torsion reinforcement must be included in the slab as well.

(4) Check the adequacy of the section dimensions. The cross-sectional dimensions need to be increased if:

$$\sqrt{\left(\frac{V_u}{b_w d}\right)^2 + \left(\frac{T_u p_h}{1.7A_{oh}^2}\right)^2} \geq \phi\left(\frac{V_c}{b_w d} + 8\sqrt{f'_c}\right) \qquad \text{(ACI 11.6.3)}$$

where:

V_u = factored shear force at the section, kips.
p_h = perimeter of centerline of outermost closed stirrups, in.
A_{oh} = area enclosed by centerline of the outermost closed stirrups, in^2
V_c = nominal shear strength provided by concrete, kips
b_w = web width, in.
d = distance from extreme compression fiber to centroid of tension reinforcement, in.

(5) Calculate the required area of stirrups for torsion:

$$\frac{A_t}{s} = \frac{T_u}{2\phi A_o f_y} \qquad \text{(ACI 11.6.3.6)}$$

where:

A_t = area of one leg of closed stirrups resisting torsion, in2
$A_o = 0.85A_{oh}$, in as defined in Step 4
s = spacing of stirrups, in.
(Note that for simplicity the angle of compression diagonals in truss analogy is assumed to be 45°)

(6) Calculate the required area of stirrups for shear:

$$V_s = \frac{V_u}{\phi} - V_c$$

$$\frac{A_v}{s} = \frac{V_s}{f_y d}$$

where:

A_v = Area of shear reinforcement within spacing s (two legs)

(7) Calculate required combined stirrups for shear and torsion (for one leg):

$$\frac{A_t}{s} = \frac{A_v}{2s}$$

Select the size and spacing of the combined stirrups to satisfy the following conditions:

a. The minimum area of stirrups A_v+2A_t (two legs) for the selected concrete with compressive strength = 4000 psi is $50b_w s/f_y$

b. The maximum spacing of transverse torsion reinforcement s is the smaller of $p_h/8$ or 12 in.

(8) Calculate the required additional longitudinal reinforcement A_l for torsion

$$A_\ell = \frac{A_t}{s} p_h \geq \frac{5\sqrt{f'_c}A_{cp}}{f_y} - \left(\frac{A_t}{s}\right)p_h \qquad \text{(ACI 11.6.3.7)}$$

A_t/s in the above equation must be taken as the actual amount calculated in Step 5 but not less than $25b_w/f_y$ (ACI 11.6.5.3). The additional longitudinal reinforcement must be distributed around the perimeter of the closed stirrups with maximum spacing of 12 in.

3.7.2.1 Example: Design of Torsion Reinforcement

Determine the required combined shear and torsion reinforcement for the E-W spandrel beam of Building #2, Alternate (1) – slab and column framing with spandrel beams.

For E-W spandrels:

Spandrel size = 12 × 20 in.
d = 20 –2.5 = 17.5 in. = 1.46 ft.
ℓ_n = 24 – (12/12) = 23.0 ft.
Beam weight = 1.2(12 × 20 × 0.150/144) = 0.30 kips/ft
w_u from slab = 1.2(136) + 1.6(50) = 243 psf

Tributary load to spandrel (1/2 panel width) = 243 × (20/2) = 2.43 kips/ft
Beam = 0.30 kips/ft
Total w_u = 2.73 kips/ft

$M_o = w_u l_2 l_n^2/8 = 0.243(24)(18.83)^2/8 = 254.5$ ft-kip

where

$$\ell_n = 20 - \frac{12+16}{2(12)} = 18.83 \text{ ft.}$$

$T_u = 0.30M_o/2 = 0.3(254.5) = 38.8$ ft-kip
T_u at distance d = 38.8 – 38.8/24(1.46) = 36.4 ft-kip
V_u (at the face of support) = 2.73(23)/2 = 31.4 kips
V_u at distance d = 31.4-2.73(1.46) = 27.4 kips

(1) Check if torsion can be neglected

A_{cp} = 20(12) + (20-8.5)(8.5) = 337.8 in²
p_{cp} = 20(2) + 12(2) + (20-8.5)(2) = 87 in

$$=0.004\left(\frac{A_{cp}^2}{p_{cp}}\right) = 0.004(337.8)^2/87 = 5.25 \text{ ft-kip} < 36.4 \text{ Torsion must be considered}$$

(2) Calculate the maximum factored torsional moment at the critical section to be considered:

$$T_u = \phi 4\sqrt{f_c'}\left(\frac{A_{cp}^2}{p_{cp}}\right) = 0.75(4)\sqrt{4000}\ (337.8)^2/87/12000 = 20.7 \text{ ft-kips} < 36.4$$

Note: The difference in torsional moments 36.4-20.7 must be redistributed (ACI 11.6.2.2). This redistribution will result in an increase of the positive midspan moment is slab.

(3) Check section adequacy:

Assume the distance to stirrups centroid = 1.75 in. (considering 1.5 in. cover and 0.5 in. stirrups diameter)

$p_h = 2(12\text{-}2 \times 1.75) + 2(20\text{-}2 \times 1.75) = 50$ in.
$A_{oh} = (12\text{-}2 \times 1.75)(20\text{-}2 \times 1.75) = 140.3$ in^2.

$$\frac{V_c}{b_w d} = 2\sqrt{4000} = 126.4 \text{ psi}$$

$$\sqrt{\left(\frac{V_u}{b_w d}\right)^2 + \left(\frac{T_u p_h}{1.7A_{oh}^2}\right)^2} = \sqrt{\left(\frac{27.4(1000)}{12(17.5)}\right)^2 + \left(\frac{20.7(12000)(50)}{1.7(140.3)^2}\right)^2} = 393.4$$

$$\phi\left(\frac{V_c}{b_w d} + 8\sqrt{f_c'}\right) = 0.75\left(126.4 + 8\sqrt{4000}\right) = 474.3 > 393.4$$

Section is adequate

(4) Required stirrups for torsion

$A_o = 0.85A_{oh} = 0.8(140.3) = 119.2$

$$\frac{A_t}{s} = \frac{T_u}{2\phi A_o f_y} = \frac{20.7(12000)}{2(0.75)(119.2)(60000)} = 0.023 \text{ in.}^2\text{/in. (one leg)}$$

(5) Required stirrups for shear

$$V_s = \frac{V_u}{\phi} - V_c = 27.4/0.75 \;\; 126.4(12)(17.5)/1000 = 10 \text{ kips}$$

$$\frac{A_v}{s} = \frac{V_s}{f_y d} = 10/60/17.5 = 0.0095 \text{ in (two legs)}$$

(6) Combined stirrups

$$\frac{A_t}{s} + \frac{A_v}{2s} = 0.023 + 0.0095/2 = 0.0325 \text{ in.}$$

maximum stirrups spacing = $p_h/8$ or 12 in. = 50/8 = 6.25
For No. 4 closed stirrups required s = 0.2/0.0325 = 6.15 in.
For No. 3 closed stirrups required s = 0.11/.0325 = 3.4 in.
Use No. 4 closed stirrups @ 6 in.

Area for 2 legs = 0.4 in.^{2}

Minimum area of stirrups:

$$50b_w s/f_y = 50(12)(6)/60000 = 0.06 \text{ in.}^2$$

(7) Required additional longitudinal reinforcement:

$$A_\ell = \frac{A_t}{s} p_h = 0.023(50) = 1.15 \text{ in.}^2 \text{ govern}$$

$$\text{Minimum} = \frac{5\sqrt{f'_c}A_{cp}}{f_y} - \left(\frac{A_t}{s}\right)p_h = \frac{5\sqrt{4000}(337.8)}{60000} - (0.023)50 = 0.63 \text{ in.}^2$$

Place the longitudinal bars around the perimeter of the closed stirrups, spaced not more than 12 in. apart. Locate one longitudinal bar in each corner of the closed stirrups (ACI 11.6.6.2). For the 20 in. deep beam, one bar is required at mid-depth on each side face, with 1/3 of the total A_ℓ required at top, mid-depth, and bottom of the closed stirrups. $A_\ell/3 = 1.15/3 = 0.38 \text{ in}^2$. Use 2-No.5 bars at mid-depth (one on each face). Longitudinal bars required at top and bottom may be combined with the flexural reinforcement.

Details of the shear and torsion reinforcement (at support) are shown in Fig. (3-11).

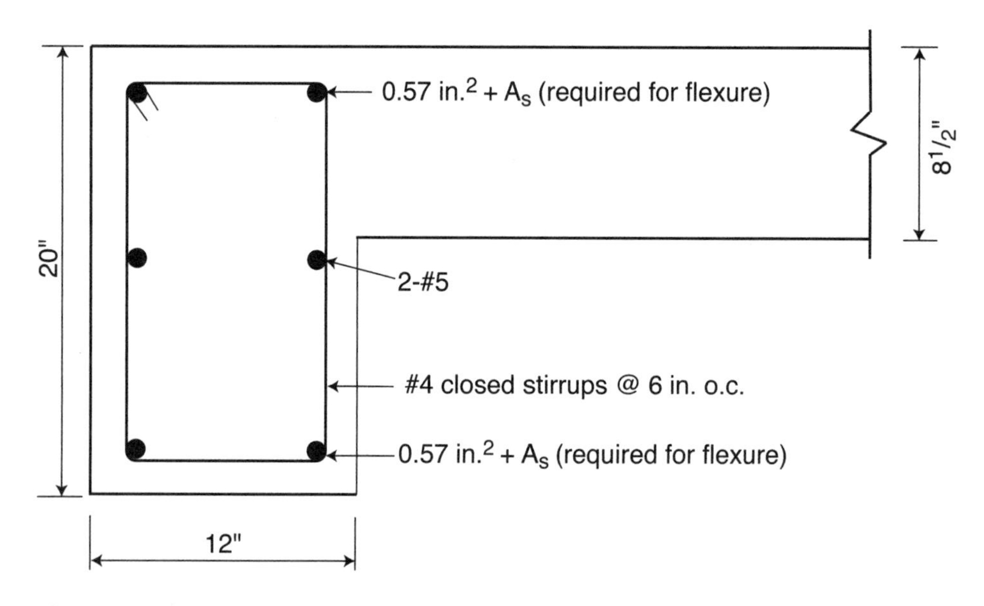

Figure 3-11 Required Shear and Torsion Reinforcement (Example 3.7.2.1)

3.8 EXAMPLES: SIMPLIFIED DESIGN FOR BEAMS AND SLABS

The following three examples illustrate the use of the simplified design data presented in Chapter 3 for proportioning beams and slabs. Typical floor members for the one-way joist floor system of Building #1 are designed.

3.8.1 Example: Design of Standard Pan Joists for Alternate (1) Floor System (Building #1)

(1) Data: $f'_c = 4000$ psi (normal weight concrete, carbonate aggregate)
$f_y = 60{,}000$ psi

Floors: LL = 60 psf
DL = 130 psf (assumed total for joists and beams + partitions + ceiling & misc.)

Required fire resistance rating = 1 hour

Floor system – Alternate (1): 30-in. wide standard pan joists
width of spandrel beams = 20 in.
width of interior beams = 36 in.

Note: The special provisions for standard joist construction in ACI 8.11 apply to the pan joist floor system of Alternate (1).

(2) Determine the factored shears and moments using the approximate coefficients of ACI 8.3.3.

Factored shears and moments for the joists of Alternate (1) are determined in Chapter 2, Section 2.3.2 (Figure 2-8).

$w_u = [1.2(130) + 1.6(60) = 252 \text{ psf}] \times 3 \text{ ft} = 756 \text{ plf}$

Note: All shear and negative moment values are at face of supporting beams.

V_u @ spandrel beams	= 10.5 kips
V_u @ first interior beams	= 12.0 kips
$-M_u$ @ spandrel beams	= 24.0 ft-kips
$+M_u$ @ end spans	= 41.1 ft-kips
$-M_u$ @ first interior beams	= 56.3 ft-kips
$+M_u$ @ interior spans	= 34.6 ft-kips
$-M_u$ @ interior beams	= 50.4 ft-kips

(3) Preliminary size of joist rib and slab thickness

From Table 3-1: depth of joist $h = \ell_n/18.5 = (27.5 \times 12)/18.5 = 17.8$ in.

where ℓ_n (end span) = 30 – 1.0 – 1.5 = 27.5 ft (governs)
ℓ_n (interior span) = 30 – 3 = 27.0 ft

From Table 10-1, required slab thickness = 3.2 in. for 1-hour fire resistance rating. Also, from ACI 8.11.6.1:

Minimum slab thickness > 30/12 = 2.5 in. > 2.0 in.

Try 16 in. pan forms * + $3^1/_2$-in. slab

h = 19.5 in. > 17.8 in. O.K.
slab thickness = 3.5 in. > 3.2 in. O.K.

* *See Table 9-3 for standard form dimensions for one-way joists.*

(4) Determine width of joist rib

(a) Code minimum (ACI 8.11.2):

$b_w > 16/3.5 = 4.6$ in. > 4.0 in.

(b) For flexural strength:

$$b_w = \frac{20M_u}{d^2} = \frac{20(57.5)}{18.25^2} = 3.45 \text{ in.}$$

where d = 19.5 – 1.25 = 18.25 in. = 1.52 ft

Check for fire resistance: from Table 10-4 for restrained members, the required cover for a fire resistance rating of 1 hr = 3/4-in. for joists.

(c) For shear strength

V_u @ distance d from support face = 12.0 – 0.76(1.52) = 10.84 kips

$\phi V_c = 1.1(0.095 b_w d) = 0.11 b_w d$*

$\therefore b_w = 10.84/(0.11 \times 18.25) = 5.40$ in.

Use 6 in. wide joists (see Table 9-3 and Fig. 3-12).

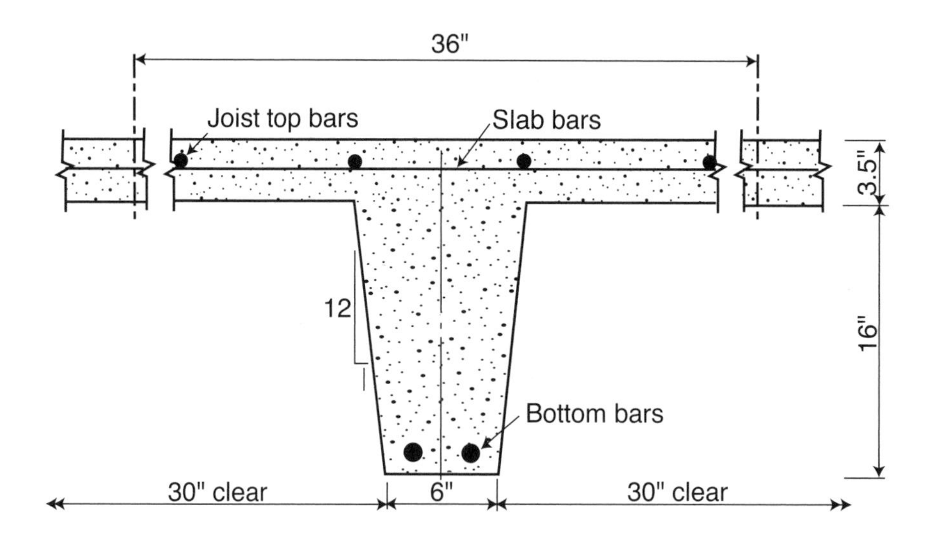

Figure 3-12 Joist Section (Example 3.8.1)

* *For standard joist ribs conforming to ACI 8.11 a 10% greater shear strength fVc is allowed. Also, minimum shear reinforcement is not required (see ACI 11.5.5).*

(5) Determine Flexural Reinforcement

(a) Top bars at spandrel beams:

$$A_s = \frac{M_u}{4d} = \frac{24.0}{4(18.25)} = 0.33 \text{ in.}^2$$

Distribute bars uniformly in top slab:

$A_s = 0.33/3 = 0.11$ in.^{2}/ft

Maximum bar spacing for an interior exposure and 3/4-in. cover will be controlled by provisions of ACI 7.6.5

$s_{max} = 3h = 3(3.5) = 10.5$ in. < 18 in.

From Table 3-6: Use No.3 @ 10 in. ($A_s = 0.13$ in.^{2}/ft)

(b) Bottom bars in end spans:

$A_s = 41.1/(4 \times 18.25) = 0.56$ in.^{2}

Check rectangular section behavior:

$a = A_s f_y / 0.85 f'_c b_e = (0.56 \times 60)/(0.85 \times 4 \times 36) = 0.27$ in. < 3.5 in. O.K.

From Table 3-5: Use 2-No.5 ($A_s = 0.62$ in.^{2})

Check $\rho = A_s/b_w d = 0.56/(6 \times 18.25) = 0.005 >$ rmin $= 0.0033$ O.K.

(c) Top bars at first interior beams:

$A_s = [56.3/ (4 \times 18.25) = 0.77$ in.$^2]/3 = 0.26$ in.^{2}/ft

From Table 3-6, with $s_{max} = 10.5$ in.:

Use No.5 @ 10 in. ($A_s = 0.37$ in.^{2}/ft)

(d) Bottom bars in interior spans:

$A_s = 34.6/(4 \times 18.25) = 0.47$ in.^{2}

From Table 3-5: Use 2-No.5 ($A_s = 0.62$ in.^{2})

(e) Top bars at interior beams:

$A_s = [50.4/(4 \times 18.25) = 0.69\ in.^2]/3 = 0.23\ in.^2/ft$

From Table 3-6, with s_{max} = 10.5 in.:

Use No.4 @ 10 in. (A_s = 0.24 in.$_2$/ft)

(f) Slab reinforcement normal to ribs (ACI 8.11.6.2):

Slab reinforcement is often located at mid-depth of slab to resist both positive and negative moments.

Use $M_u = w_u\ell_n^2/12$ (see Fig. 2-5)

$= 0.19(2.5)^2/12 = 0.10$ ft-kips

where w_u = 1.2(44 + 30) + 1.6(60) = 185 psf = 0.19 kips/ft
ℓ_n= 30 in. = 2.5 ft (ignore rib taper)

With bars on slab centerline, d = 3.5/2 = 1.75 in.

$A_s = 0.10/4(1.75) = 0.015\ in.^2/ft$ (but not less than required temperature and shrinkage reinforcement)

From Table 3-4, for a 3$^1/_4$-in. slab: Use No.3 @ 16 in.

Note: For slab reinforcement normal to ribs, space bars per ACI 7.12.2.2 at 5h or 18 in.
Check for fire resistance: From Table 10-3, required cover for fire resistance rating of 1 hour = $^3/_4$-in. O.K.

(6) Reinforcement details shown in Fig. 3-13 are determined directly from Fig. 8-5*. Note that one of the No.5 bars at the bottom must be continuous or be spliced over the support with a Class A tension splice, and be terminated with a standard hook at the non-continuous supports for structural integrity (ACI 7.13.2.1).

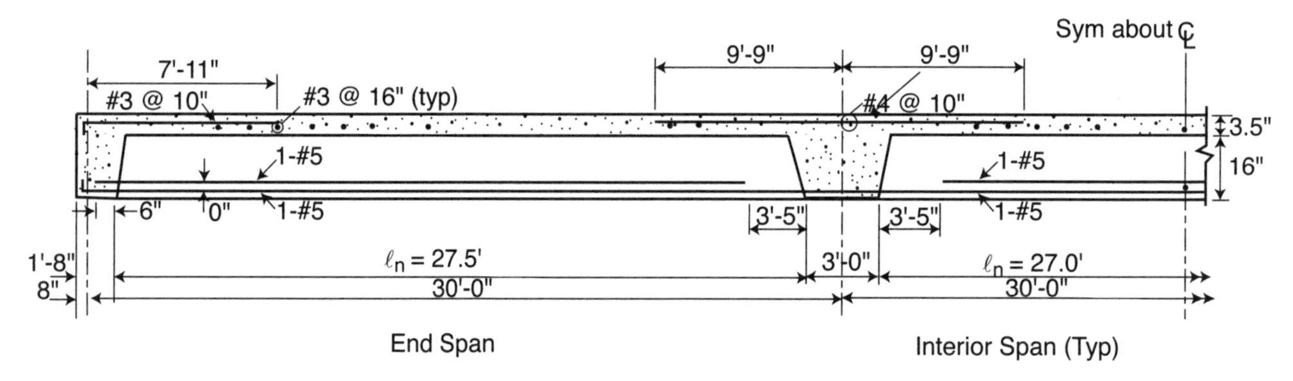

Figure 3-13 Reinforcement Details for 30 in. Standard Joist Floor System [Building #1—Alternate (1)]

* *The bar cut-off points shown in Fig. 8-5 are recommended for one-way joist construction. The reader may consider determining actual bar lengths using the provisions in ACI 12.10.*

3.8.2 Example: Design of Wide-Module Joists for Alternate (2) Floor System (Building #1)

(1) Data: f'_c = 4000 psi (normal weight concrete, carbonate aggregate)
f_y = 60,000 psi

Floors: LL = 60 psf
DL = 130 psf (assumed total for joists and beams + partitions + ceiling & misc.)

Fire resistance rating = assume 2 hours

Floor system – Alternate (2): Assume joists on 6 ft – 3 in. centers (53" standard forms*)
width of spandrel beams = 20 in.
width of interior beams = 36 in.

Note: The provisions for standard joist construction in ACI 8.11 do not apply for the wide-module joist system (clear spacing between joists > 30 in.). Wide-module joists are designed as beams and one-way slabs (ACI 8.11.4).

(2) Determine factored shears and moments using the approximate coefficients of ACI 8.3.3 (see Figs. 2-3, 2-4, and 2-7).

w_u = 1.2(130) + 1.6(60)** = 252 psf × 6.25 = 1575 plf, say 1600 plf

Note: All shear and negative moment values are at face of supporting beams.

V_u @ spandrel beams = 1.6(27.5)/2 = 22.0 kips
V_u @ first interior beams = 1.15(22.0) = 25.3 kips
V_u @ interior beams = 1.6(27.0)/2 = 21.6 kips
$-M_u$ @ spandrel beams = 1.6(27.5)2/24 = 50.4 ft-kips
$+M_u$ @ end spans = 1.6(27.5)2/14 = 86.4 ft-kips
$-M_u$ @ first interior beams = 1.6(27.25)2/10 = 118.8 ft-kips
$+M_u$ @ interior spans = 1.6(27)2/16 = 72.9 ft-kips
$-M_u$ @ interior beams = 1.6(27)2/11 = 106 ft-kips

(3) Preliminary size of joists (beams) and slab thickness

From Table 3-1: depth of joist h = ℓ_n /18.5 = (27.5 × 12)/18.5 = 17.8 in.

where ℓ_n (end span) = 30 – 1.0 – 1.5 = 27.5 ft (governs)
ℓ_n (interior span) = 30 – 3 = 27.0 ft

Check for fire resistance: from Table 10-1, required slab thickness = 4.6 in. for 2-hour rating.

* *See Table 9-3 for standard form dimensions.*

** *No live load reduction permitted: AI = 12.5 × 30 = 375 sq ft < 400 sq ft (see Table 2-1).*

Try 16 in. pan forms + $4^1/_2$-in. slab

h = 20.5 in. > 17.8 in. O.K.

(4) Determine width of joist rib

(a) For moment strength:

$$b_w = \frac{20M_u}{d^2} = \frac{20(118.8)}{18.0^2} = 7.33 \text{ in.}$$

where d = 20.5 – 2.5 = 18.0 in. = 1.50 ft

Check for fire resistance: for joists designed as beams, minimum cover per ACI 7.7 = 1.5 in. From Table 10-4, the required cover for restrained beams for a fire resistance rating of 2 hr = $^1/_4$-in. < 1.5 in. O.K.

(b) For shear strength

V_u @ distance d from support face = 25.3 – 1.6(1.50) = 22.9 kips

$\phi V_c/2 = 0.048 b_w d$*

$b_w = 22.9/(0.048 \times 18.0) = 26.5$ in.

Use 9 in.-wide joists (standard width) and provide stirrups where required. A typical cross-section of the joist is shown in Fig. 3-14.

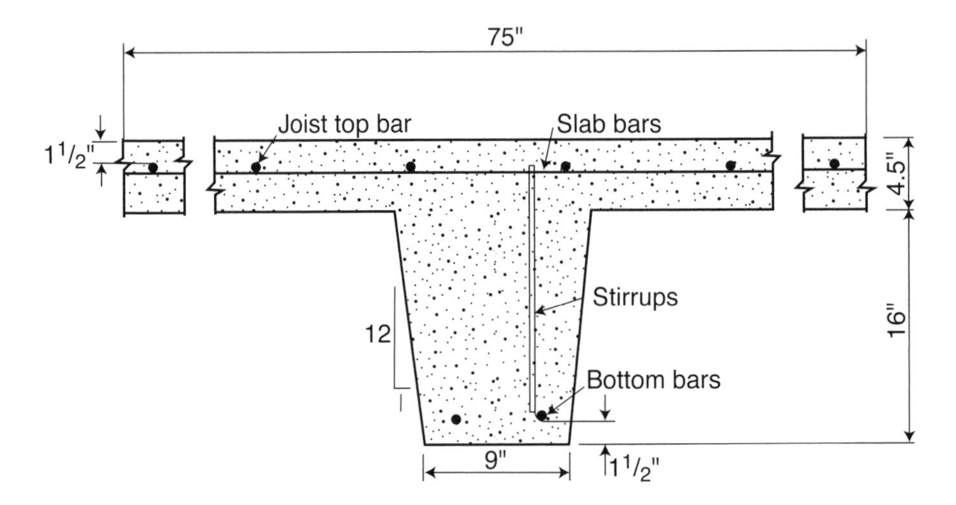

Figure 3-14 Joist Section

* *For joists designed as beams, the 10% increase in ϕV_c is not permitted. Also, minimum shear reinforcement is required when $V_u > \phi V_c/2$.*

(5) Determine Flexural Reinforcement

(a) Top bars at spandrel beams:

$$A_s = \frac{M_u}{4d} = \frac{50.4}{4(18.0)} = 0.70 \text{ in.}^2$$

Distribute bars uniformly in top slab according to ACI 10.6.6.

Effective flange width: $(30 \times 12)/10 = 36$ in. (governs

$6.25 \times 12 = 75$ in.

$9 + 2(8 \times 4.5) = 81$ in.

$A_s = 0.70/3 = 0.23 \text{ in.}^2/\text{ft}$

maximum bar spacing for No.4 bars,

$s_{max} = 3h = 3(4.5) = 13.5 \text{ in.} < 18 \text{ in.}$

From Table 3-6: Use No. 4 @ 9 in. ($A_s = 0.27 \text{ in.}^2/\text{ft}$)

(b) Bottom bars in end spans:

$A_s = 86.4/(4 \times 18.0) = 1.2 \text{ in.}^2$

$a = A_s f_y/0.85 f'_c b_e = (1.2 \times 60)/(0.85 \times 4 \times 36) = 0.59 \text{ in.} < 4.5 \text{ in.}$ O.K.

From Table 3-5: Use 2-No. 7 ($A_s = 1.20 \text{ in.}^2$)

Check $\rho = A_s/b_w d = 1.2/(9 \times 18.0) = 0.0074 > \rho_{min} = 0.0033$ O.K.

The 2-No.7 bars satisfy all requirements for minimum and maximum number of bars in a single layer.

(c) Top bars at first interior beams:

$A_s = [118.8/(4 \times 18.0) = 1.65 \text{ in.}^2]/3 = 0.55 \text{ in.}^2/\text{ft}$

From Table 3-6, assuming No.7 bars ($s_{max} = 10.8$ in. from Table 3-6),

Use No. 7 @ 10 in. ($A_s = 0.72 \text{ in.}^2/\text{ft}$)

(d) Bottom bars in interior spans:

$A_s = 72.9/(4 \times 18.0) = 1.01 \text{ in.}^2$

From Table 3-5: Use 2 No. 7 ($A_s = 1.20 \text{ in.}^2$)

(e) Top bars at interior beams:

$A_s = [106/(4 \times 18) = 1.47 \text{ in.}^2]/3 \text{ ft} = 0.50 \text{ in.}^2/\text{ft}$

From Table 3-6, with $s_{max} = 11.6$ in.

Use No. 6 @ 10 in. ($A_s = 0.59 \text{ in.}^2/\text{ft}$)

(f) Slab reinforcement normal to joists:

Use $M_u = w_u \ell_n^2/12$ (see Fig. 2-5)

$= 0.20(6.25)^2/12 = 0.65$ ft-kips

where $w_u = 1.2(56 + 30) + 1.6(60) = 200$ psf $= 0.20$ kips/ft

Place bars on slab centerline: $d = 4.5/2 = 2.25$ in.

$A_s = 0.65/4(2.25) = 0.07 \text{ in.}^2/\text{ft}$
(but not less than required temperature reinforcement).

From Table 3-4, for a 4½ in. slab: Use No. 3 @ 13 in. ($A_s = 0.10 \text{ in.}^2/\text{ft}$)

Check for fire resistance: from Table 10-3, for restrained members, required cover for fire resistance rating of 2 hours = ¾ in. O.K.

(6) Reinforcement details shown in Fig. 3-15 are determined directly from Fig. 8-3(a).* For structural integrity (ACI 7.13.2.3), one of the No. 7 and No. 8 bars at the bottom must be spliced over the support with a Class A tension splice; the No. 8 bar must be terminated with a standard hook at the non-continuous supports.

(7) Design of Shear Reinforcement

(a) End spans:

V_u at face of interior beam = 25.3 kips
V_u at distance d from support face = 25.3 – 1.6(1.50) = 22.9 kips

Use average web width for shear strength calculations

$b_w = 9 + (20.5/12) = 10.7$ in.

$(\phi V_c + \phi V_s)_{max} = 0.48\, b_w d = 0.48(10.7)(18.0) = 92.4$ kips

$\phi V_c = 0.095\, b_w d = 0.095(10.7)(18.0) = 18.3$ kips

$\phi V_c/2 = 0.048\, b_w d = 9.3$ kips

* *The bar cut-off points shown in Fig. 8-3(a) are recommended for beams without closed stirrups. The reader may consider determining actual bar lengths using the provisions in ACI 12.10.*

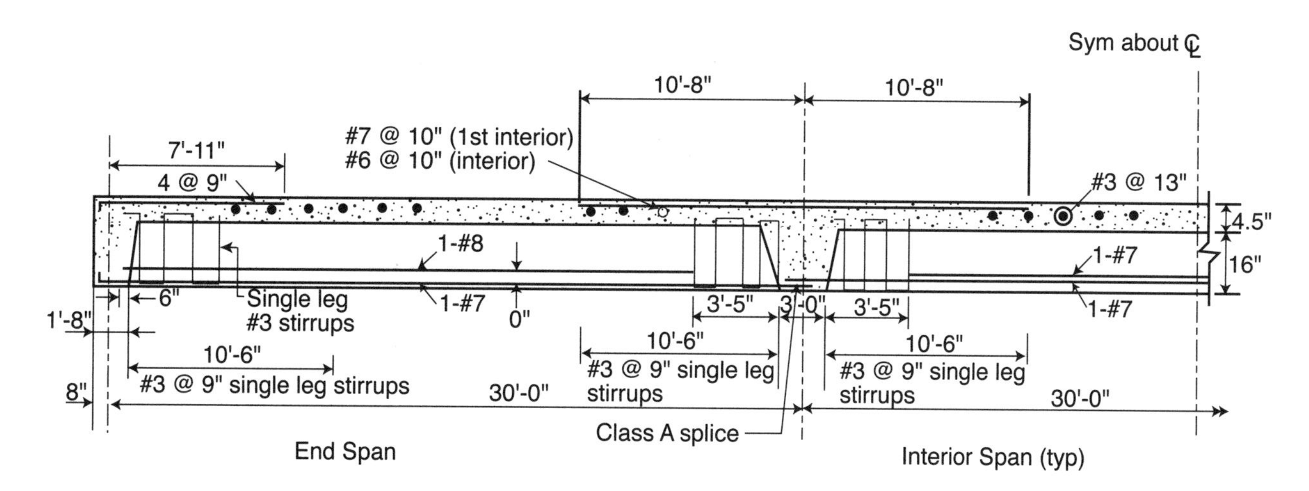

Figure 3-15 Reinforcement Details for 6 ft-3 in. Wide-Module Joist Floor System [Building #1—Alternate (2)]

Beam size is adequate for shear strength since 92.3 kips > 22.9 kips.

Since $V_u > \phi V_c$, more than minimum shear reinforcement is required. Due to the sloping face of the joist rib and the narrow widths commonly used, shear reinforcement is generally a one-legged stirrup rather than the usual two. The type commonly used is a continuous bar located near the joist centerline and bent into the configuration shown in Fig. 3-16. The stirrups are attached to the joist bottom bars.

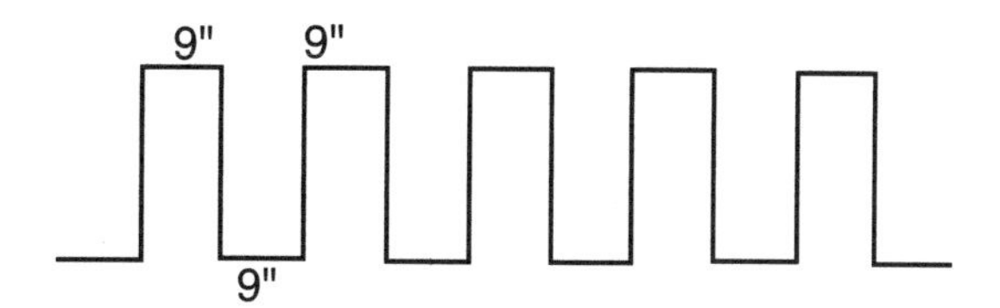

Figure 3-16 Stirrup Detail

Single leg No. 3 @ d/2 is adequate for full length where stirrups are required. Length over which stirrups are required: (25.3 – 9.3)/1.8 = 8.9 ft. Stirrup spacing s = d/2 = 18.0/2 = 9.0 in.

Check $A_{v(min)} = 50\ b_w s/f_y = 50 \times 10.7 \times 9.0/60{,}000 = 0.08\ \text{in.}^2$

Single leg No.3 stirrup O.K.

Use 14-No. 3 single leg stirrups @ 9 in. Use the same stirrup detail at each end of all joists.

3.8.3 Example: Design of the Support Beams for the Standard Pan Joist Floor Along a Typical N-S Interior Column Line (Building #1)

(1) Data: $f'_c = 4000$ psi (normal weight concrete, carbonate aggregate)

$f_y = 60{,}000$ psi

Floors: LL = 60 psf
DL = 130 psf (assumed total for joists and beams + partitions + ceiling & misc.)

Required fire resistance rating = 1 hour (2 hours for Alternate (2)).

Preliminary member sizes
Columns interior = 18 × 18 in.
exterior = 16 × 16 in.

width of interior beams = 36 in. 2 × depth – 2 × 19.5 = 39.0 in.

The most economical solution for a pan joist floor is making the depth of the supporting beams equal to the depth of the joists. In other words, the soffits of the beams and joists should be on a common plane. This reduces formwork costs sufficiently to override the savings in materials that may be accomplished by using a deeper beam. See Chapter 9 for a discussion on design considerations for economical formwork. The beams are often made about twice as wide as they are deep. Overall joist floor depth = 16 in. + 3.5 in. = 19.5 in. Check deflection control for the 19.5 in. beam depth. From Table 3-1:

$h = 19.5 \text{ in.} > /18.5 = (28.58 \times 12)/18.5 = 18.5 \text{ in.}$ O.K.

where ℓ_n (end span) = 30 – 0.67 – 0.75 = 28.58 ft (governs)
ℓ_n (interior span) = 30 – 1.50 = 28.50 ft.

(2) Determine factored shears and moments from the gravity loads using the approximate coefficients (see Figs. 2-3, 2-4, and 2-7).

Check live load reduction. For interior beams:

$K_{LL}A_T = 2(30 \times 30) = 1800 \text{ sq ft} > 400 \text{ sq ft}$

$L = 60(0.25 + 15/\sqrt{1800}) = 60(0.604)^* = 36.2 \text{ psf} > 50\% \, L_o$

$DL = 130 \times 30 \times \frac{1}{1000} = 3.9 \text{ klf}$

$LL = 36.2 \times 30 \times \frac{1}{1000} = 1.09 \text{ klf}$

$w_u = [1.20(130) + 1.6(36.2) = 214 \text{ psf}] \times 30 \text{ ft} = 6.4 \text{ klf}$

Note: All shear and negative moment values are at face of supporting beams.

Vu @ exterior columns	= 6.4(28.58)/2 = 91.5 kips
Vu @ first interior columns	= 1.15(91.5) = 105.3 kips
Vu @ interior columns	= 6.4(28.5)/2 = 91.2 kips
-Mu @ exterior columns	= $6.4(28.58)^2/16$ = 326.7 ft-kips
+Mu @ end spans	= $6.4(28.58)^2/14$ = 373.4 ft-kips
-Mu @ first interior columns	= $6.4(28.58)^2/10$ = 522.8 ft-kips
+Mu @ interior span	= $6.4(28.50)^2/16$ = 324.9 ft-kips

(3) Design of the column line beams also includes consideration of moments and shears due to wind. The wind load analysis for Building #1 is summarized in Fig. 2-13.

Note: The reduced load factor (0.5) permitted for load combinations including the wind effect (ACI 9-3 and 9-4) is in most cases, sufficient to accommodate the wind forces and moments without an increase in the required beam size or reinforcement (i.e., the load combination for gravity load only will usually govern for proportioning the beam).

(4) Check beam size for moment strength

Preliminary beam size = 19.5 in. × 36 in.

For negative moment section:

$$b_w = \frac{20M_u}{d^2} = \frac{20(522.8)}{17^2} = 36.2 \text{ in.} > 36 \text{ in}$$

where d = 19.5 – 2.5 = 17.0 in. = 1.42 ft

For positive moment section:

$$b_w = 20\,(373.4)/17^2 = 25.8 \text{ in.} < 36 \text{ in.}$$

Check minimum size permitted with $\rho = 0.0206$:

$$b_w = 14.6(522.8)/17^2 = 26.4 \text{ in.} < 36 \text{ in.} \quad \text{O.K.}$$

Use 36 in. wide beam and provide slightly higher percentage of reinforcement ($\rho > 0.5\ \rho_{max}$) at interior columns.

Check for fire resistance: from Table 10-4, required cover for fire resistance rating of 4 hours or less = ¾ in. < provided cover. O.K.

(5) Determine flexural reinforcement for the beams at the 1st floor level

(a) Top bars at exterior columns

Check governing load combination:

- gravity loads

  M_u = 326.7 ft-kips ACI Eq. (9-2)

- gravity + wind load

  $M_u = 1.2(3.9)(28.58)^2/16 + 0.8(99.56$ 317.9 ft-kips ACI Eq. (9-3)

---

* *For members supporting one floor only, maximum reduction = 0.5 (see Table 2-1).*

or

$M_u = 1.2(3.9)(28.58)^2/16 + 0.5(1.09)(28.58)^2/16 + 1.6(99.56)$ = 426.1 ft-kips ACI Eq. (9-4)

- also check for possible moment reversal due to wind moment:

$M_u = 0.9(3.9)(28.58)^2/16 \pm 1.6(99.56)$ = 338.5 klf, 19.9 ft-kips ACI Eq. (9-6)

$$A_s = \frac{M_u}{4d} = \frac{426.1}{4(17)} = 6.27 \text{ in.}^2$$

From Table 3-5: Use 8-No. 8 bars ($A_s = 6.32$ in.^{2})

minimum $n = 36[1.5 + 0.5 + (1.0/2)]^2/57.4 = 4$ bars < 8 O.K.

Check $\rho = A_s/bd = 6.32/(36 \times 17) = 0.0103 > \rho_{min} = 0.0033$ O.K.

(b) Bottom bars in end spans:

$A_s = 373.4/4(17) = 5.49$ in.^{2}

Use 8-No. 8 bars ($A_s = 6.32$ in.^{2})

(c) Top bars at interior columns:

Check governing load combination:

- gravity load only

M_u = 522.8 ft-kips ACI Eq. (9-2)

- gravity + wind loads:

$M_u = 1.2(3.9)(28.54)^2/10 + 0.8(99.56)$ = 460.8 ft-kips ACI Eq. (9-3)

or

$M_u = 1.2(3.9)(28.54)^2/10 + 0.5(1.09)(28.54)^2/10 + 1.6(99.56)$ = 584.9 ft-kips ACI Eq. (9-4)

$A_s = 584.9/4(17) = 8.6$ in.^{2}

Use 11-No. 8 bars ($A_s = 8.6$ in.^{2})

(d) Bottom bars in interior span:

$A_s = 324.9/4(17) = 4.78$ in.^{2}

Use 7-No. 8 bars ($A_s = 5.53$ in.^{2})

(6) Reinforcement details shown in Fig. 3-17 are determined directly from Fig. 8-3(a).* Provided 2-No. 5 top bars within the center portion of all spans to account for any variations in required bar lengths due to wind effects.

Since the column line beams are part of the primary wind-force resisting system. ACI 12.11.2 requires at least one-fourth the positive moment reinforcement to be extended into the supporting columns and be anchored to develop full fy at face of support. For the end spans: $A_s/4 = 8/4 = 2$ bars. Extend 2-No. 8 center bars anchorage distance into the supports:

- At the exterior columns, provide a 90° standard end-hook (general use). From Table 8-5, for No. 8 bar:

  $\ell_n = 14$ in. $= 16 - 2 = 14$ in. O.K.

- At the interior columns, provide a Class A tension splice (ACI 13.2.4). Clear space between No. 8 bars = 3.4 in. = $3.4d_b$. From Table 8-2, length of splice = $1.0 \times 30 = 30$ in. (ACI 12.15).

(7) Design of Shear Reinforcement

Design shear reinforcement for the end span at the interior column and use the same stirrup requirements for all three spans.

Check governing load combination:

- gravity load only

  V_u = at interior column = 105.5 kips (governs)

- gravity + wind loads:

  $V_u = 1.2(1.15)(3.9)(28.54)/2 + 0.8(6.64) = 82.1$ kips

  or

  $V_u = 1.2(1.15)(3.9)(28.54)/2 + 0.5(1.15)(1.09)(28.54)/2 + 1.6(6.64) = 95.7$ kips

- wind only at span center

  $V_u = 1.6(6.64) = 10.6$ kips

  V_u @ face of column = 105.5 kips

  V_u at distance d from column face $= 105.5 - 6.4(1.42) = 96.4$ kips

  $(\phi V_c + \phi V_s)_{max} = 0.48\ b_w d = 0.48(36)17 = 293.8$ kips > 96.4 kips O.K.

  $\phi V_c = 0.095\ b_w d = 0.095(36)17 = 58.1$ kips

  $\phi V_c/2 = 29.1$ kips

* *The bar cut-off points shown in Fig. 8-3(a) are recommended for beams without closed stirrups. The reader may consider determining actual bar lengths using the provisions in ACI 12.10.*

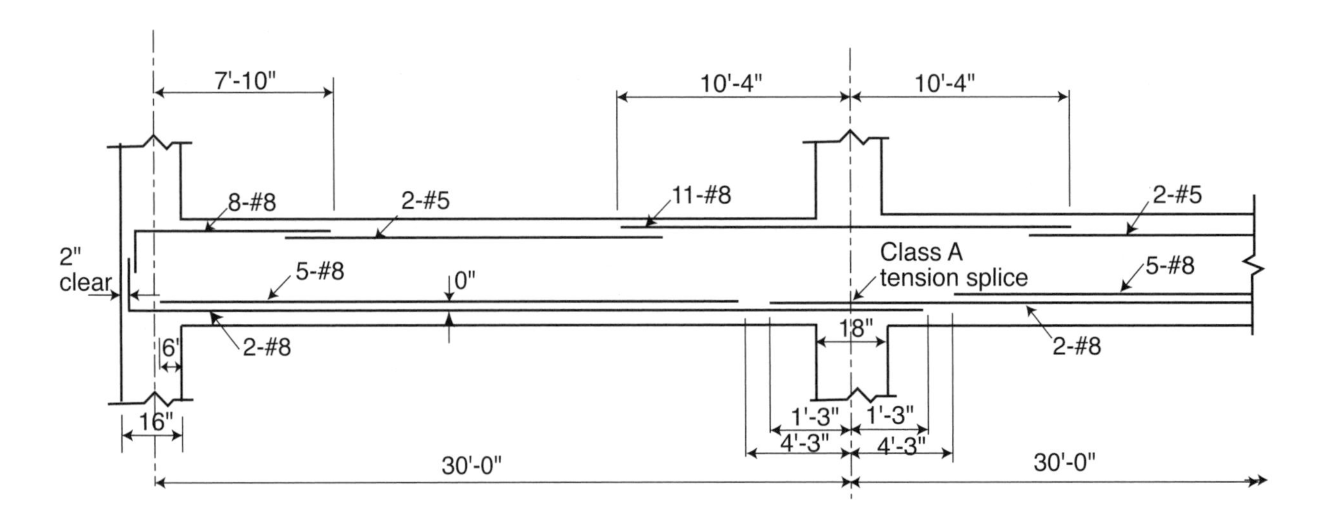

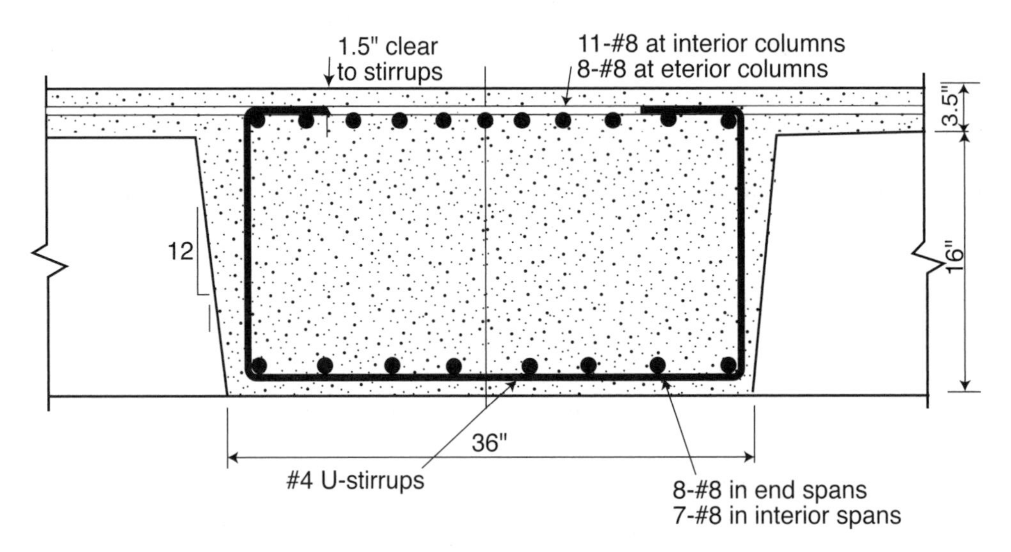

Figure 3-17 Reinforcement Details for Support Beams along N-S Interior Column Line

Length over which stirrups are required: (105.5 – 29.1)/6.4 = 11.9 ft

$$\phi V_s \text{ (required)} = 96.4 - 58.1 = 38.3 \text{ kips}$$

Try No. 4 U-stirrups

From Fig. 3-4, use No.4 @ 8 in. over the entire length where stirrups are required (see Fig. 3-18).

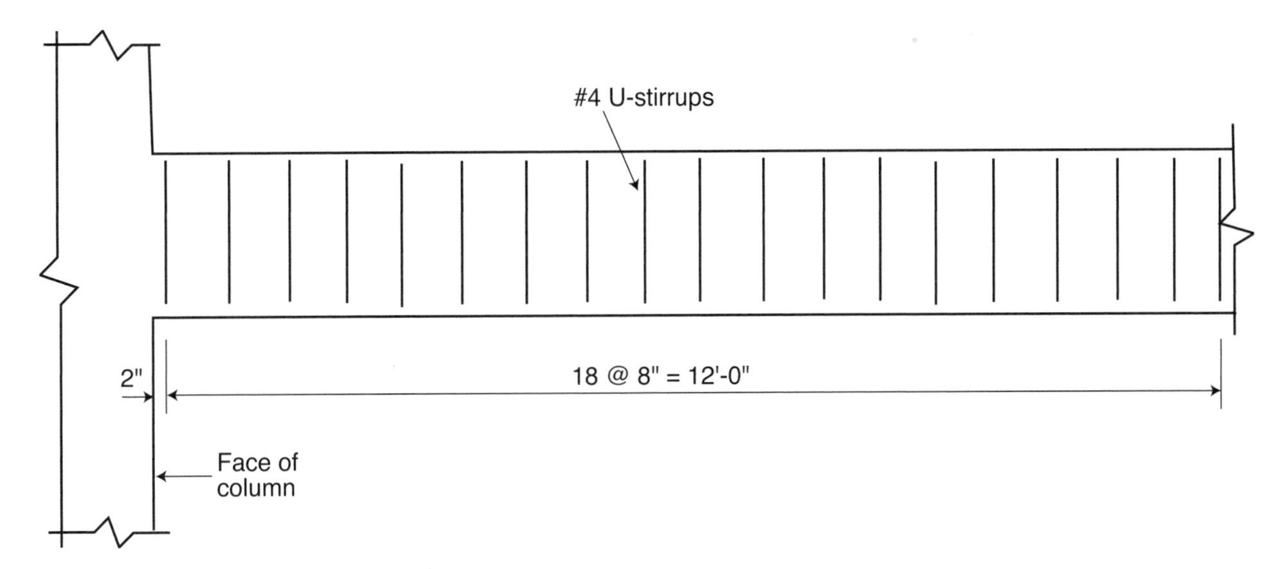

Figure 3-18 Stirrup Spacing Layout

References

3.1 Fling. R.S., "Using ACI 318 the Easy Way," *Concrete International,* Vol. 1, No. 1, January 1979.

3.2 Pickett, C., *Notes on ACI 318-77, Appendix A – Notes on Simplified Design,* 3rd Edition, Portland Cement Association, Skokie, Illinois.

3.3 Rogers, P., "Simplified Method of Stirrup Spacing," *Concrete International,* Vol. 1, No. 1, January 1979.

3.4 Fanella D. A, "Time-Saving Design Aids for Reinforced Concrete – Part 1", Structural Engineer, August 2001, (PCA RP410).

3.5 Fanella D. A, "Time-Saving Design Aids for Reinforced Concrete - Part 2 Two-Way Slabs", *Structural Engineer,* October 2001, (PCA RP410).

3.6 Fanella D. A, "Time-Saving Design Aids for Reinforced Concrete - Part 3 Columns & Walls", *Structural Engineer,* November 2001, (PCA RP410).

3.7 *PSI – Product Services and Information,* Concrete Reinforcing Steel Institute, Schaumburg, Illinois.

(a) Bulletin 7901A *Selection of Stirrups in Flexural Members for Economy*

(b) Bulletin 7701A *Reinforcing Bars Required – Minimum vs. Maximum*

(c) Bulletin 7702A *Serviceability Requirements with Grade 60 Bars*

3.8 *Notes on ACI 318-02,* Chapter 6, "General Principles of Strength Design," EB702, Portland Cement Association, Skokie, Illinois, 2002.

3.9 *Design Handbook in Accordance with the Strength Design Method of ACI 318-89: Vol. 1–Beams, Slabs, Brackets, Footings, and Pile Caps,* SP-17(91), American Concrete Institute, Detroit, 1991.

3.10 *CRSI Handbook,* Concrete Reinforcing Steel Institute, Schaumburg, Illinois, 9th Edition, 2002, (PCA LT265).

3.11 *ACI Detailing Manual* – American Concrete Institute, Detroit, 1994, (PCA LT185).

Chapter 4

Simplified Design for Two-Way Slabs

4.1 INTRODUCTION

Figure 4-1 shows the various types of two-way reinforced concrete slab systems in use at the present time.

A solid slab supported on beams on all four sides [Fig. 4-1(a)] was the original slab system in reinforced concrete. With this system, if the ratio of the long to the short side of a slab panel is two or more, load transfer is predominantly by bending in the short direction and the panel essentially acts as a one-way slab. As the ratio of the sides of a slab panel approaches unity (square panel), significant load is transferred by bending in both orthogonal directions, and the panel should be treated as a two-way rather than a one-way slab.

As time progressed and technology evolved, the column-line beams gradually began to disappear. The resulting slab system, consisting of solid slabs supported directly on columns, is called a flat plate [Fig. 4-1(b)]. The flat plate is very efficient and economical and is currently the most widely used slab system for multistory residential and institutional construction, such as motels, hotels, dormitories, apartment buildings, and hospitals. In comparison to other concrete floor/roof systems, flat plates can be constructed in less time and with minimum labor costs because the system utilizes the simplest possible formwork and reinforcing steel layout. The use of flat plate construction also has other significant economic advantages. For instance, because of the shallow thickness of the floor system, story heights are automatically reduced resulting in smaller overall heights of exterior walls and utility shafts, shorter floor to ceiling partitions, reductions in plumbing, sprinkler and duct risers, and a multitude of other items of construction. In cities like Washington, D.C., where the maximum height of buildings is restricted, the thin flat plate permits the construction of the maximum number of stories on a given plan area. Flat plates also provide for the most flexibility in the layout of columns, partitions, small openings, etc. Where job conditions allow direct application of the ceiling finish to the flat plate soffit, (thus eliminating the need for suspended ceilings), additional cost and construction time savings are possible as compared to other structural systems.

The principal limitation on the use of flat plate construction is imposed by punching shear around the columns (Section 4.4). For heavy loads or long spans, the flat plate is often thickened locally around the columns creating what are known as drop panels. When a flat plate is equipped with drop panels, it is called a flat slab [Fig. 4-1(c)]. Also, for reasons of shear capacity around the columns, the column tops are sometimes flared, creating column capitals. For purposes of design, a column capital is part of the column, whereas a drop panel is part of the slab.

Waffle slab construction [Fig. 4-1(d)] consists of rows of concrete joists at right angles to each other with solid heads at the columns (to increase shear resistance). The joists are commonly formed by using standard square dome forms. The domes are omitted around the columns to form the solid heads acting as drop panels. Waffle slab construction allows a considerable reduction in dead load as compared to conventional flat slab construc-

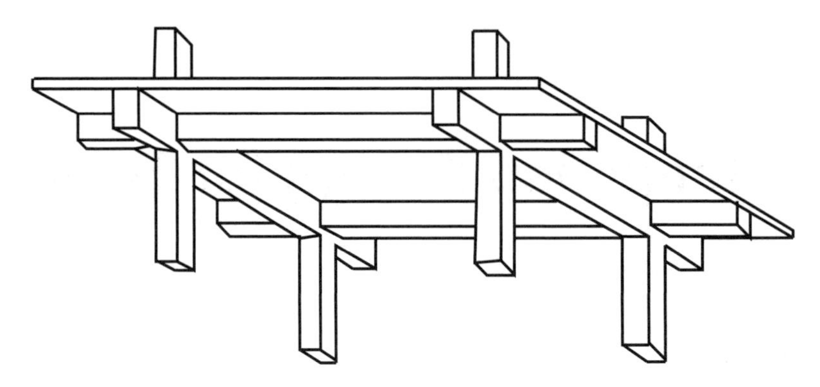

(a) Two-Way Beam Supported Slab

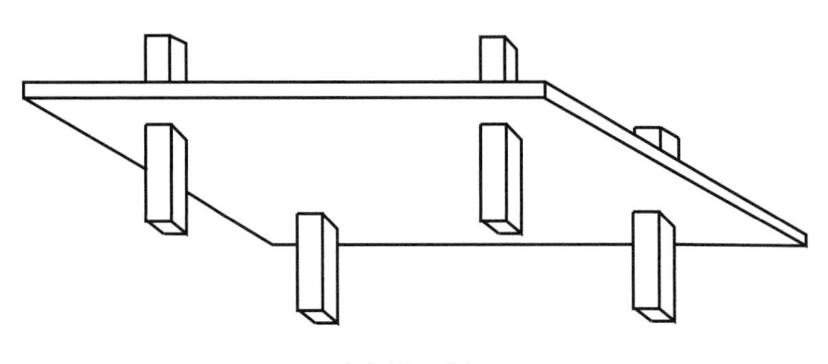

(b) Flat Plate

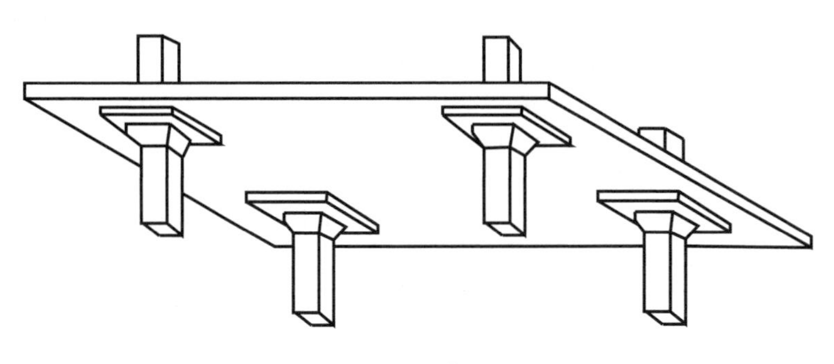

(c) Flat Slab

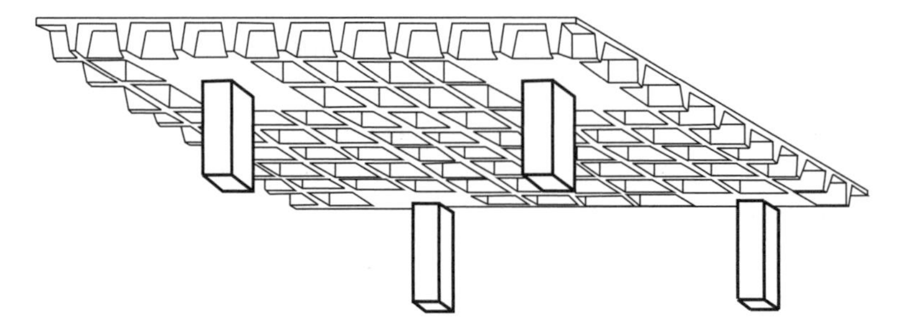

(d) Waffle Slab (Two-Way Joist Slab)

Figure 4-1 Types of Two-Way Slab Systems

tion. Thus, it is particularly advantageous where the use of long span and/or heavy loads is desired without the use of deepened drop panels or support beams. The geometric shape formed by the joist ribs is often architecturally desirable.

Discussion in this chapter is limited largely to flat plates and flat slabs subjected only to gravity loads.

4.2 DEFLECTION CONTROL—MINIMUM SLAB THICKNESS

Minimum thickness/span ratios enable the designer to avoid extremely complex deflection calculations in routine designs. Deflections of two-way slab systems need not be computed if the overall slab thickness meets the minimum requirements specified in ACI 9.5.3. Minimum slab thicknesses for flat plates, flat slabs (and waffle slabs), and two-way slabs, based on the provisions in ACI 9.5.3, are summarized in Table 4-1, where ℓ_n is the clear span length in the long direction of a two-way slab panel. The tabulated values are the controlling minimum thicknesses governed by interior or exterior panels assuming a constant slab thickness for all panels making up a slab system.[4.1] Practical spandrel beam sizes will usually provide beam-to-slab stiffness ratios, a, greater than the minimum specified value of 0.8; if this is not the case, the spandrel beams must be ignored in computing minimum slab thickness. A standard size drop panel that would allow a 10% reduction in the minimum required thickness of a flat slab floor system is illustrated in Fig. 4-2. Note that a larger size and depth drop may be used if required for shear strength; however, a corresponding lesser slab thickness is not permitted unless deflections are computed.

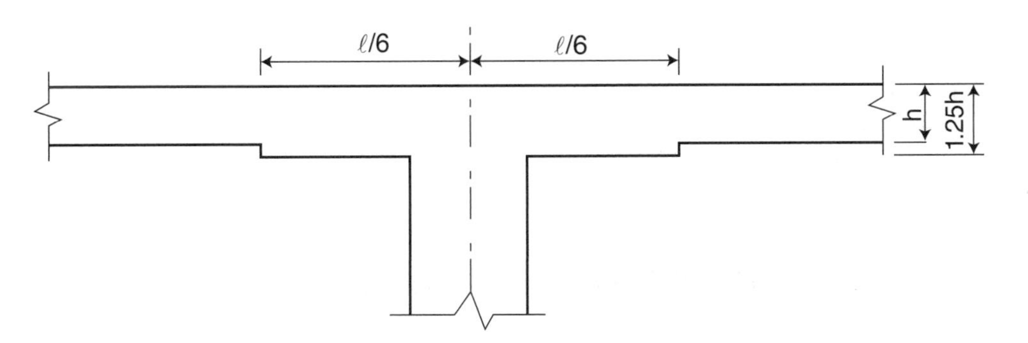

Figure 4-2 Drop Panel Details (ACI 13-3-7)

Table 4-1 gives the minimum slab thickness h based on the requirements given in ACI 9-5.3; α_m is the average value of α (ratio of flexural stiffness of beam to flexural stiffness of slab) for all beams on the edges of a panel, and β is the ratio of clear spans in long to short direction.

For design convenience, minimum thicknesses for the six types of two-way slab systems listed in Table 4-1 are plotted in Fig. 4-3.

Table 4-1 Minimum Thickness for Two-Way Slab Systems

Two-Way Slab System		α_m	β	Minimum h
Flat Plate		—	≤ 2	$\ell_n/30$
Flat Plate with Spandrel Beams[1]	[Min. h = 5 in.]	—	≤ 2	$\ell_n/33$
Flat Slab[2]		—	≤ 2	$\ell_n/33$
Flat Slab[2] with Spandrel Beams[1]	[Min. h = 4 in.]	—	≤ 2	$\ell_n/36$
Two-Way Beam-Supported Slab[3]		≤ 0.2	≤ 2	$\ell_n/30$
		1.0	1	$\ell_n/33$
			2	$\ell_n/36$
		≥ 2.0	1	$\ell_n/37$
			2	$\ell_n/44$
Two-Way Beam-Supported Slab[1,3]		≤ 0.2	≤ 2	$\ell_n/33$
		1.0	1	$\ell_n/36$
			2	$\ell_n/40$
		≥ 2.0	1	$\ell_n/41$
			2	$\ell_n/49$

[1] Spandrel beam-to-slab stiffness ratio $\alpha \geq 0.8$ (ACI 9.5.3.3)

[2] Drop panel length $\geq \ell/3$, depth $\geq 1.25h$ (ACI 13.3.7)

[3] Min. h = 5 in. for $\alpha_m \leq 2.0$; min. h = 3.5 in. for $\alpha_m > 2.0$ (ACI 9.5.3.3)

4.3 TWO-WAY SLAB ANALYSIS BY COEFFICIENTS

For gravity loads, ACI Chapter 13 provides two analysis methods for two-way slab systems: the Direct Design Method (ACI 13.6) and the Equivalent Frame Method (ACI 13.7). The Equivalent Frame Method, using member stiffnesses and complex analytical procedures, is not suitable for hand calculations. Only the Direct Design Method, using moment coefficients, will be presented in this Chapter.

The Direct Design Method applies when all of the conditions illustrated in Fig. 4-4 are satisfied (ACI 13.6.1):

- There shall be three or more continuous spans in each direction.
- Slab panels shall be rectangular with a ratio of longer to shorter span (c/c of supports) not greater than 2.
- Successive span lengths (c/c of supports) in each direction shall not differ by more than one-third of the longer span.
- Columns shall not be offset more than 10% of the span (in direction of offset) from either axis between centerlines of successive columns.

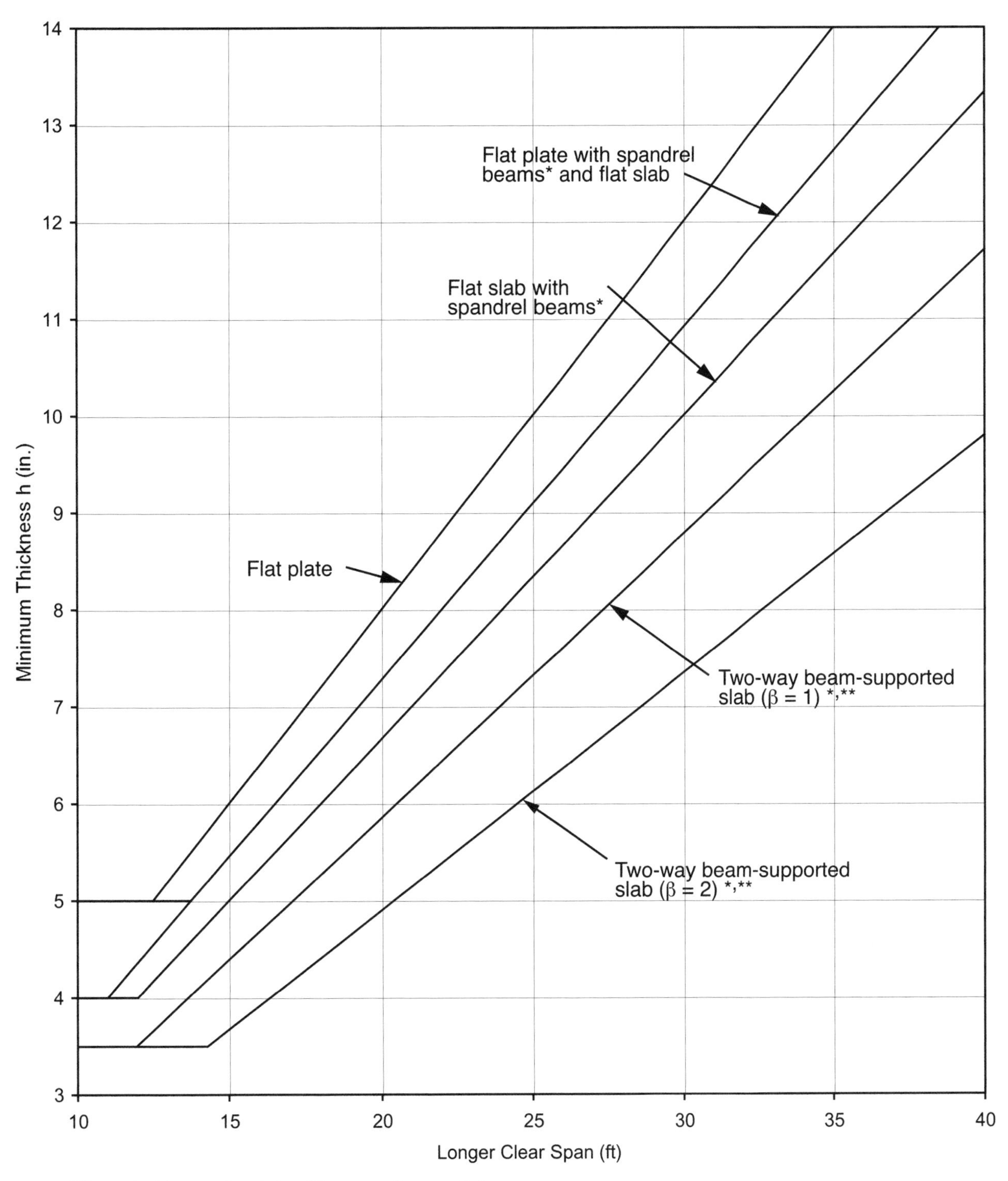

*Spandrel beam-to-slab stiffness ratio $\alpha \geq 0.8$

**$\alpha_m > 2.0$

Figure 4-3 Minimum Slab Thickness for Two-Way Slab Systems

- Loads must be due to gravity only and must be uniformly distributed over the entire panel.
- The live load shall not be more than 2 times the dead load ($L/D \leq 2$).
- For two-way slabs, relative stiffnesses of beams in two perpendicular directions must satisfy the minimum and maximum requirements given in ACI 13.6.1.6.
- Redistribution of moments by ACI 8.4 shall not be permitted.

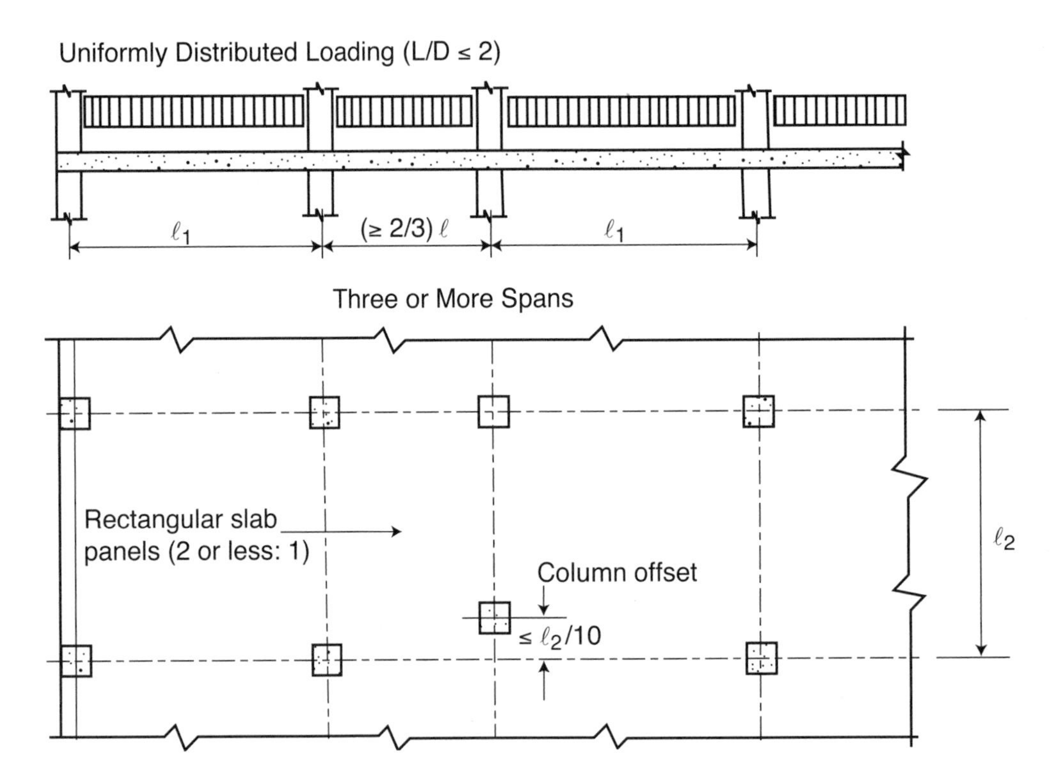

Figure 4-4 Conditions for Analysis for Coefficients

In essence, the Direct Design Method is a three-step analysis procedure. The first step is the calculation of the total design moment M_o for a given panel. The second step involves the distribution of the total moment to the negative and positive moment sections. The third step involves the assignment of the negative and positive moments to the column strips and middle strips.

For uniform loading, the total design moment M_o for a panel is calculated by the simple static moment expression, ACI Eq. (13-3):

$$M_o = w_u \ell_2 \ell_n^2 / 8$$

where w_u is the factored combination of dead and live loads (psf), $w_u = 1.2\, w_d + 1.6\, w_\ell$. The clear span ℓ_n is defined in a straightforward manner for slabs supported on columns or other supporting elements of rectangular cross section (ACI 13.6.2.5). Circular or regular polygon shaped supports shall be treated as square supports with the same area (see ACI Fig. R13.6.2.5). The clear span starts at the face of support and shall not be taken less than 65% of the span center-to-center of supports (ACI 13.6.2.5). The span ℓ_2 is simply the span transverse to ℓ_n; however, when the panel adjacent and parallel to an edge is being considered, the distance from edge of slab to panel centerline is used for ℓ_2 in calculation of M_o (ACI 13.6.2.4).

Division of the total panel moment M_o into negative and positive moments, and then into column and middle strip moments, involves direct application of moment coefficients to the total moment M_o. The moment coefficients are a function of span (interior or exterior) and slab support conditions (type of two-way slab system). For design convenience, moment coefficients for typical two-way slab systems are given in Tables 4-2 through 4-6. Tables 4-2 through 4-5 apply to flat plates or flat slabs with various end support conditions. Table 4-6 applies to two-way slabs supported on beams on all four sides. Final moments for the column strip and middle strip are computed directly using the tabulated values. All coefficients were determined using the appropriate distribution factors in ACI 13.6.3 through 13.6.6.

Table 4-2 Flat Plate or Flat Slab Supported Directly on Columns

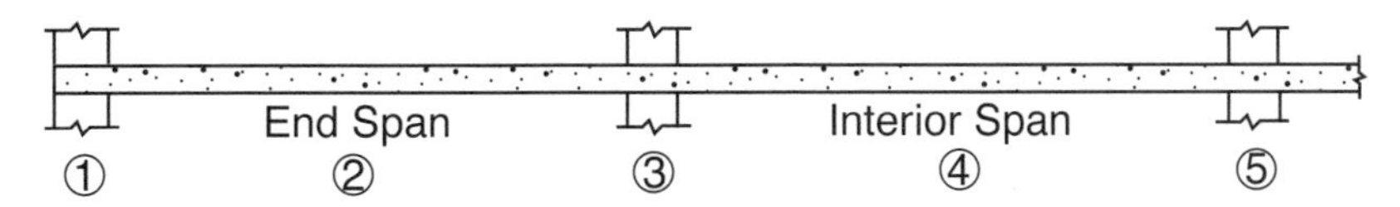

Slab Moments	End Span 1 Exterior Negative	End Span 2 Positive	End Span 3 First Interior Negative	Interior Span 4 Positive	Interior Span 5 Interior Negative
Total Moment	$0.26\ M_o$	$0.52\ M_o$	$0.70\ M_o$	$0.35\ M_o$	$0.65\ M_o$
Column Strip	$0.26\ M_o$	$0.31\ M_o$	$0.53\ M_o$	$0.21\ M_o$	$0.49\ M_o$
Middle Strip	0	$0.21\ M_o$	$0.17\ M_o$	$0.14\ M_o$	$0.16\ M_o$

Note: All negative moments are at face of support.

Table 4-3 Flat Plate or Flat Slab with Spandrel Beams

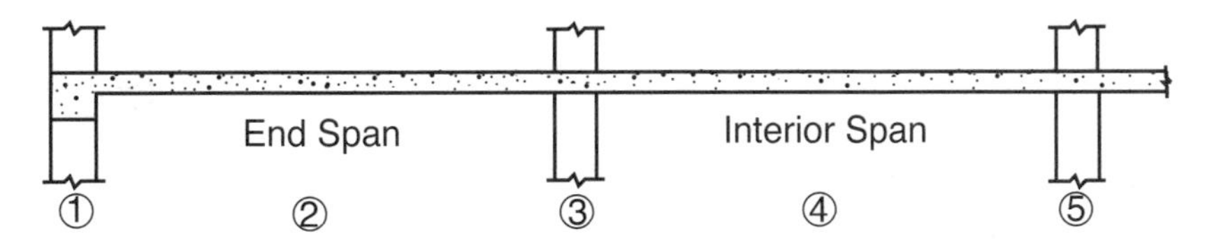

Slab Moments	End Span 1 Exterior Negative	End Span 2 Positive	End Span 3 First Interior Negative	Interior Span 4 Positive	Interior Span 5 Interior Negative
Total Moment	$0.30\ M_o$	$0.50\ M_o$	$0.70\ M_o$	$0.35\ M_o$	$0.65\ M_o$
Column Strip	$0.23\ M_o$	$0.30\ M_o$	$0.53\ M_o$	$0.21\ M_o$	$0.49\ M_o$
Middle Strip	$0.07\ M_o$	$0.20\ M_o$	$0.17\ M_o$	$0.14\ M_o$	$0.16\ M_o$

Note: (1) All negative moments are at face of support.
(2) Torsional stiffness of spandrel beams $\beta_t \geq 2.5$. For values of β_t less than 2.5, exterior negative column strip moment increases to $(0.30 - 0.03\ \beta_t)\ M_o$

Table 4-4 Flat Plate or Flat Slab with End Span Integral with Wall

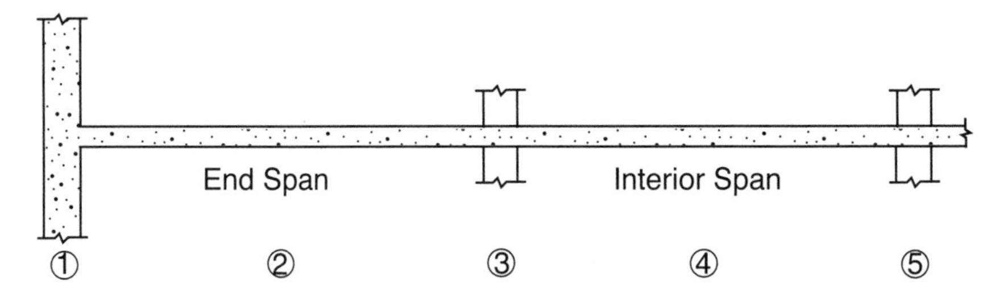

Slab Moments	End Span			Interior Span	
	1 Exterior Negative	2 Positive	3 First Interior Negative	4 Positive	5 Interior Negative
Total Moment	0.65 M_o	0.35 M_o	0.65 M_o	0.35 M_o	0.65 M_o
Column Strip	0.49 M_o	0.21 M_o	0.49 M_o	0.21 M_o	0.49 M_o
Middle Strip	0.16 M_o	0.14 M_o	0.16 M_o	0.14 M_o	0.16 M_o

Note: All negative moments are at face of support.

Table 4-5 Flat Plate or Flat Slab with End Span Simply Supported on Wall

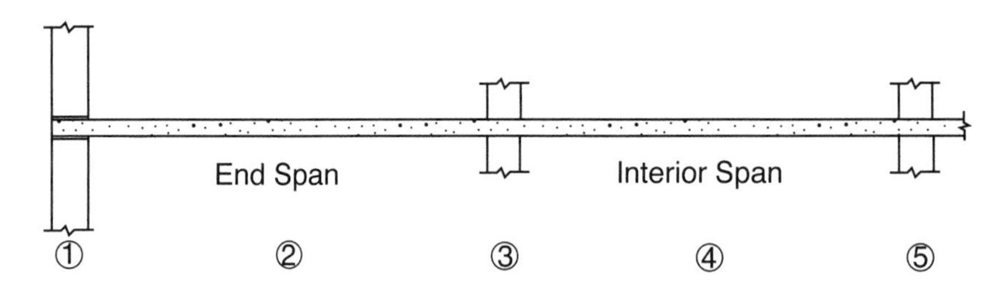

Slab Moments	End Span			Interior Span	
	1 Exterior Negative	2 Positive	3 First Interior Negative	4 Positive	5 Interior Negative
Total Moment	0	0.63 M_o	0.75 M_o	0.35 M_o	0.65 M_o
Column Strip	0	0.38 M_o	0.56 M_o	0.21 M_o	0.49 M_o
Middle Strip	0	0.25 M_o	0.19 M_o	0.14 M_o	0.16 M_o

Note: All negative moments are at face of support.

Table 4-6 Two-Way Beam-Supported Slab

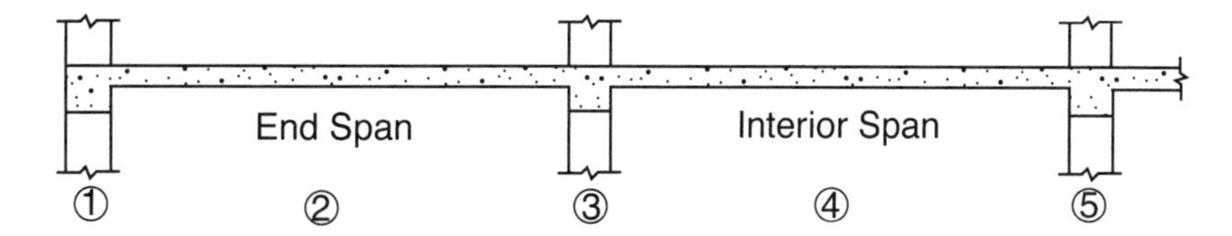

Span ratio	Slab Moments		End Span			Interior Span	
			1 Exterior Negative	2 Positive	3 First Interior Negative	4 Positive	5 Interior Negative
ℓ_2/ℓ_1	Total Moment		0.16 M_o	0.57 M_o	0.70 M_o	0.35 M_o	0.65 M_o
0.5	Column Strip	Beam	0.12 M_o	0.43 M_o	0.54 M_o	0.27 M_o	0.50 M_o
		Slab	0.02 M_o	0.08 M_o	0.09 M_o	0.05 M_o	0.09 M_o
	Middle Strip		0.02 M_o	0.06 M_o	0.07 M_o	0.03 M_o	0.06 M_o
1.0	Column Strip	Beam	0.10 M_o	0.37 M_o	0.45 M_o	0.22 M_o	0.42 M_o
		Slab	0.02 M_o	0.06 M_o	0.08 M_o	0.04 M_o	0.07 M_o
	Middle Strip		0.04 M_o	0.14 M_o	0.17 M_o	0.09 M_o	0.16 M_o
2.0	Column Strip	Beam	0.06 M_o	0.22 M_o	0.27 M_o	0.14 M_o	0.25 M_o
		Slab	0.01 M_o	0.04 M_o	0.05 M_o	0.02 M_o	0.04 M_o
	Middle Strip		0.09 M_o	0.31 M_o	0.38 M_o	0.19 M_o	0.36 M_o

Note: (1) Beams and slab satisfy stiffness criteria: $\alpha_1 \ell_2/\ell_1 \geq 1.0$ and $\beta_t \geq 2.5$.

(2) Interpolated between values shown for different ℓ_2/ℓ_1 ratios.

(3) All negative moments are at face of support.

(4) Concentrated loads applied directly to beams must be accounted for separately.

The moment coefficients of Table 4-3 (flat plate with spandrel beams) are valid for $\beta_t \geq 2.5$, the coefficients of Table 4-6 (two-way beam-supported slabs), are applicable when $\alpha_1\ell_2/\ell_1 \geq 1.0$ and $\beta_t \geq 2.5$ (β_t, and α_1 are stiffness parameters defined below). Many practical beam sizes will provide beam-to-slab stiffness ratios such that $\alpha_1\ell_2/\ell_1$ and β_t would be greater than these limits, allowing moment coefficients to be taken directly from the tables. However, if beams are present, the two stiffness parameters α_1 and β_t will need to be evaluated. For two-way slabs, the stiffness parameter α_1 is simply the ratio of the moments of inertia of the effective beam and slab sections in the direction of analysis, $\alpha_1 = I_b/I_s$, as illustrated in Fig. 4-5. Figures 4-6 and 4-7 can be used to determine α.

Relative stiffness provided by a spandrel beam is reflected by the parameter $\beta_t = C/2I_s$, where I_s is the moment of inertia of the effective slab spanning in the direction of ℓ_1 and having a width equal to ℓ_2, i.e., $I_s = \ell_2 h^3/12$. The constant C pertains to the torsional stiffness of the effective spandrel beam cross section. It is found by dividing the beam section into its component rectangles, each having smaller dimension x and larger dimension y, and summing the contribution of all the parts by means of the equation.

$$C = \sum\left(1 - 0.63\frac{x}{y}\right)\frac{x^3y}{3}$$

The subdivision can be done in such a way as to maximize C. Figure 4-8 can be used to determine the torsional constant C.

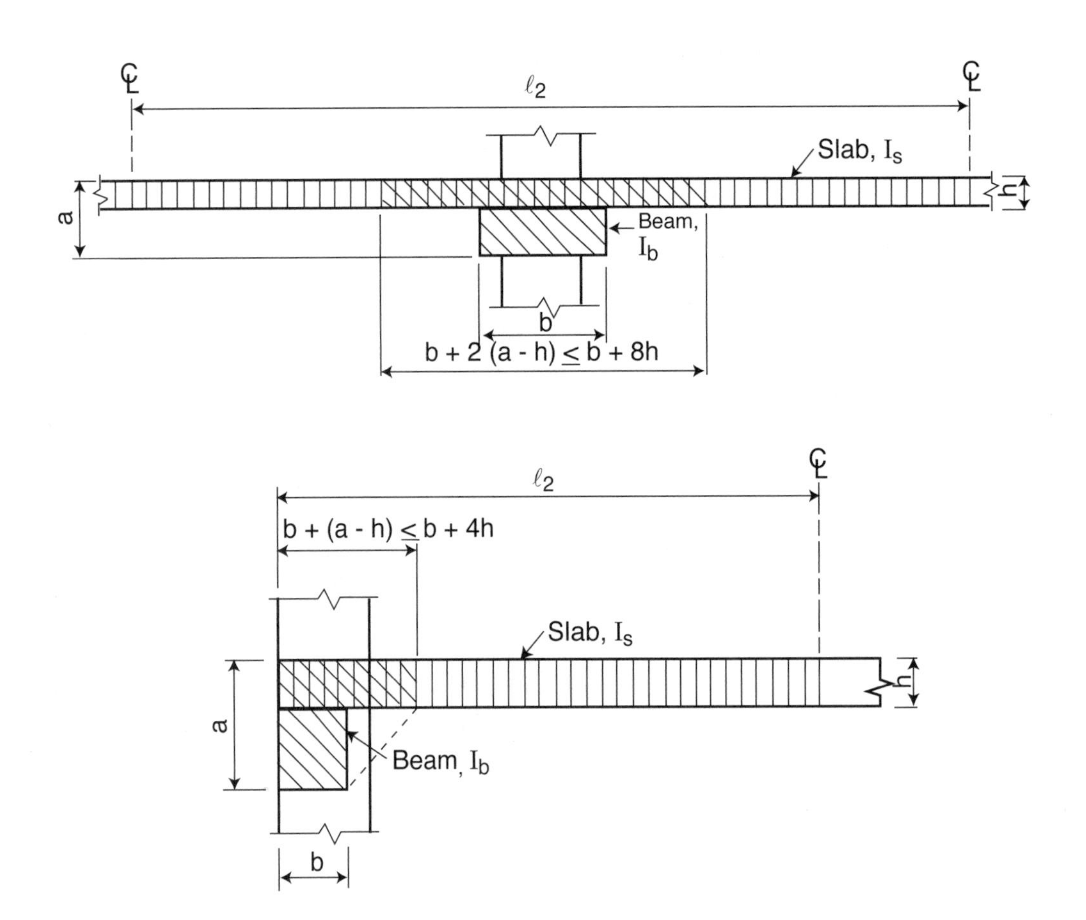

Figure 4-5 Effective Beam and Slab Sections for Stiffness Ratio α (ACI 13.2.4)

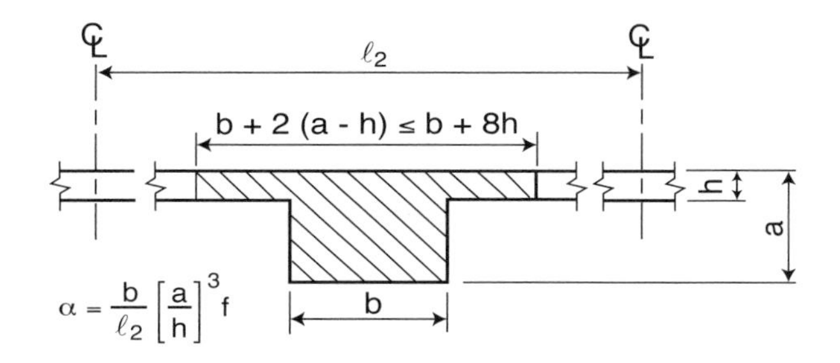

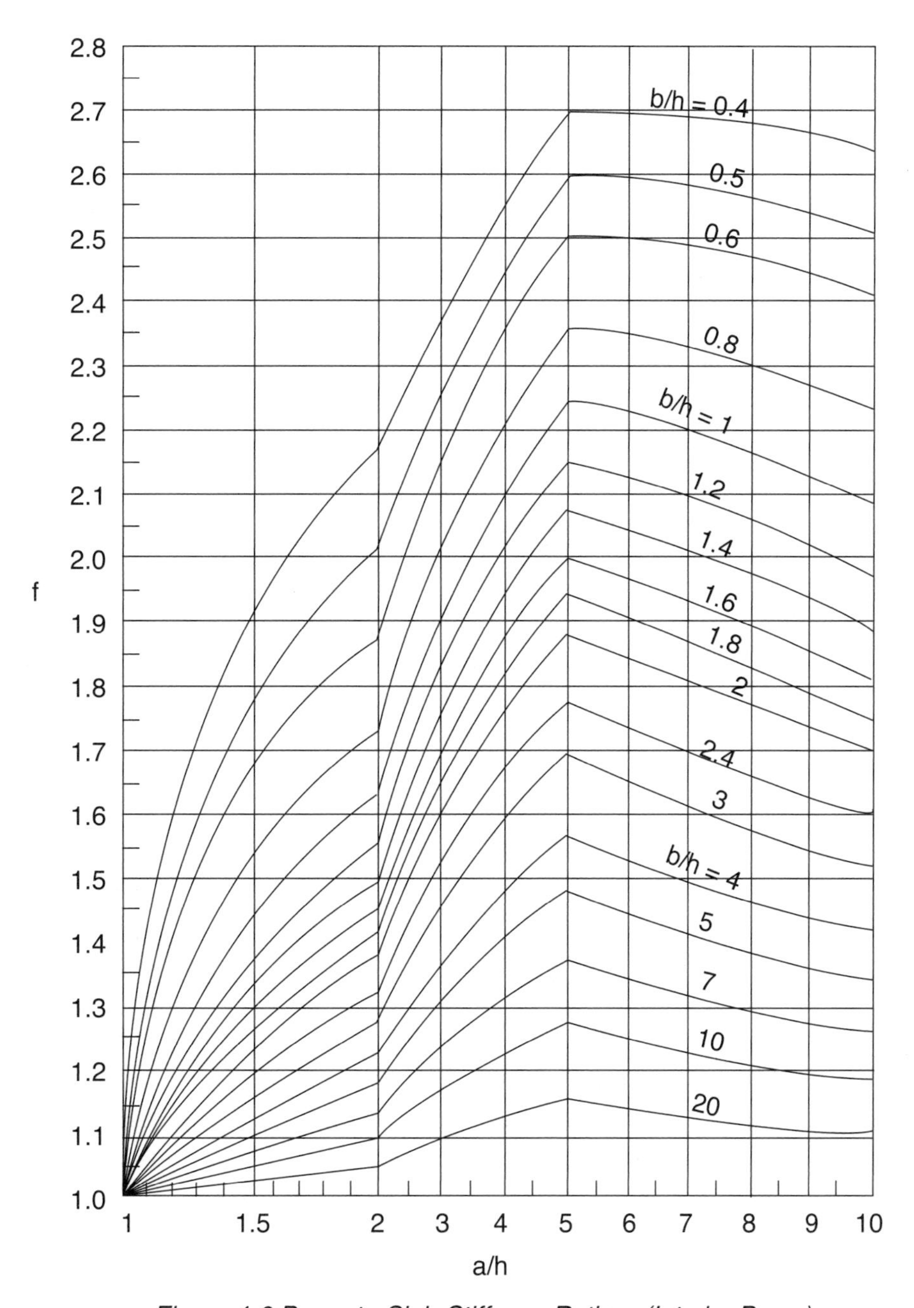

Figure 4-6 Beam to Slab Stiffness Ratio α (Interior Beam)

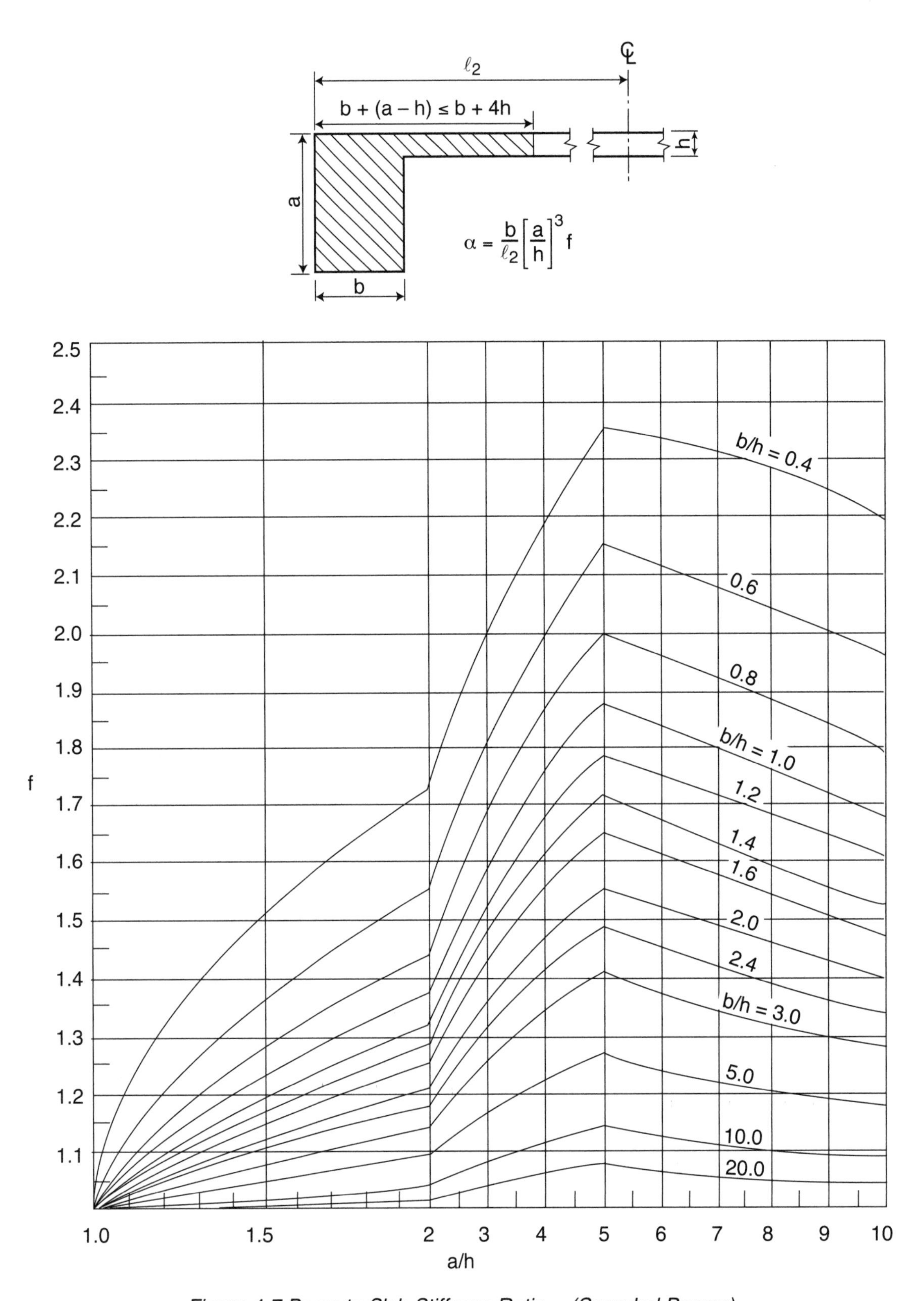

Figure 4-7 Beam to Slab Stiffness Ratio α (Spandrel Beams)

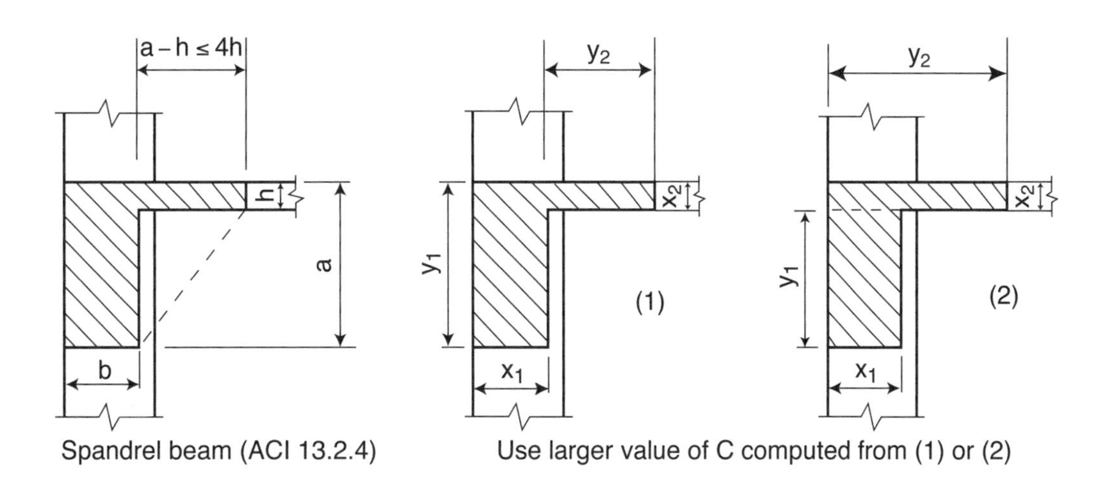

Values of torsion constant, $C = (1-0.63\ x/y)(x^3\ y/3)$

y \ x*	4	5	6	7	8	9	10	12	14	16
12	202	369	592	868	1,188	1,538	1,900	2,557	—	—
14	245	452	736	1,096	1,529	2,024	2,567	3,709	4,738	—
16	288	535	880	1,325	1,871	2,510	3,233	4,861	6,567	8,083
18	330	619	1,024	1,554	2,212	2,996	3,900	6,013	8,397	10,813
20	373	702	1,168	1,782	2,553	3,482	4,567	7,165	10,226	13,544
22	416	785	1,312	2,011	2,895	3,968	5,233	8,317	12,055	16,275
24	458	869	1,456	2,240	3,236	4,454	5,900	9,469	13,885	19,005
27	522	994	1,672	2,583	3,748	5,183	6,900	11,197	16,629	23,101
30	586	1,119	1,888	2,926	4,260	5,912	7,900	12,925	19,373	27,197
33	650	1,244	2,104	3,269	4,772	6,641	8,900	14,653	22,117	31,293
36	714	1,369	2,320	3,612	5,284	7,370	9,900	16,381	24,861	35,389
42	842	1,619	2,752	4,298	6,308	8,828	11,900	19,837	30,349	43,581
48	970	1,869	3,184	4,984	7,332	10,286	13,900	23,293	35,837	51,773
54	1,098	2,119	3,616	5,670	8,356	11,744	15,900	26,749	41,325	59,965
60	1,226	2,369	4,048	6,356	9,380	13,202	17,900	30,205	46,813	68,157

*Small side of a rectangular cross section with dimensions x and y.

Figure 4-8 Design Aid for Computing torsional Section Constant C

The column strip and middle strip moments are distributed over an effective slab width as illustrated in Fig. 4-9. The column strip is defined as having a width equal to one-half the transverse or longitudinal span; whichever is smaller (ACI 13.2.1). The middle strip is bounded by two column strips.

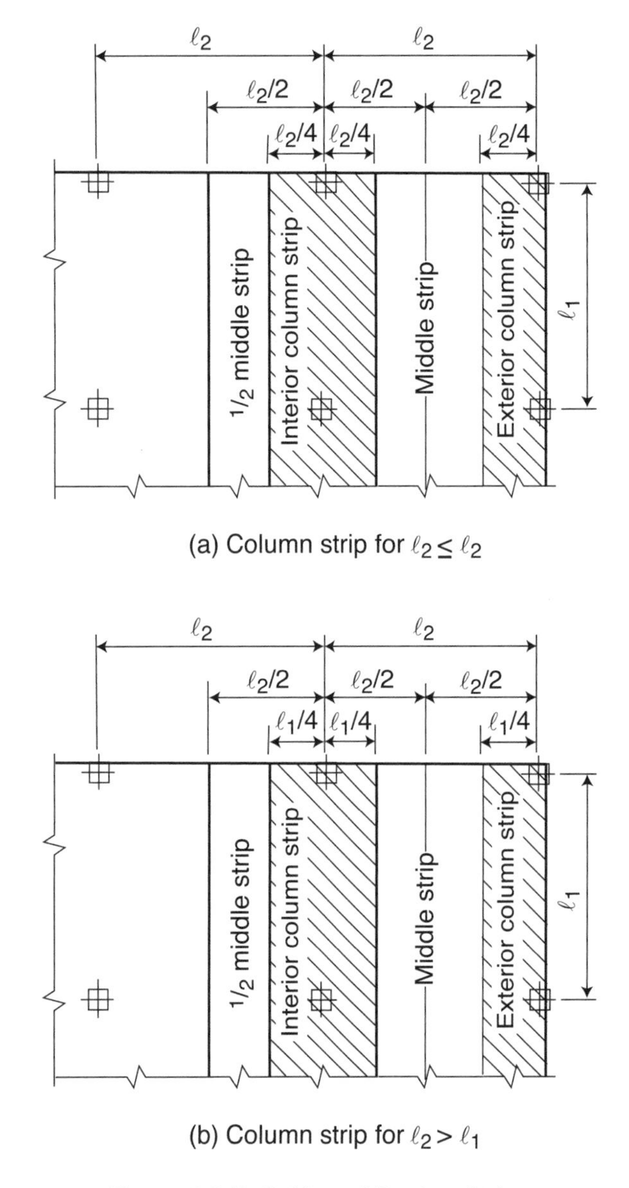

(a) Column strip for $\ell_2 \leq \ell_2$

(b) Column strip for $\ell_2 > \ell_1$

Figure 4-9 Definition of Design Strips

Once the slab and beam (if any) moments are determined, design of the slab and beam sections follows the simplified design approach presented in Chapter 3. Slab reinforcement must not be less than that given in Table 3-5, with a maximum spacing of 2h or 18 in. (ACI 13.3).

4.4 SHEAR IN TWO-WAY SLAB SYSTEMS

When two-way slab systems are supported by beams or walls, the shear capacity of the slab is seldom a critical factor in design, as the shear force due to the factored loads is generally well below the capacity of the concrete.

In contrast, when two-way slabs are supported directly by columns (as in flat plates and flat slabs), shear near the columns is of critical importance. Shear strength at an exterior slab-column connection (without spandrel beams) is especially critical because the total exterior negative slab moment must be transferred directly to the column. This aspect of two-way slab design should not be taken lightly by the designer. Two-way slab systems will normally be found to be fairly "forgiving" if an error in the distribution or even in the amount of flexural reinforcement is made, but there will be no forgiveness if a critical lapse occurs in providing the required shear strength.

For slab systems supported directly by columns, it is advisable at an early stage in the design to check the shear strength of the slab in the vicinity of columns as illustrated in Fig. 4-10.

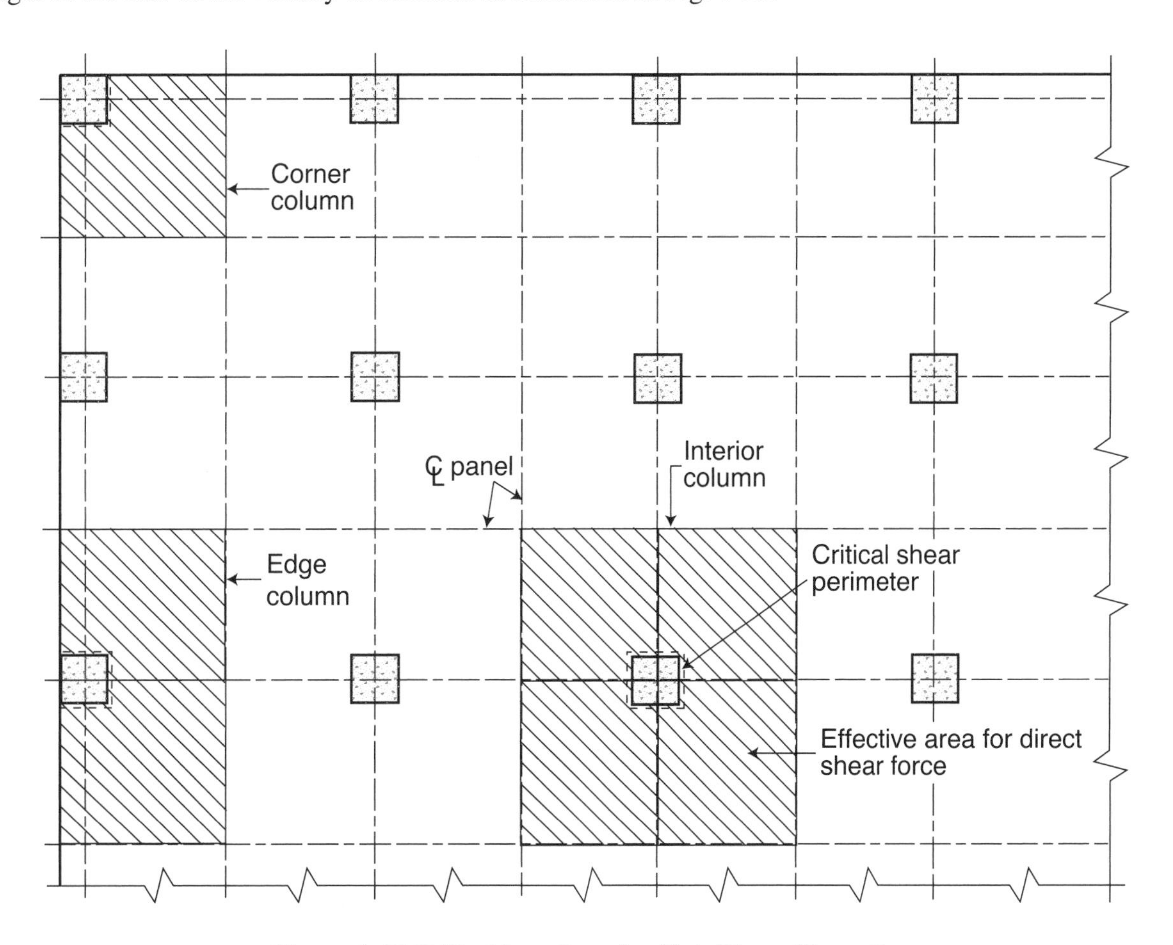

Figure 4-10 Critical Locations for Slab Shear Strength

4.4.1 Shear in Flat Plate and Flat Slab Floor Systems

Two types of shear need to be considered in the design of flat plates or flat slabs supported directly on columns. The first is the familiar one-way or beam-type shear, which may be critical in long narrow slabs. Analysis for beam shear considers the slab to act as a wide beam spanning between the columns. The critical section is taken a distance d from the face of the column. Design against beam shear consists of checking the requirement indicated in Fig. 4-11(a). Beam shear in slabs is seldom a critical factor in design, as the shear force is usually well below the shear capacity of the concrete.

Two-way or "punching" shear is generally the more critical of the two types of shear in slab systems supported directly on columns. Punching shear considers failure along the surface of a truncated cone or inverted pyramid around a column. The critical section is taken perpendicular to the slab at a distance d/2 from the perimeter of a column. The shear force V_u to be resisted can be easily calculated as the total factored load on the area bounded by panel centerlines around the column, less the load applied within the area defined by the critical shear perimeter (see Fig. 4-10). In the absence of a significant moment transfer from the slab to the column, design against punching shear consists of ensuring that the requirement in Fig. 4-11(b) is satisfied. Figures 4-12 through 4-14 can be used to determine ϕV_c for interior, edge and corner columns, respectively.

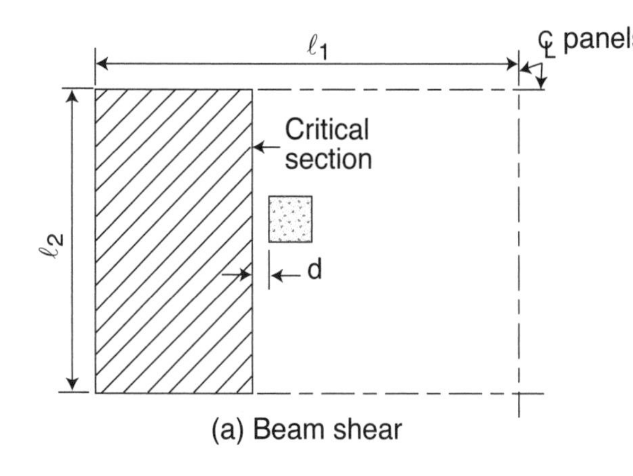

$$V_u \le \phi V_c$$
$$\le \phi 2\sqrt{f'_c}\,\ell_2 d$$
$$\le 0.095\,\ell_2 d \quad (f'_c = 4000 \text{ psi})$$

where V_u is factored shear force (total factored load on shaded area). V_u is in kips and ℓ_2 and d are in inches.

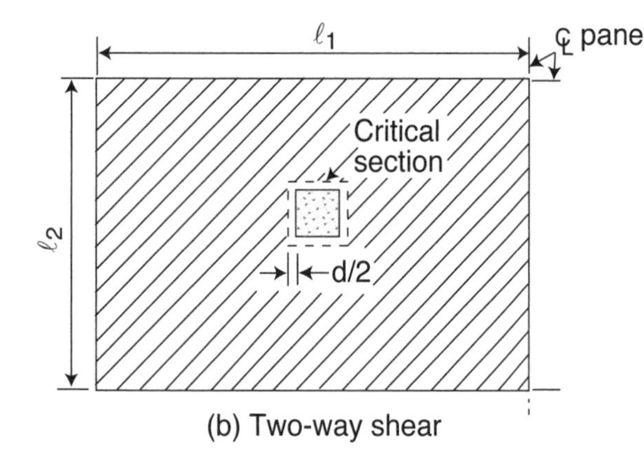

$$V_u \le \phi V_c$$

where:

$$\phi V_c = \text{least of} \begin{cases} \phi\left(2+\dfrac{4}{\beta_c}\right)\sqrt{f'_c}\,b_o d = 0.048\left(2+\dfrac{4}{\beta_c}\right)b_o d \\ \phi\left(\dfrac{\alpha_s d}{b_o}+2\right)\sqrt{f'_c}\,b_o d = 0.048\left(\dfrac{\alpha_s d}{b_o}+2\right)b_o d \\ \phi 4\sqrt{f'_c}\,b_o d = 0.19\,b_o d \end{cases}$$

V_u = factored shear force (total factored load on shaded area), kips

b_o = perimeter of critical section, in.

β_c = long side/short side of reaction area

α_s = constant (ACI 11.12.2.1 (b))

Figure 4-11 Direct Shear at an Interior Slab-Column Support

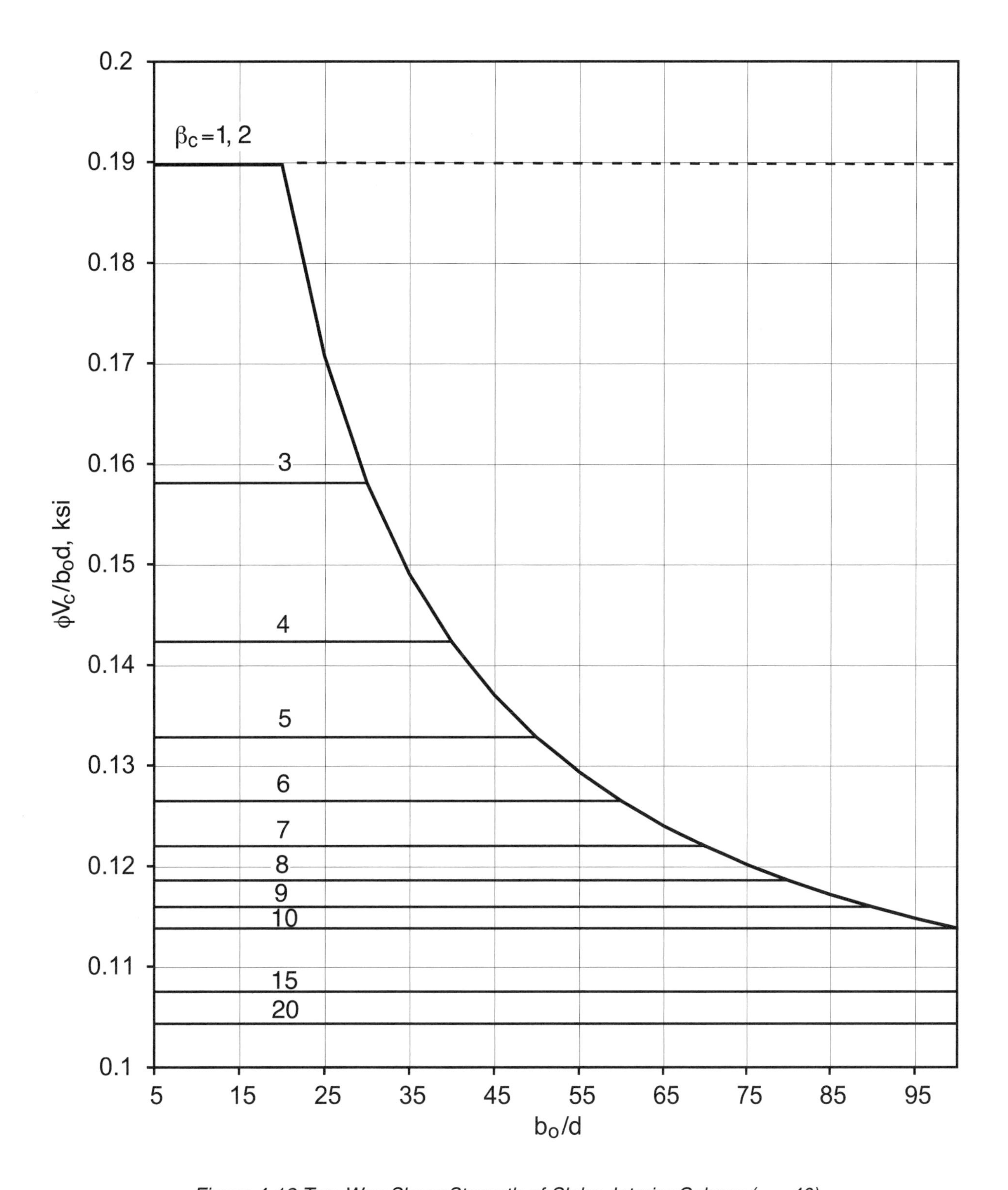

Figure 4-12 Two-Way Shear Strength of Slabs, Interior Column (α = 40)

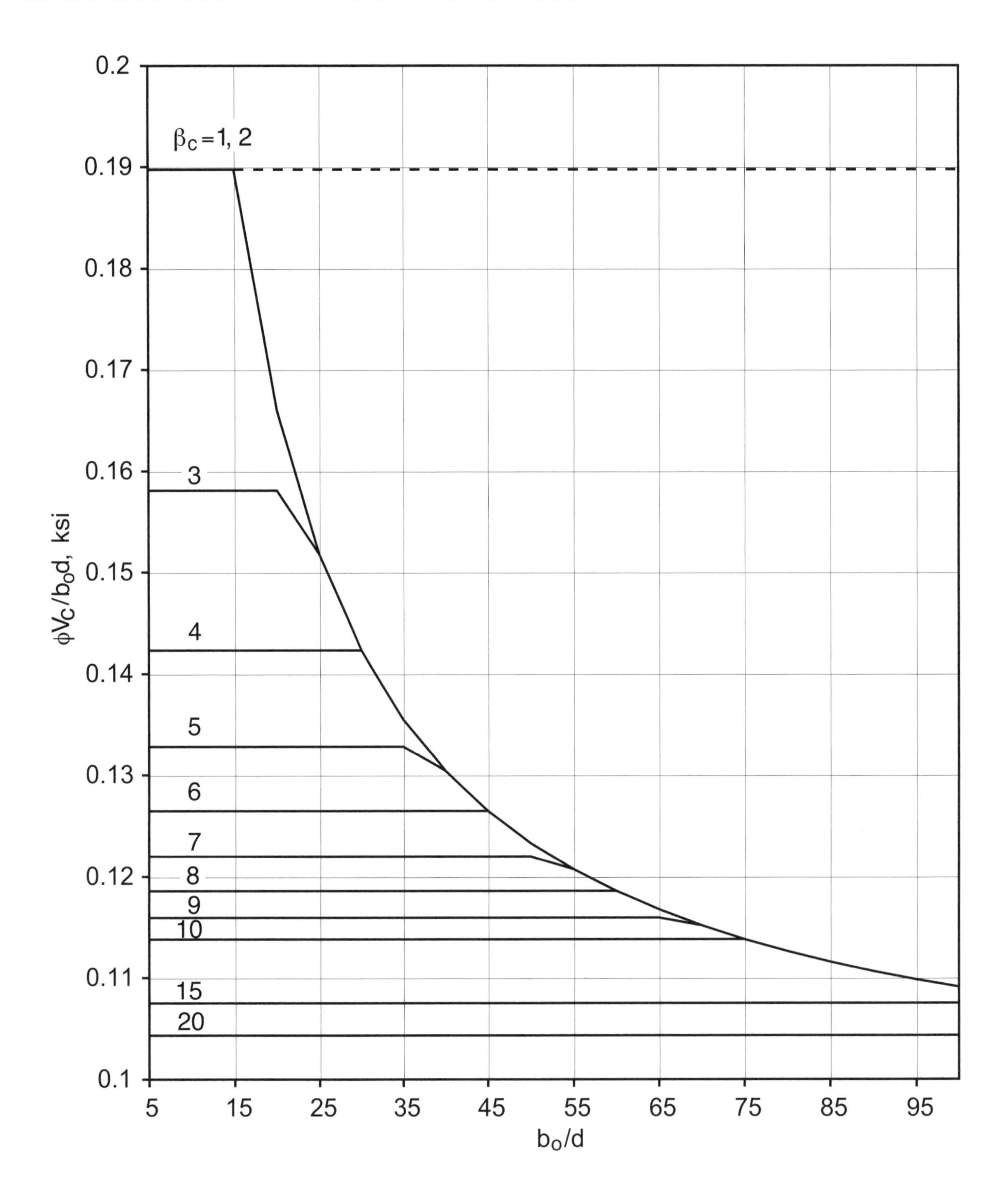

Figure 4-13 Two-Way Shear Strength of Slabs, Edge Column (α_s = 30)

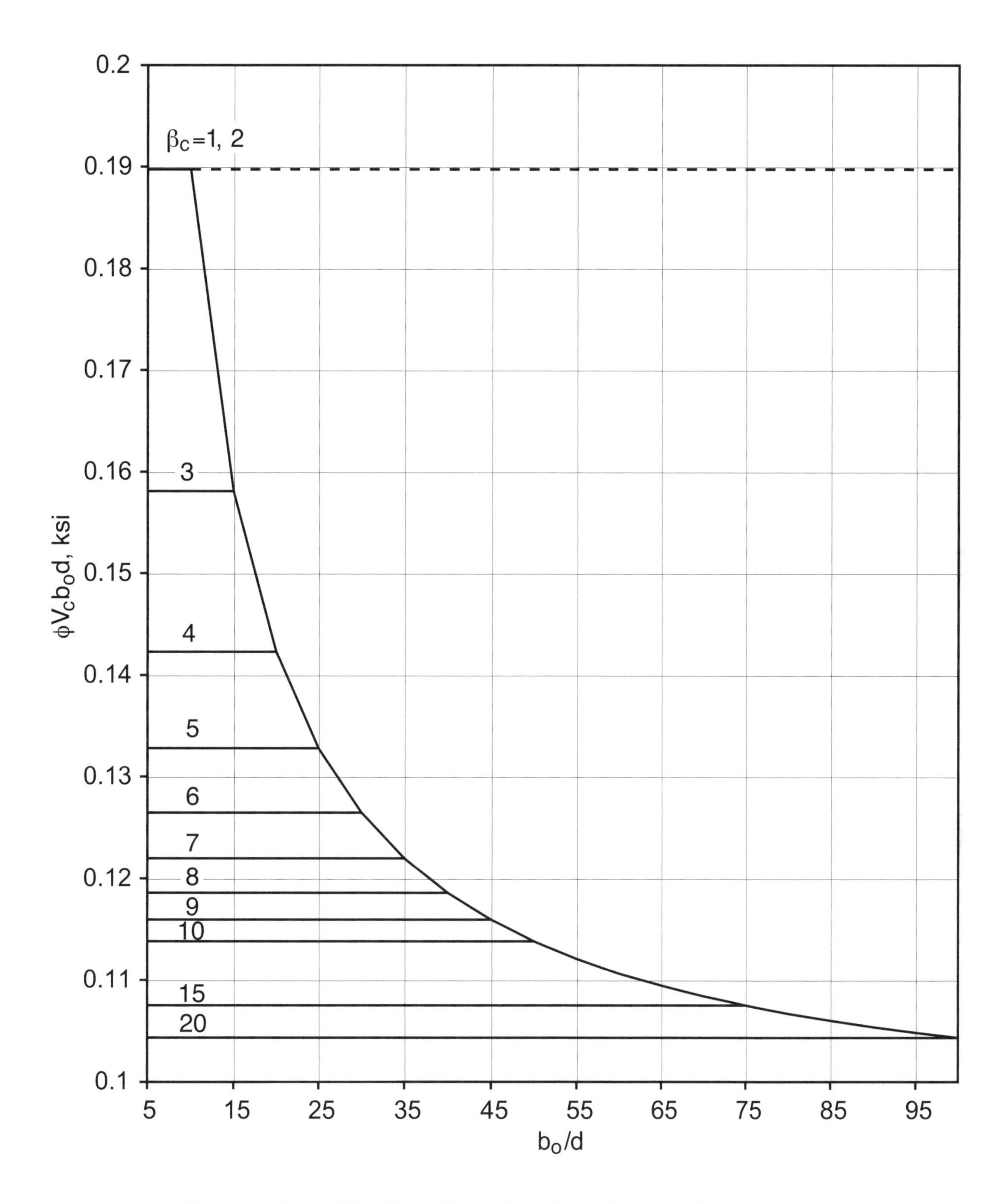

Figure 4-14 Two-Way Shear Strength of Slabs, Corner Column (α_s = 20)

For practical design, only direct shear (uniformly distributed around the perimeter b_o) occurs around interior slab-column supports where no (or insignificant) moment is to be transferred from the slab to the column. Significant moments may have to be carried when unbalanced gravity loads on either side of an interior column or horizontal loading due to wind must be transferred from the slab to the column. At exterior slab-column supports, the total exterior slab moment from gravity loads (plus any wind moments) must be transferred directly to the column.

Transfer of unbalanced moment between a slab and a column takes place by a combination of flexure (ACI 13.5.3) and eccentricity of shear (ACI 11.12.6). Shear due to moment transfer is assumed to act on a critical section at a distance d/2 from the face of the column, the same critical section around the column as that used for direct shear transfer [Fig. 4-11(b)]. The portion of the moment transferred by flexure is assumed to be transferred over a width of slab equal to the transverse column width c_2, plus 1.5 times the slab or drop panel thickness (1.5h) on each side of the column. Concentration of negative reinforcement is to be used to resist moment on this effective slab width. The combined shear stress due to direct shear and moment transfer often governs the design, especially at the exterior slab-columns supports.

The portions of the total moment to be transferred by eccentricity of shear and by flexure are given by ACI Eqs. (11-39) and (13-1), respectively. For square interior or corner columns, 40% of the moments is considered transferred by eccentricity of the shear ($\gamma_v M_u = 0.40\ M_u$), and 60% by flexure ($\gamma_f M_u = 0.60\ M_u$), where M_u is the transfer moment at the centroid of the critical section. The moment M_u at an exterior slab-column support will generally not be computed at the centroid of the critical transfer section in the frame analysis. In the Direct Design Method, moments are computed at the face of the support. Considering the approximate nature of the procedure used to evaluate the stress distribution due to moment transfer, it seems unwarranted to consider a change in moment to the transfer centroid; use of the moment values at the faces of the supports would usually be accurate enough.

The factored shear stress on the critical transfer section is the sum of the direct shear and the shear caused by moment transfer,

$$v_u = V_u/A_c + \gamma_v M_u\, c/J$$

or

$$v_u = V_u/A_c - \gamma_v M_u\, c'/J$$

Computation of the combined shear stress involves the following properties of the critical transfer section:

A_c = area of critical section, in.^{2}

c or c' = distance from centroid of critical section to the face of section where stress is being computed, in.

J_c = property of critical section analogous to polar moment of inertia, in.^{4}

The above properties are given in terms of formulas in Tables 4-7 through 4-10 (located at the end of this chapter) for the four cases that can arise with a rectangular column section: interior column (Table 4-7), edge column with bending parallel to the edge (Table 4-8), edge column with bending perpendicular to the edge (Table 4-9), and corner column (Table 4-10). Numerical values of the above parameters for various combinations of square column sizes and slab thicknesses are also given in these tables. Properties of the critical shear transfer section for circular interior columns can be found in Reference 4.2. Note that in the case of flat slabs, two different critical sections need to be considered in punching shear calculations as shown in Fig. 4-15. Tables 4-7 through 4-10 can be used in both cases. Also, Fig. 4-16 can be used to determine γ_v and γ_f given b_1 and b_2.

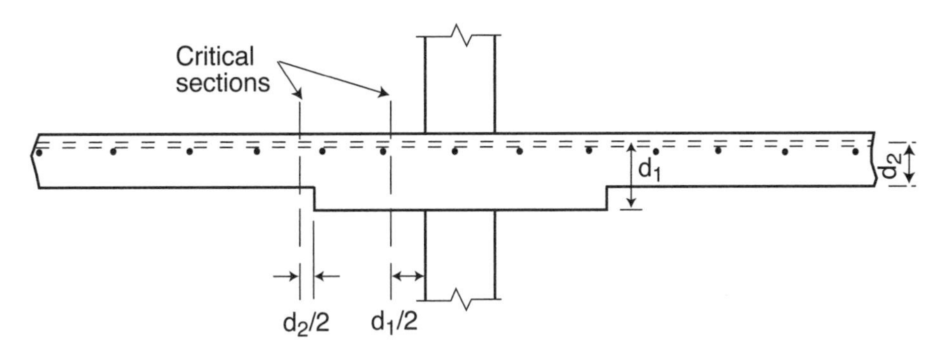

Figure 4-15 Critical Shear-Transfer Sections for Flat Slabs

Unbalanced moment transfer between slab and an edge column (without spandrel beams) requires special consideration when slabs are analyzed by the Direct Design Method for gravity loads. To assure adequate shear strength when using the approximate end-span moment coefficient, the moment 0.30 M_o must be used in determining the fraction of unbalanced moment transferred by eccentricity of shear ($\gamma_v M_u = \gamma_v 0.30 M_o$) according to ACI 13.6.3.6. For end spans without spandrel beams, the column strip is proportioned to resist the total exterior negative factored moment (Table 4-2). The above requirement is illustrated in Fig. 4-17. The total reinforcement provided in the column strip includes the additional reinforcement concentrated over the column to resist the fraction of unbalanced moment transferred by flexure $\gamma_f M_u = \gamma_f (0.26 M_o)$, where the moment coefficient (0.26) is from Table 4-2. Application of this special design requirement is illustrated in Section 4.7.

4.5 COLUMN MOMENTS DUE TO GRAVITY LOADS

Supporting columns (and walls) must resist any negative moments transferred from the slab system. For interior columns, the approximate ACI Eq. (13-4) may be used for unbalanced moment transfer due to gravity loading, unless an analysis is made considering the effects of pattern loading and unequal adjacent spans. The transfer moment is computed directly as a function of the span length and gravity loading. For the more usual case with equal transverse and longitudinal spans, ACI Eq. (13-4) simplifies to:

$$M_u = 0.07(0.05\, w_\ell \ell_2 \ell_n^2) = 0.035\, w_\ell \ell_2 \ell_n^2$$

where w_ℓ = factored live load, psf

ℓ_2 = span length transverse to

ℓ_n = clear span length in direction M_u is being determined

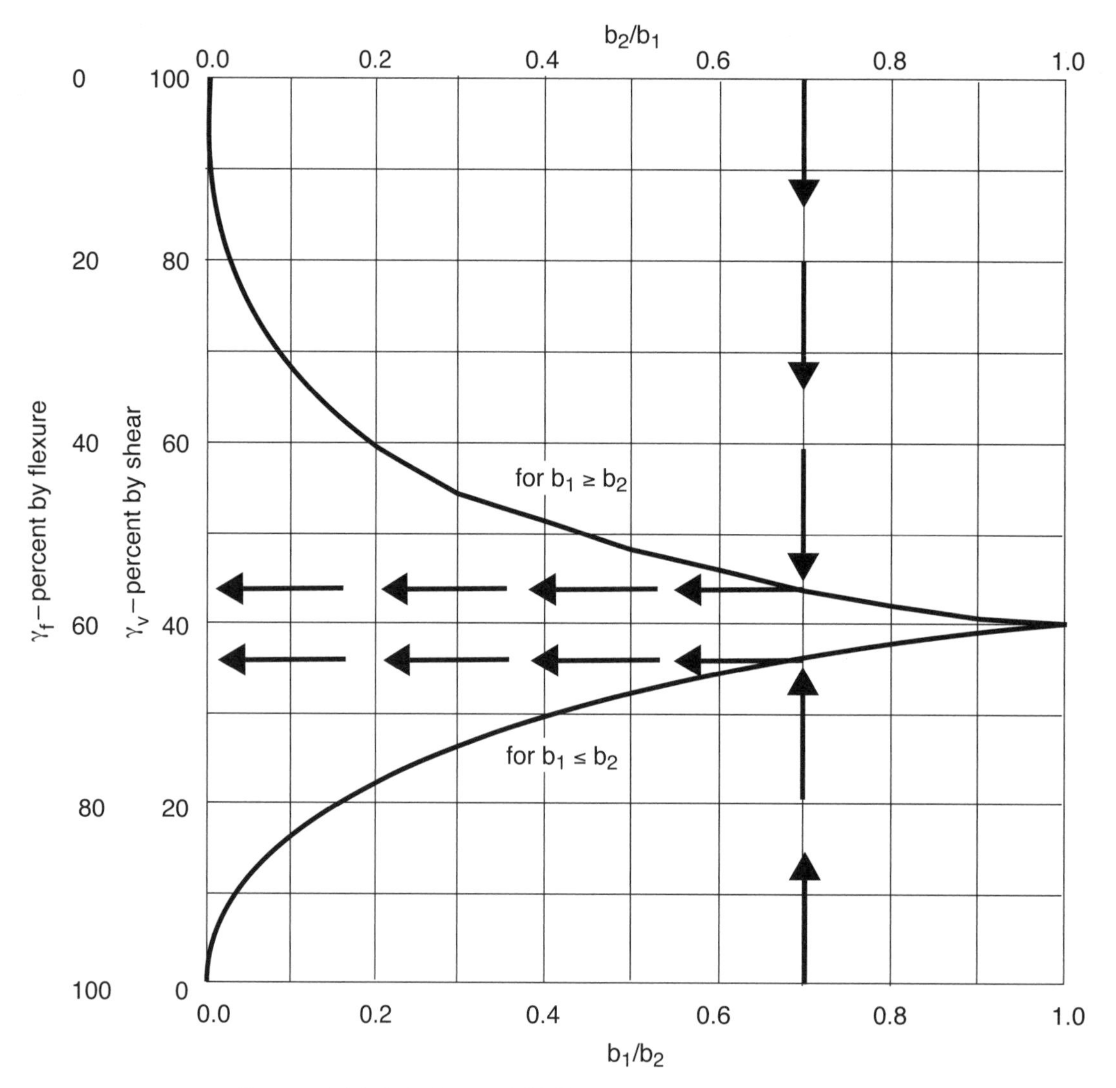

Figure 4-16 Solution to ACI Equations (11-42) and (13-1)

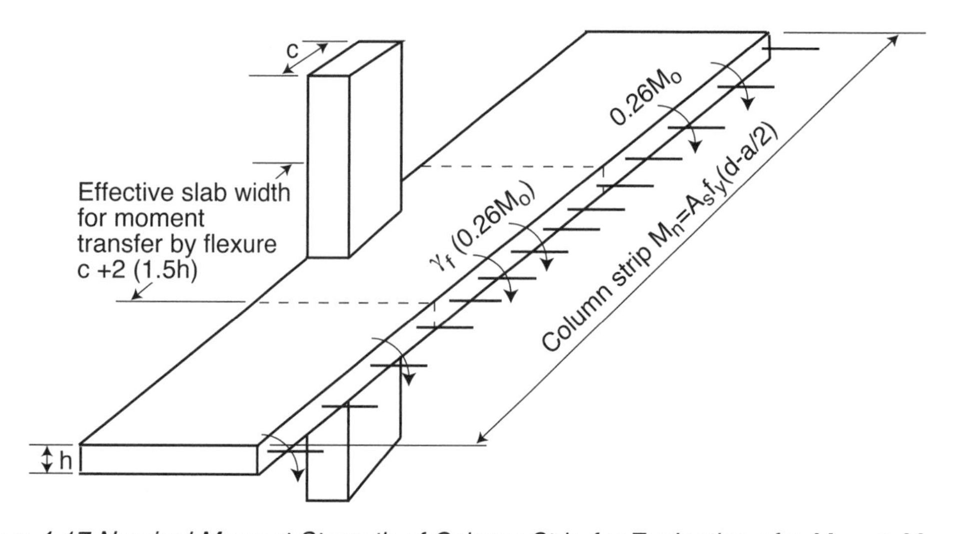

Figure 4-17 Nominal Moment Strength of Column Strip for Evaluation of $\gamma_v M_u = \gamma_v M_n$

At an exterior column, the total exterior negative moment from the slab system is transferred directly to the column. Due to the approximate nature of the moment coefficients of the Direct Design Method, it seems unwarranted to consider the change in moment from face of support to centerline of support; use of the exterior negative slab moment directly would usually be accurate enough.

Columns above and below the slab must resist portions of the support moment based on the relative column stiffnesses (generally, in proportion to column lengths above and below the slab). Again, due to the approximate nature of the moment coefficients, the refinement of considering the change in moment from centerline of slab to top or bottom of slab seems unwarranted.

4.6 REINFORCEMENT DETAILING

In computing required steel areas and selecting bar sizes, the following will ensure conformance to the Code and a practical design.

1. Minimum reinforcement area = 0.0018 bh (b = slab width, h = total thickness) for Grade 60 bars for either top or bottom steel. These minima apply separately in each direction (ACI 13.3.1).

2. Maximum bar spacing is 2h, but not more than 18 in. (ACI 13.3.2).

3. Maximum top bar spacing at all interior locations subject to construction traffic should be limited. Not less than No. 4 @ 12 in. is recommended to provide adequate rigidity and to avoid displacement of top bars with standard bar support layouts under ordinary foot traffic.

4. Maximum $\rho = A_s/bd$ is limited to 0.0206 however, $\rho \leq 0.50\ \rho_{max}$ is recommended to provide deformability, to avoid overly flexible systems subject to objectionable vibration or deflection, and for a practical balance to achieve overall economy of materials, construction and design time.

5. Generally, the largest size of bars that will satisfy the maximum limits on spacing will provide overall economy. Critical dimensions that limit size are the thickness of slab available for hooks and the distances from the critical design sections to edges of slab.

4.7 EXAMPLES: SIMPLIFIED DESIGN FOR TWO-WAY SLABS

The following two examples illustrate the use of the simplified design data presented in this chapter for the analysis and design of two-way slab systems. The two-way slab system for Building #2 is used to illustrate the simplified design.

4.7.1 Example: Interior Strip (N-S Direction) of Building #2, Alternate (2)

The slab and column framing will be designed for gravity loads only; the structural walls will carry the total wind forces.

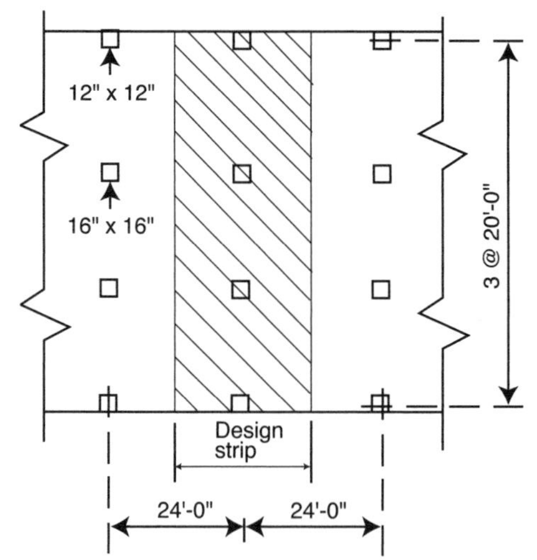

1. Data: f'_c = 4000 psi (carbonate aggregate)
   f_y = 60,000 psi

   Floors: LL = 50 psf
   DL = 142 psf (9 in. slab + 20 psf partitions + 10 psf ceiling and misc.)

   Required fire resistance rating = 2 hours

   Preliminary slab thickness:

   Determine preliminary h based on two-way shear at an interior column (see Fig. 1-8).

   w_u = 1.2(142) + 1.6(29.5*) = 218 psf
   A_2 = 24 × 20 = 480 ft²
   c_1^2 = 16 × 16 = 256 in.² = 1.8 ft²
   A/c_1^2 = 480/1.8 = 267

   From Fig. 1-8, required $d/c_1 \cong 0.38$

   Required d = 0.38 × 16 = 6.1 in.

   h = 6.10 + 1.25 = 7.35 in.

   To account for moment transfer at the edge columns, increase h by 20%

   Try preliminary h = 9 in.

2. Check the preliminary slab thickness for deflection control and shear strength.

   (a) Deflection control:

   From Table 4-1 (flat plate): $h = \ell_n/30 = (22.67 \times 12)/30 = 9.07$ in.

where $\ell_n = 24 - (16/12) = 22.67$ ft

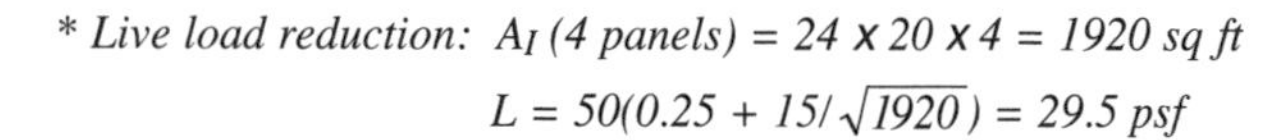

* *Live load reduction:* A_I *(4 panels) = 24 x 20 x 4 = 1920 sq ft*
$L = 50(0.25 + 15/\sqrt{1920}) = 29.5$ psf

(b) Shear Strength:

From Fig. 4-12: Check two-way shear strength at interior slab column support for h = 9 in.

From Table 4-7: $A_c = 736.3$ in.^{2} for 9 in. slab with 16 × 16 in column.

$b_1 = b_2 = 2(11.88) = 23.76$ in. = 1.98 ft

$V_u = 0.218(24 \times 20 - 1.98^*) = 103.8$ kips

From Fig. 4-12, with $\beta_c = 1$ and $b_o/d = 4(23.76)/(9 - 1.25) = 12.3$:

$\phi V_c = 0.19\ A_c = 0.19(736.3) = 139.9$ kips > 118.5 kips O.K.

Check for fire resistance: From Table 10-1, for fire resistance rating of 2 hours, required slab thickness = 4.6 in. < 9.0 in. O.K.

Use 9 in. slab.

3. Check limitations for slab analysis by coefficients (ACI 13.6.1)

- 3 continuous spans in one direction, 5 in the other
- rectangular panels with long-to-short span ratio = 24/20 = 1.2 < 2
- successive span lengths in each direction are equal
- LL/DL = 50/142 = 0.35 < 2
- slab system is without beams

Since all requirements are satisfied, the Direct Design Method can be used to determine the moments.

4. Factored moments in slab (N-S direction)

(a) Total panel moment M_o:

$M_o = w_u \ell_2 \ell_n^2/8$

$= 0.245 \times 24 \times 18.83^2/8 = 260.6$ ft-kips

where $w_u = 1.2(142) + 1.6(46.5^*) = 245$ psf

$\ell_2 = 24$ ft

ℓ_n (interior span) = 20 – 1.33 = 18.67 ft

ℓ_n (end span) = 20 – 0.67 – 0.50 = 18.83 ft

---

* *Live load reduction:* A_I *(1 panel) = 24 x 20 = 480 sq ft*

$L = 50(0.25 + 15/\sqrt{480}) = 46.5$ *psf*

Use larger value of ℓ_n for both spans.

(b) Negative and positive factored moments:

Division of the total panel moment M_o into negative and positive moments, and then, column and middle strip moments, involves direct application of the moment coefficients in Table 4-2.

Slab Moments (ft-kips)	End Spans Exterior Negative	End Spans Positive	End Spans Interior Negative	Interior Span Positive
Total Moment	67.8	135.5	182.4	91.2
Column Strip	67.8	80.8	138.1	54.7
Middle Strip	0	54.7	44.3	36.5

Note: All negative moments are at face of columns.

5. Slab Reinforcement

Required slab reinforcement is easily determined using a tabular form as follows:

Span Location		M_u (ft-kips)	b^1 (in.)	b^2 (in.)	$A_s = M_u/4d$ (in.^{2})	A_s^3 (min) (in.^{2})	No. of #4 Bars	No. of #5 Bars
END SPAN								
Column Strip	Ext. Negative	67.8	120	7.75	2.19	1.94	11	8
	Positive	80.8	120	7.75	2.61	1.94	14	9
	Int. Negative	138.1	120	7.75	4.45	1.94	23	15
Middle Strip	Ext. Negative	0	168	7.75	0.00	2.72	14	9
	Positive	54.7	168	7.75	1.76	2.72	14	9
	Int. Negative	44.3	168	7.75	1.43	2.72	14	9
INTERIOR SPAN								
Column Strip	Positive	54.7	120	7.75	1.76	1.94	10	7
Middle Strip	Positive	36.5	168	7.75	1.18	2.72	14	9

[1]Column strip = 0.5(20 x 12) = 120 in. (see Fig. 4-9b)
Middle Strip = (24 x 12) – 120 = 168 in.
[2]Use average d = 9 – 1.25 = 7.75 in.
[3]$A_{s(min)}$ = 0.0018 bh = 0.0162b
s_{max} = 2h < 18 in = 2(9) = 18 in.

6. Check slab reinforcement at exterior column (12 × 12 in.) for moment transfer between slab and column. For a slab without spandrel beams, the total exterior negative slab moment is resisted by the column strip (i.e., M_u = 67.8 ft-kips).

Fraction transferred by flexure using ACI Eq. (13-1):

$$b_1 = 12 + (7.75/2) = 15.88 \text{ in.}$$

$$b_2 = 12 + 7.75 = 19.75 \text{ in.}$$

From Fig. 4-16, $\gamma_f \cong 0.62$ with $b_1/b_2 = 0.8$

$M_u = 0.62\ (67.8) = 42.0$ ft-kips

$A_s = M_u/4d = 42/(4 \times 7.75) = 1.35$ in.^{2}

No. of No. 4 bars = 1.35/0.20 = 6.75 bars, say 7 bars

Must provide 7-No. 4 bars within an effective slab width (ACI 13.5.3.3) = $3h + c_2 = 3(9) + 12 = 39$ in.

Provide the required 7-No. 4 bars by concentrating 7 of the column strip bars (13-No. 4) within the 3 ft-3 in. slab width over the column. For symmetry, add one column strip bar to the remaining 5 bars so that 3 bars will be on each side of the 3 ft-3 in. strip. Check bar spacing:

For 7-No. 4 within 39 in. width: 39.8 = 4.9 in.

For 6-No. 4 within (120 – 39) = 81 in. width: 81/6 = 13.5 in. < 18 in. O.K.

No additional bars are required for moment transfer.

7. Reinforcement details are shown in Figs. 4-18, 4-19, and 4-21. Bar lengths are determined directly from Fig. 8-6. Note that for structural integrity, all bottom bars within the column strip, in each direction must be continuous or spliced with class A splices. Splices must be located as shown in the figure. At least two of the column strip bottom bars must pass within the column core and must be anchored at exterior supports.

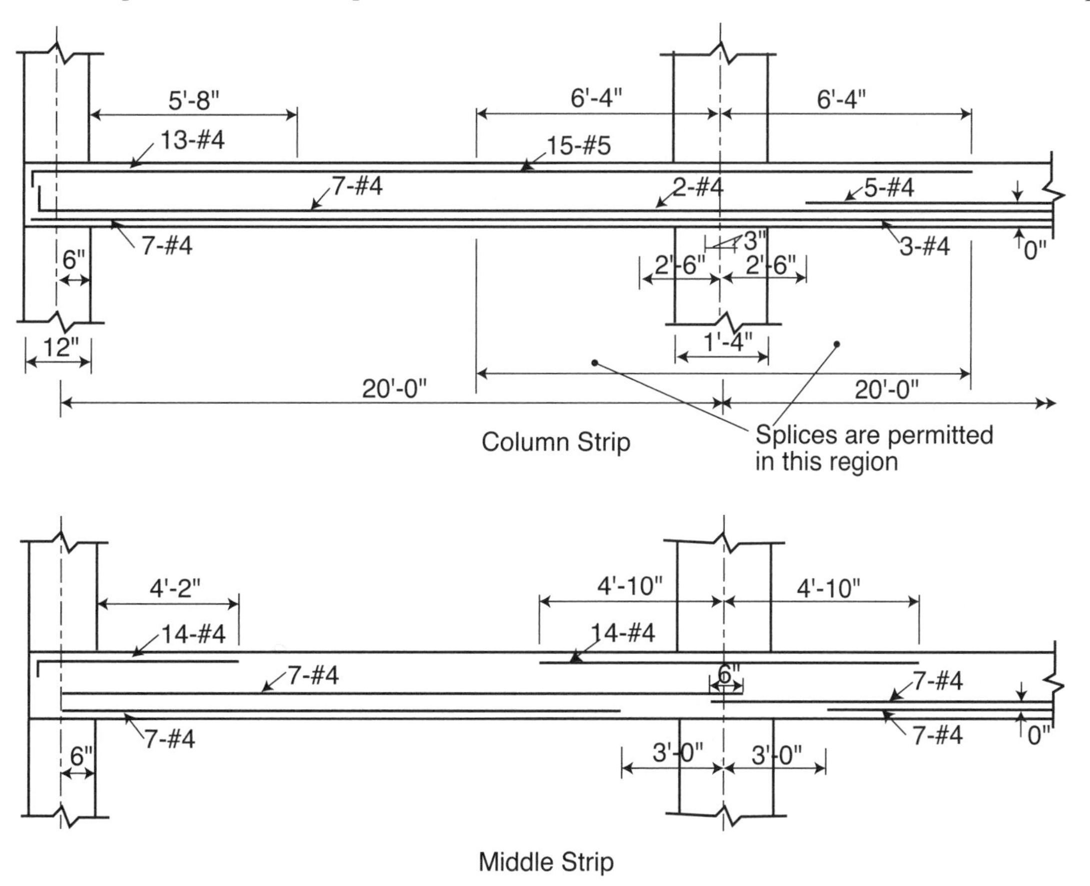

Figure 4-18 Reinforcement Details for Flat Plate of Building #2—Alternate (2) Interior Slab Panel (N-S Direction)

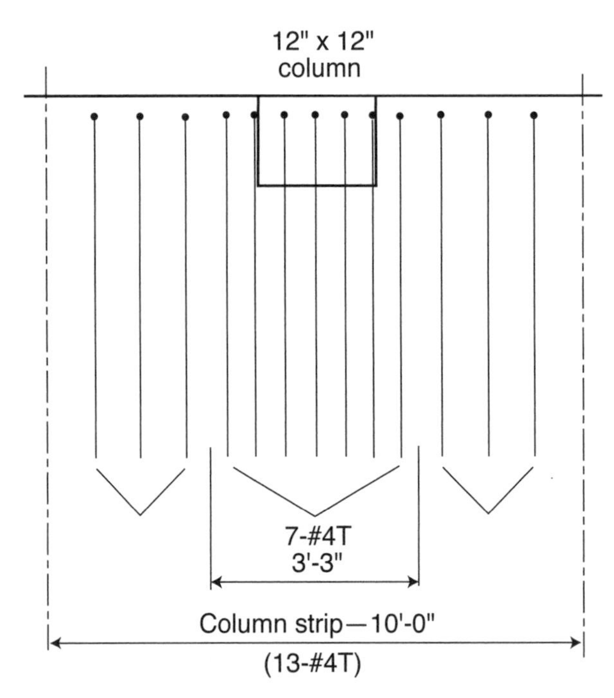

Figure 4-19 Bar Layout Detail for 13-#4 Top Bars at Exterior Columns

8. Check slab shear strength at edge column for gravity load shear and moment transfer (see Fig. 4-20).

(a) Direct shear from gravity loads:

live load reduction:

$K_{LL}A_T$ (2 panels) = $24 \times 20 \times 2 = 960$ sq ft

$L = 50(0.25 + 15/\sqrt{960}) = 36.7$ psf

$w_u = 1.2(142) + 1.6(36.7) = 229$ psf

$V_u = 0.229[(24)(10.5) - (1.32)(1.65)] = 57.2$ kips

(b) Moment transfer from gravity loads:

When slab moments are determined using the approximate moment coefficients, the special provisions of ACI 13.6.3.6 apply for moment transfer between slab and an edge column. The fraction of unbalanced moment transferred by eccentricity of shear (ACI 11.12.6.1) must be based on 0.3 M_o.

(c) Combined shear stress at inside face of critical transfer section:

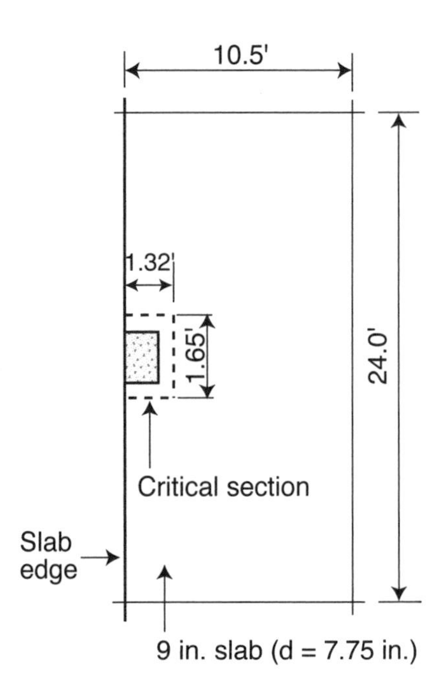

Figure 4-20 Critical Section for Edge Column

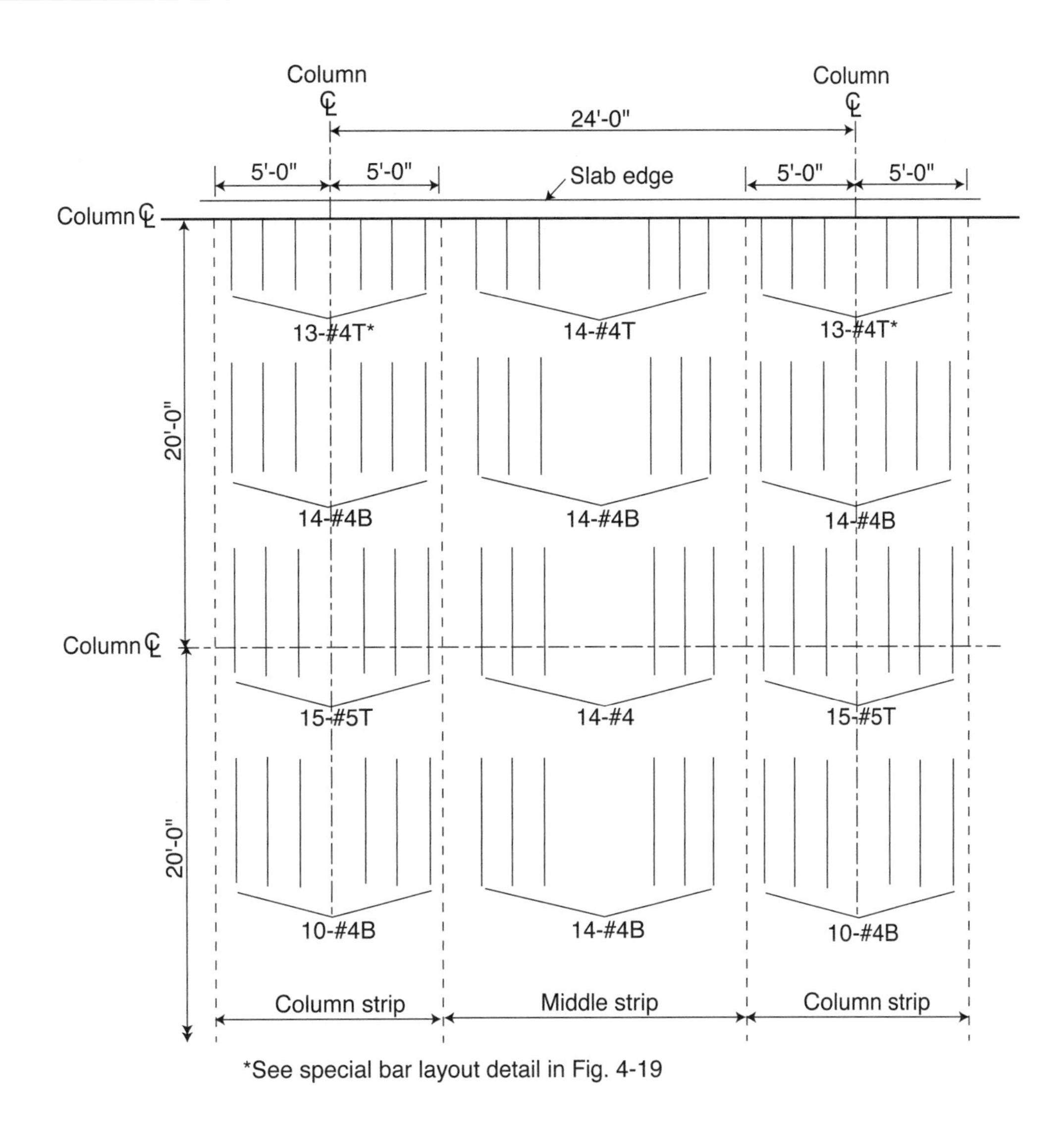

Figure 4-21 Bar Layout—Space Bars Uniformly within Each Column Strip and Middle Strip

From Table 4-9, for 9 in. slab with 12 × 12 in. column:

$$A_c = 399.1 \text{ in.}^2$$

$$J/c = 2523 \text{ in.}^3$$

From Fig. 4-16, with $b_1/b_2 = 15.88/19.75 = 0.8$, $\gamma_v \cong 0.38$

$$v_u = V_u/A_c + \gamma_v M\, c/J$$

$$= (57{,}200/399.1) + (0.38 \times 0.3 \times 260.6 \times 12{,}000/2523)$$

$$= 143.3 + 139.4 = 284.6 \text{ psi} >> \phi 4\sqrt{f'_c} = 215 \text{ psi*}$$

* $b_o/d = [(2 \times 15.88) + 19.75]/7.75 = 6.7$; *from Fig. 4-13 with* $\beta_c = 1$, $\phi V_c/b_o d = 215$ *psi*

The 9 in. slab is not adequate for shear and unbalanced moment transfer at the edge columns. Increase shear strength by providing drop panels at edge columns. Calculations not shown here.

4.7.2 Example: Interior Strip (N-S Direction) of Building #2, Alternate (1)

The slab and column framing will be designed for both gravity and wind loads. Design an interior strip for the 1st-floor level (greatest wind load effects).

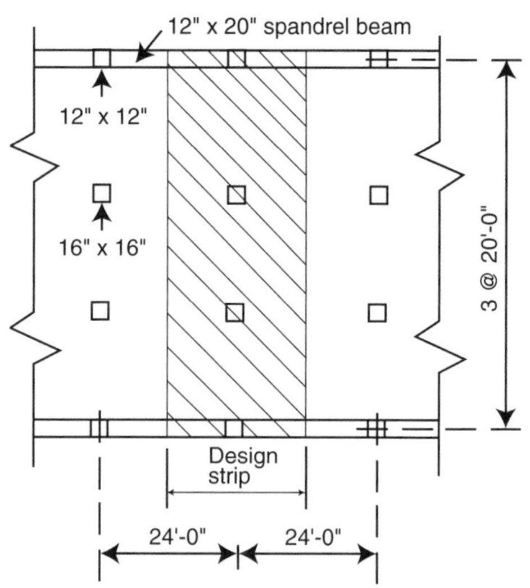

1. Data: f'_c = 4000 psi (carbonate aggregate)

   f_y = 60,000 psi

   Floors: LL = 50 psf

   DL = 136 psf (8½ in. slab + 20 psf partitions + 10 psf ceiling & misc.)

   Preliminary sizing: Slab = 8½ in. thick

   Columns interior = 16 × 16 in.

   Exterior = 12 × 12 in.

   Spandrel beams = 12 × 20 in.

   Required fire resistance rating = 2 hours

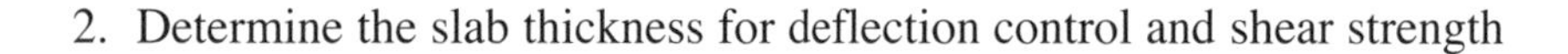

2. Determine the slab thickness for deflection control and shear strength

   (a) Deflection control:

   From Table 4-1 (flat plate with spandrel beams, $\alpha \geq 0.8$):

   $$h = \ell_n/33 = (22.67 \times 12)/33 = 8.24 \text{ in.}$$

   where ℓ_n = 24 – (16/12) = 22.67 ft

   (b) Shear strength. Check shear strength for an 8½ in. slab:

   With the slab and column framing designed for both gravity and wind loads, slab shear strength needs to be checked for the combination of direct shear from gravity loads plus moment transfer from wind loads. Wind load analysis for Building #2 is summarized in Fig. 2-15. Moment transfer between slab and column is greatest at the 1st-floor level where wind moment is the largest. Transfer moment (unfactored) at 1st-floor level due to wind, M_w = 55.78 + 55.78 = 111.56 ft-kips.

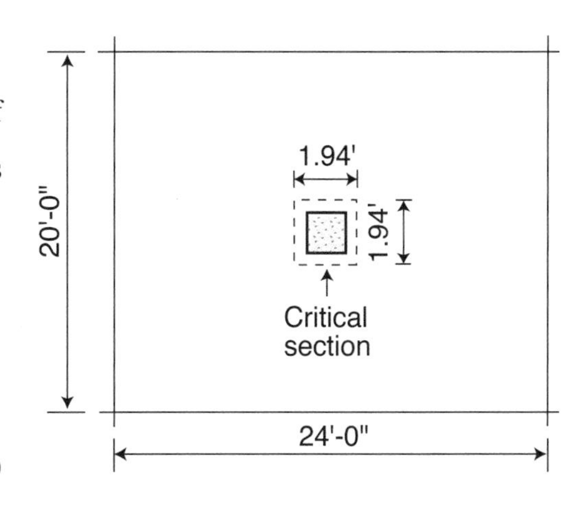

Direct shear from gravity loads:

$$w_u = 1.20(136) + 1.60(29.5^*) = 210.4 \text{ psf}$$

$$V_u = 0.210(24 \times 20 - 1.94^2) = 100.2 \text{ kips}$$

where d = 8.50 – 1.25 = 7.25 in.

and $b_1 = b_2 = (16 + 7.25)/12 = 1.94$ ft

Gravity + wind load:

Two load combinations have to be considered ACI Eq. (9-3) and Eq. (9-4).

ACI Eq. (9-3) [1.2D + 0.8W or 12.D + 0.5L]

$$V_u = [1.2(136) + 0.5(29.5)] \times [24 \times 20 - 1.94^2] = 84.7 \text{ kips}$$

$$M_u = 0.8 \times 111.56 = 89.25 \text{ ft-kips}$$

ACI Eq. (9-4) [1.2D + 0.5L + 1.6W]

$$V_u = [1.2(136) + 0.5(29.5)] \times [24 \times 20 - 1.94^2] = 84.75 \text{ kips}$$

$$M_u = 1.6 \times 111.56 = 178.5 \text{ ft-kips}$$

From Table 4-7, for 8½ in. slab with 16 x 16 in. columns:

$$A_c = 674.3 \text{ in.}^2$$

$$J/c = 5352 \text{ in.}^3$$

Shear stress at critical transfer section:

$$v_u = V_u/A_c + \gamma_v M_u \, c/J$$

$$= (84{,}700/674.3) + (0.4 \times 178.5 \times 12{,}000/5352)$$

$$= 125.6 + 160.1 = 285.7 \text{ psi} > \phi 4\sqrt{f'_c} = 215 \text{ psi}$$

The 8½ in. slab is not adequate for gravity plus wind load transfer at the interior columns.

Increase shear strength by providing drop panels at interior columns. Minimum slab thickness at drop panel = 1.25(5.5) = 10.63 in. (see Fig. 4-2). Dimension drop to actual lumber dimensions for economy of formwork. Try 2¼ in. drop (see Table 9-1).

$$h = 8.5 + 2.25 = 10.75 \text{ in.} > 10.63 \text{ in.}$$

$$d = 7.25 + 2.25 = 9.5 \text{ in.}$$

* *Live load reduction:* A_I *(4 panels) = 24 x 20 x 4 = 1920 sq ft*

$$L = 50(0.25 + 15/\sqrt{1920}) = 29.5 \text{ psf}$$

Refer to Table 4-7:

$b_1 = b_2 = 16 + 9.5 = 25.5$ in. $= 2.13$ ft

$A_c = 4(25.5) \times 9.5 = 969$ in.^{2}

$J/c = [25.5 \times 9.5(25.5 \times 4) + 9.5^3]/3 = 8522$ in.^{3}

$V_u = (84{,}700/969) + (0.4 \times 178.5 \times 12{,}000/8522)$

$= 87.4 + 100.5 = 187.9$ psi < 215 psi

Note that the shear stress around the drop panel is much less than the allowable stress (calculations not shown here).

With drop panels, a lesser slab thickness for deflection control is permitted. From Table 4-1 (flat slab with spandrel beams): $h = \ell_n/36 = (22.67 \times 12)/36 = 7.56$ in. could possibly reduce slab thickness from 8½ to 8 in.; however, shear strength may not be adequate with the lesser slab thickness. For this example hold the slab thickness at 8½ in. Note that the drop panels may not be required in the upper stories where the transfer moment due to wind become substantially less (see Fig. 2-15).

Use 8½ in. slab with 2¼ in. drop panels at interior columns of 1st story floor slab. Drop panel dimensions $= \ell/3 = 24/3 = 8$ ft. Use same dimension in both directions for economy of formwork.

Check for fire resistance: From Table 10-1 for fire resistance rating of 2 hours, required slab thickness $= 4.6$ in. < 8.5 in. O.K.

3. Factored moments in slab due to gravity load (N-S direction).

   (a) Evaluate spandrel beam-to-slab stiffness ratio α and β_t:

Referring to Fig. 4-7:

$\ell_2 = (20 \times 12)/2 = 120$ in.

$a = 20$ in.

$b = 12$ in.

$h = 8.5$ in.

$a/h = 20/8.5 = 2.4$

$b/h = 12/8.5 = 1.4$

$f \cong 1.37$

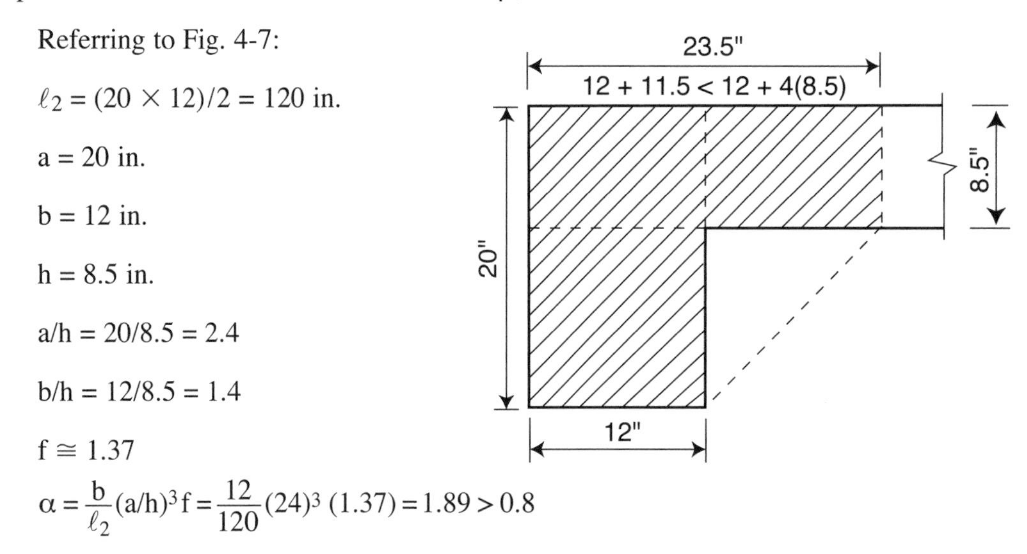

$$\alpha = \frac{b}{\ell_2}(a/h)^3 f = \frac{12}{120}(24)^3\,(1.37) = 1.89 > 0.8$$

Note that the original assumption that the minimum $h = \ell_n/33$ is O.K. since a > 0.8 (see Table 4-1).

$$\beta_t \frac{C}{2I_s} = \frac{8425}{2(14{,}740)} = 0.29 < 2.5$$

where $I_s = (24 \times 12)(8.5)^3/12 = 14{,}740$ in.^{4}

C = larger value computed for the spandrel beam section (see Fig. 4-8).

$x_1 = 8.5$ $y_1 = 23.5$ $C_1 = 3714$	$x_2 = 11.5$ $y_2 = 12$ $C_2 = 2410$	$x_1 = 12$ $y_1 = 20$ $C_1 = 7165$	$x_2 = 8.5$ $y_2 = 11.5$ $C_2 = 1260$

$\Sigma C = 3714 + 2410 = 6124$ $\qquad$ $\Sigma C = 7165 + 1260 = 8425$ (governs)

(b) Total panel moment M_o:

$$M_o = w_u \ell_2 \ell_n^2/8$$

where $w_u = 1.2(136) + 1.6(46.5^*) = 238$ psf

$\ell_2 = 24$ ft

ℓ_n (interior span) $= 20 - 1.33 = 18.67$ ft

ℓ_n (end span) $= 20 - 0.67 - 0.50 = 18.83$ ft

Use larger value for both spans.

(c) Negative and positive factored gravity load moments:

Division of the total panel moment M_o into negative and positive moments, and the, column strip and middle strip moments involves direct application of the moment coefficients of Table 4-3. Note that the moment coefficients for the exterior negative column and middle strip moments need to be modified for β_t less than 2.5. For $\beta_t = 0.29$:

$$\text{Column strip moment} = (0.30 - 0.03 \times 0.29)M_o = 0.29M_o$$

$$\text{Middle strip moment} = 0.30M_o - 0.29M_o = 0.01M_o$$

	End Spans			Interior Span
Slab Moments (ft-kips)	Exterior Negative	Positive	Interior Negative	Positive
Total Moment	75.9	126.5	177.2	88.6
Column Strip	73.4	75.9	134.1	53.2
Middle Strip	2.5	50.6	43.1	38.1

Note: All negative moments are at faces of columns.

* *Live load reduction:* A_I *(4 panels)* $= 24 \times 20 = 480$ *sq ft*

$L = 50(0.25 + 15/\sqrt{480}) = 46.5$ *psf*

4. Check negative moment sections for combined gravity plus wind load moments

   (a) Exterior Negative:

Consider wind load moments resisted by column strip as defined in Fig. 4-9. Column strip width $= 0.5(20 \times 12) = 120$ in.

gravity loads only:

$M_u = 73.4$ ft-kips

gravity + wind loads:

$$M_u = \frac{1.2(136)+0.5(46.5)}{238} \times 73.4 + 1.6 \times 55.78 = 146.75 \text{ ft-kips (govern)}$$

ACI Eq. (9-4)

Also check for possible moment reversal due to wind moments:

$M_u = 0.9(42) \pm 1.6(55.78) = 127.0$ ft-kips, -51.4 ft-kips (reversal)

ACI Eq. (9-6)

where $w_d = 136$ psf

$M_d = 0.29(0.136 \times 24 \times 18.83^2/8) = 42$ ft-kips

   (b) Interior Negative:

gravity loads only:

$M_u = 134.1$ ft-kips

gravity + wind loads:

$$M_u = \frac{1.2(136)+0.5(46.5)}{238} \times 134.1 + 1.6 \times 55.78 = 194.3 \text{ (governs)}$$

and $M_u = 0.9(76.8) \pm 1.6(55.78) = 158.4$ ft-kips, -20.1 ft-kips (reversal)

where $M_d = 42(0.53/0.29) = 76.8$ ft-kips

5. Check slab section for moment strength

   (a) At exterior negative support section:

   $b = 20\, M_u/d^2 = 20 \times 146.75/7.25^2 = 55.8$ in. < 120 in. O.K.

where $d = 8.5 - 1.25 = 7.25$ in.

   (b) At interior negative support section:

   $b = 20 \times 194.3/9.50^2 = 43.1$ in. < 120 in. O.K.

where d = 7.25 + 2.25 = 9.50 in.

6. Slab Reinforcement

Required slab reinforcement is easily determined using a tabular form as follows:

Span Location		M_u (ft-kips)	b^1 (in.)	b^2 (in.)	$A_s = M_u/4d$ (in.^{2})	$A_s{}^3$ (min) (in.^{2})	No. of #4 Bars	No. of #5 Bars
END SPAN								
Column Strip	Ext. Negative	146.75	120	7.25	5.06	1.84	26	17
		-51.4	120	7.25	1.77	0.00	9	9
	Positive	75.9	120	7.25	2.62	1.84	14	9
	Int. Negative	194.3	120	9.5	5.11	2.32	26	17
		-20.1	120	7.25	0.69	0.00	9	9
Middle Strip	Ext. Negative	2.5	168	7.25	0.09	2.57	13	9
	Positive	50.6	168	7.25	1.74	2.57	13	9
	Int. Negative	43.1	168	7.25	1.49	2.57	13	9
INTERIOR SPAN								
Column Strip	Positive	53.2	120	7.25	1.83	1.84	10	9
Middle Strip	Positive	40	168	7.25	1.38	2.57	13	9

[1]Column strip = 0.5(20 x 12) = 120 in. (see Fig. 4-9b)
Middle Strip = (24 x 12) – 120 = 168 in.
[2]Use average d = 8.5 – 1.25 = 7.25 in.
At drop panel, d = 7.25 + 2.25 = 9.50 in. (negative moment only)
[3]$A_{s(min)} = 0.0018\ bh$
$s_{max} = 2h < 18$ in = 2(8.5) = 17 in.

7. Check slab reinforcement at interior columns for moment transfer between slab and column. Shear strength of slab already checked for direct shear and moment transfer in Step (2)(b). Transfer moment (unfactored) at 1st-story due to wind, M_w = 111.56 ft-kips.

Fraction transferred by flexure using ACI Eqs. (13-1) and (9-6):

$$M_u = 0.60(1.6 \times 111.56) = 107.1 \text{ ft-kips}$$

$$A_s = M_u/4d = 107.1/(4 \times 9.50) = 2.82 \text{ in.}^2$$

For No. 5 bars, 2.82/0.31 = 9.1 bars, say 10-No. 5 bars

Must provide 10-No. 5 bars within an effective slab width = $3h + c_2$ = 3(10.75) + 16 = 48.3 in.

Provide the required 10-No. 5 bars by concentrating 10 of the column strip bars (18-No. 5) within the 4-ft slab width over the column. Distribute the other 8 column strip bars (4 on each side) in the remaining column strip width. Check bar spacing:

48/9 spaces = ± 5.3 in.

(120 – 48)/7 spaces = ± 10.3 in. < 18 in. O.K.

Reinforcement details for the interior slab are shown in Figs. 4-22 and 4-23. Bar lengths for the middle strip are taken directly from Fig. 8-6. For the column strip, the bar lengths given in Fig. 8-6 (with drop panels) need to be modified to account for wind moment effects. In lieu of a rigorous analysis to determine bar cut-offs based on a combination of gravity plus wind moment variations, provide bar length details as follows:

For bars near the top face of the slab, cut off one-half of the bars at $0.2\ell_n$ from supports and extend the remaining half the full span length, with a Class B splice near the center of span. Referring to Table 8-1, splice length = $1.3 \times 23.7 = 30.8$ in. ≅ 2.5 ft. At the exterior columns, provide a 90° standard hook with 2 in. minimum cover to edge of slab. From Table 8-2, for No. 5 bars, $\ell_{dh} = 9$ in. < 12 – 2 = 10 in. O.K. (For easier bar placement, alternate equal bar lengths at interior column supports.)

At the exterior columns, provide a 90° end-hook with 2 in. minimum cover to edge of slab for all bottom bars. At least 2 of the bottom bars in the column strip must pass within the column core and anchored at the exterior support.

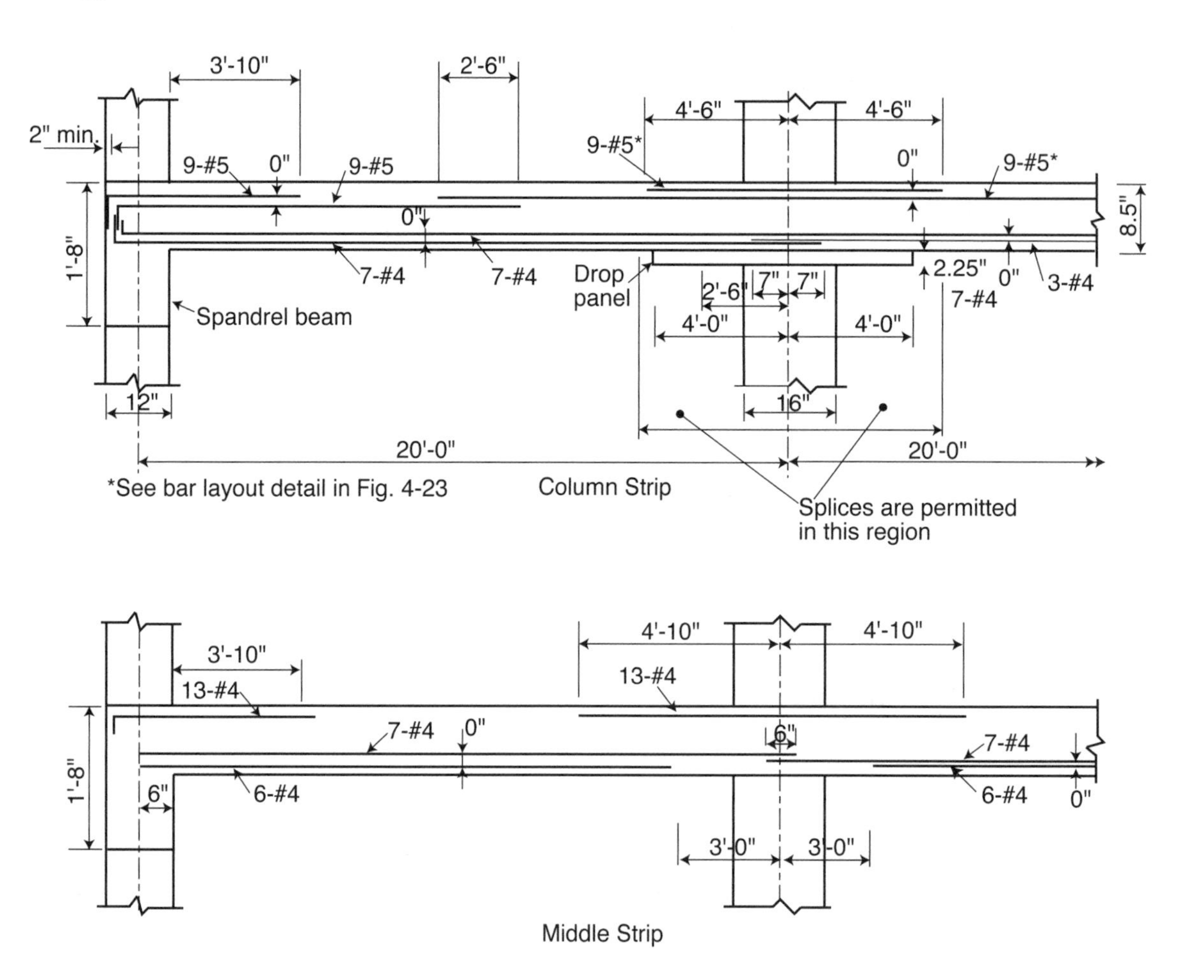

Figurer 4-22 Reinforcement Details for Flat Slab of Building #2 Alternate (1)—1st Floor Interior Slab Panel (N-S Direction)

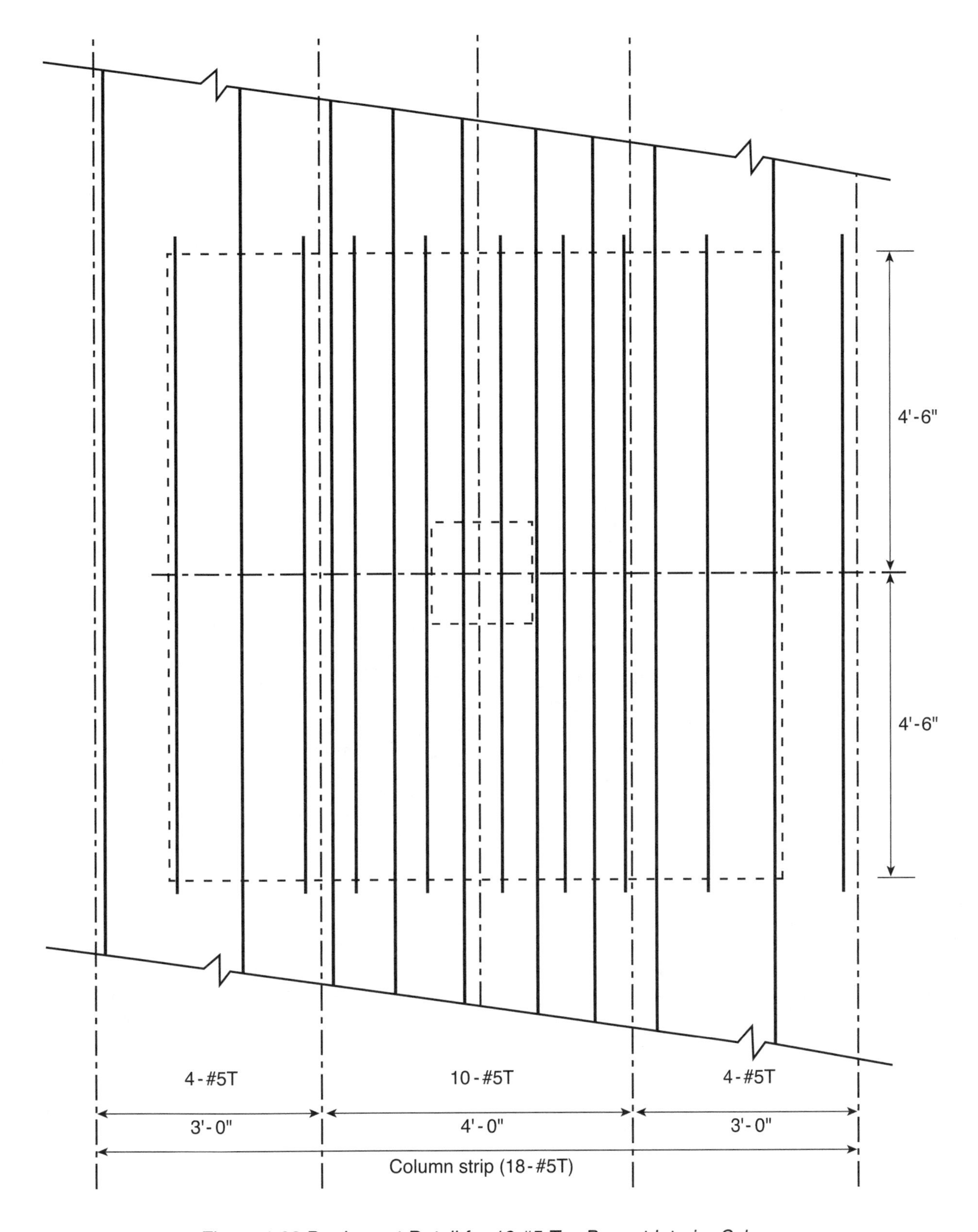

Figure 4-23 Bar Layout Detail for 18-#5 Top Bars at Interior Columns

Table 4-7 Properties of Critical Transfer Section-Interior Column

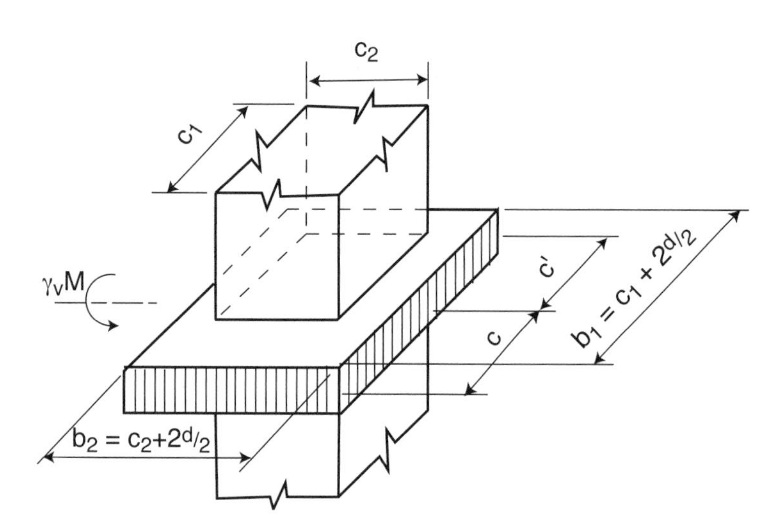

Concrete area of critical section:

$A_c = 2(b_1 + b_2)d$

Modulus of critical section:

$\frac{J}{c} = \frac{J}{c'} = [b_1 d(b_1 + 3b_2) + d^3]/3$

where:

$c = c' = b_1/2$

COL. SIZE	h = 5 in., d = 3 3/4 in.			h = 5 1/2 in., d = 4 1/4 in.			h = 6 in., d = 4 3/4 in.			h = 6 1/2 in., d = 5 1/4 in.		
	A_c in.^{2}	J/c = J/c' in.^{3}	c = c' in.	A_c in.^{2}	J/c = J/c' in.^{3}	c = c' in.	A_c in.^{2}	J/c = J/c' in.^{3}	c = c' in.	A_c in.^{2}	J/c = J/c' in.^{3}	c = c' in.
10x10	206.3	963	6.88	242.3	1176	7.13	280.3	1414	7.38	320.3	1676	7.63
12x12	236.3	1258	7.88	276.3	1522	8.13	318.3	1813	8.38	362.3	2131	8.63
14x14	266.3	1593	8.88	310.3	1913	9.13	356.3	2262	9.38	404.3	2642	9.63
16x16	296.3	1968	9.88	344.3	2349	10.13	394.3	2763	10.38	446.3	3209	10.63
18x18	326.3	2383	10.88	378.3	2831	11.13	432.3	3314	11.38	488.3	3832	11.63
20x20	356.3	2838	11.88	412.3	3358	12.13	470.3	3915	12.38	530.3	4511	12.63
22x22	386.3	3333	12.88	446.3	3930	13.13	508.3	4568	13.38	572.3	5246	13.63
24x24	416.3	3868	13.88	480.3	4548	14.13	546.3	5271	14.38	614.3	6037	14.63

COL. SIZE	h = 7 in., d = 5 3/4 in.			h = 7 1/2 in., d = 6 1/4 in.			h = 8 in., d = 6 3/4 in.			h = 8 1/2 in., d = 7 1/4 in.		
	A_c in.^{2}	J/c = J/c' in.^{3}	c = c' in.	A_c in.^{2}	J/c = J/c' in.^{3}	c = c' in.	A_c in.^{2}	J/c = J/c' in.^{3}	c = c' in.	A_c in.^{2}	J/c = J/c' in.^{3}	c = c' in.
10x10	362.3	1965	7.88	406.3	2282	8.13	452.3	2628	8.38	500.3	3003	8.63
12x12	408.3	2479	8.88	456.3	2857	9.13	506.3	3267	9.38	558.3	3709	9.63
14x14	454.3	3054	9.88	506.3	3499	10.13	560.3	3978	10.38	616.3	4492	10.63
16x16	500.3	3690	10.88	556.3	4207	11.13	614.3	4761	11.38	674.3	5352	11.63
18x18	546.3	4388	11.88	606.3	4982	12.13	668.3	5616	12.38	732.3	6290	12.63
20x20	592.3	5147	12.88	656.3	5824	13.13	722.3	6543	13.38	790.3	7305	13.63
22x22	638.3	5967	13.88	706.3	6732	14.13	776.3	7542	14.38	848.3	8397	14.63
24x24	684.3	6849	14.88	756.3	7707	15.13	830.3	8613	15.38	906.3	9567	15.63

Table 4-7 continued

COL. SIZE	h = 9 in., d = 7 3/4 in. A_c in.^{2}	J/c = J/c' in.^{3}	c = c' in.	h = 9 1/2 in., d = 8 1/4 in. A_c in.^{2}	J/c = J/c' in.^{3}	c = c' in.	h = 10 in., d = 8 3/4 in. A_c in.^{2}	J/c = J/c' in.^{3}	c = c' in.
10x10	550.3	3411	8.88	602.3	3851	9.13	656.3	4325	9.38
12x12	612.3	4186	9.88	668.3	4698	10.13	726.3	5247	10.38
14x14	674.3	5043	10.88	734.3	5633	11.13	796.3	6262	11.38
16x16	736.3	5984	11.88	800.3	6656	12.13	866.3	7370	12.38
18x18	798.3	7007	12.88	866.3	7767	13.13	936.3	8572	13.38
20x20	860.3	8112	13.88	932.3	8966	14.13	1006.3	9867	14.38
22x22	922.3	9301	14.88	998.3	10253	15.13	1076.3	11255	15.38
24x24	984.3	10572	15.88	1064.3	11628	16.13	1146.3	12737	16.38

Table 4-8 Properties of Critical Transfer Section—Edge Column—Bending Parallel to Edge

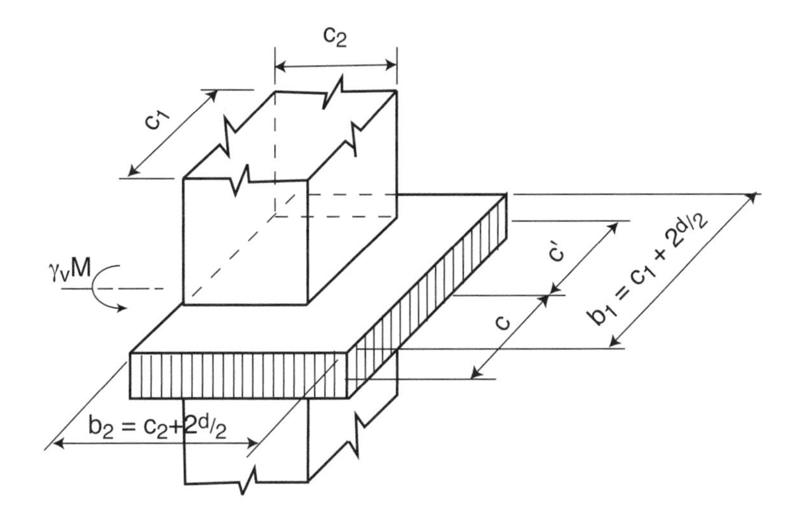

Concrete area of critical section:

$A_c = (b_1 + 2b_2)d$

Modulus of critical section:

$\frac{J}{c} = \frac{J}{c'} = [b_1 d(b_1 + 6b_2) + d^3]/6$

where:

$c = c' = b_1/2$

COL. SIZE	h = 5 in., d = 3 3/4 in. A_c in.^{2}	J/c = J/c' in.^{3}	c = c' in.	h = 5 1/2 in., d = 4 1/4 in. A_c in.^{2}	J/c = J/c' in.^{3}	c = c' in.	h = 6 in., d = 4 3/4 in. A_c in.^{2}	J/c = J/c' in.^{3}	c = c' in.	h = 6 1/2 in., d = 5 1/4 in. A_c in.^{2}	J/c = J/c' in.^{3}	c = c' in.
10x10	140.6	739	6.88	163.6	891	7.13	187.6	1057	7.38	212.6	1238	7.63
12x12	163.1	983	7.88	189.1	1175	8.13	216.1	1384	8.38	244.1	1609	8.63
14x14	185.6	1262	8.88	214.6	1499	9.13	244.6	1755	9.38	275.6	2029	9.63
16x16	208.1	1576	9.88	240.1	1863	10.13	273.1	2170	10.38	307.1	2497	10.63
18x18	230.6	1926	10.88	265.6	2267	11.13	301.6	2629	11.38	338.6	3015	11.63
20x20	253.1	2310	11.88	291.1	2710	12.13	330.1	3133	12.38	370.1	3581	12.63
22x22	275.6	2729	12.88	316.6	3192	13.13	358.6	3681	13.38	401.6	4197	13.63
24x24	298.1	3183	13.88	342.1	3715	14.13	387.1	4274	14.38	433.1	4861	14.63

Table 4-8 continued

COL. SIZE	h = 7 in., d = 5 3/4 in. A_c in.^{2}	J/c = J/c' in.^{3}	c = c' in.	h = 7 1/2 in., d = 6 1/4 in. A_c in.^{2}	J/c = J/c' in.^{3}	c = c' in.	h = 8 in., d = 6 3/4 in. A_c in.^{2}	J/c = J/c' in.^{3}	c = c' in.	h = 8 1/2 in., d = 7 1/4 in. A_c in.^{2}	J/c = J/c' in.^{3}	c = c' in.
10x10	238.6	1435	7.88	265.6	1649	8.13	293.6	1879	8.38	322.6	2127	8.63
12x12	273.1	1852	8.88	303.1	2113	9.13	334.1	2393	9.38	366.1	2692	9.63
14x14	307.6	2322	9.88	340.6	2635	10.13	374.6	2969	10.38	409.6	3325	10.63
16x16	342.1	2846	10.88	378.1	3216	11.13	415.1	3609	11.38	453.1	4025	11.63
18x18	376.6	3423	11.88	415.6	3855	12.13	455.6	4311	12.38	496.6	4793	12.63
20x20	411.1	4054	12.88	453.1	4552	13.13	496.1	5077	13.38	540.1	5628	13.63
22x22	445.6	4739	13.88	490.6	5308	14.13	536.6	5905	14.38	583.6	6531	14.63
24x24	480.1	5477	14.88	528.1	6122	15.13	577.1	6797	15.38	627.1	7502	15.63

COL. SIZE	h = 9 in., d = 7 3/4 in. A_c in.^{2}	J/c = J/c' in.^{3}	c = c' in.	h = 9 1/2 in., d = 8 1/4 in. A_c in.^{2}	J/c = J/c' in.^{3}	c = c' in.	h = 10 in., d = 8 3/4 in. A_c in.^{2}	J/c = J/c' in.^{3}	c = c' in.
10x10	352.6	2393	8.88	383.6	2678	9.13	415.6	2983	9.38
12x12	399.1	3011	9.88	433.1	3351	10.13	468.1	3713	10.38
14x14	445.6	3702	10.88	482.6	4101	11.13	520.6	4524	11.38
16x16	492.1	4464	11.88	532.1	4928	12.13	573.1	5417	12.38
18x18	538.6	5299	12.88	581.6	5832	13.13	625.6	6392	13.38
20x20	585.1	6207	13.88	613.1	6814	14.13	678.1	7449	14.38
22x22	631.6	7187	14.88	680.6	7872	15.13	730.6	8587	15.38
24x24	678.1	8239	15.88	730.1	9007	16.13	783.1	9807	16.38

Table 4-9 Properties of Critical Transfer Section—Edge Column—Bending Perpendicular to Edge

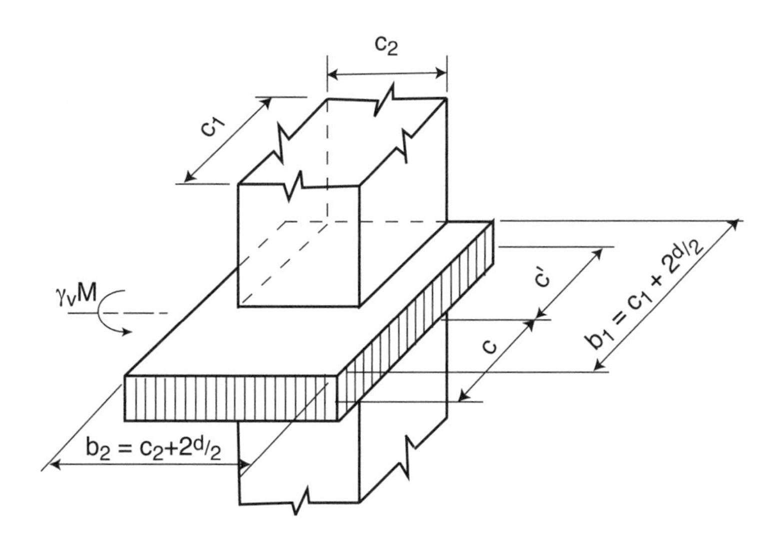

Concrete area of critical section:

$A_c = (2b_1 + b_2)d$

Modulus of critical section:

$\frac{J}{c} = [2b_1d(b_1 + 2b_2) + d^3(2b_1 + b_2)/b_1]/6$

$\frac{J}{c'} = [2b_1^2d(b_1 + 2b_2) + d^3(2b_1 + b_2)]/6(b_1 + b_2)$

where:

$c = b_1^2/(2b_1 + b_2)$

$c' = b_1(b_1 + b_2)/(2b_1 + b_2)$

	h = 5 in., d = 3³/4 in.					h = 5¹/2 in., d = 4¹/4 in.					h = 6 in., d = 4³/4 in.				
COL. SIZE	A_c in.²	J/c in.³	J/c' in.³	c in.	c' in.	A_c in.²	J/c in.³	J/c' in.³	c in.	c' in.	A_c in.²	J/c in.³	J/c' in.³	c in.	c' in.
10x10	140.6	612	284	3.76	8.11	163.6	738	339	3.82	8.31	187.6	878	400	3.88	8.50
12x12	163.1	815	381	4.43	9.45	189.1	973	453	4.48	9.64	216.1	1146	529	4.54	9.83
14x14	185.6	1047	494	5.09	10.78	214.6	1242	583	5.15	10.98	244.6	1453	677	5.21	11.17
16x16	208.1	1309	622	5.76	12.12	240.1	1545	730	5.81	12.31	273.1	1798	844	5.87	12.50
18x18	230.6	1602	765	6.42	13.45	265.6	1882	894	6.48	13.64	301.6	2181	1030	6.54	13.84
20x20	253.1	1924	923	7.09	14.79	291.1	2253	1075	7.15	14.98	330.1	2602	1235	7.20	15.17
22x22	275.6	2277	1095	7.76	16.12	316.6	2658	1273	7.81	16.31	358.6	3061	1459	7.87	16.51
24x24	298.1	2659	1283	8.42	17.45	342.1	3097	1488	8.48	17.65	387.1	3558	1702	8.54	17.84

	h = 6 ¹/2 in., d = 5¹/4 in.					h = 7 in., d = 5³/4 in.					h = 7¹/2 in., d = 6¹/4 in.				
COL. SIZE	A_c in.²	J/c in.³	J/c' in.³	c in.	c' in.	A_c in.²	J/c in.³	J/c' in.³	c in.	c' in.	A_c in.²	J/c in.³	J/c' in.³	c in.	c' in.
10x10	212.6	1030	467	3.94	8.69	238.6	1197	538	3.99	8.88	265.6	1379	616	4.05	9.07
12x12	244.1	1334	612	4.60	10.03	273.1	1537	701	4.66	10.22	303.1	1757	796	4.72	10.41
14x14	275.6	1680	779	5.26	11.36	307.6	1924	886	5.32	11.55	340.6	2185	1001	5.38	11.74
16x16	307.1	2068	966	5.93	12.70	342.1	23.56	1095	5.99	12.89	378.1	2664	1231	6.05	13.08
18x18	338.6	2498	1174	6.60	14.03	376.6	2835	1326	6.65	14.22	415.6	3192	1486	6.71	14.41
20x20	370.1	2970	1404	7.26	15.36	411.1	3360	1581	7.32	15.56	453.1	3771	1766	7.38	15.75
22x22	401.6	3485	1654	7.93	16.70	445.6	3931	1858	7.98	16.89	490.6	4400	2071	8.04	17.08
24x24	433.1	4041	1926	8.59	18.03	480.1	4548	2158	8.65	18.23	528.1	5078	2401	8.71	18.42

Table 4-9 continued

COL. SIZE	h = 8 in., d = 6 3/4 in.					h = 8 1/2 in., d = 7 1/4 in.					h = 9 in., d = 7 3/4 in.				
	A_c in.^{2}	J/c in.^{3}	J/c' in.^{3}	c in.	c' in.	A_c in.^{2}	J/c in.^{3}	J/c' in.^{3}	c in.	c' in.	A_c in.^{2}	J/c in.^{3}	J/c' in.^{3}	c in.	c' in.
10x10	293.6	1577	700	4.11	9.26	322.6	1792	791	4.17	9.45	352.6	2024	888	4.23	9.64
12x12	334.1	1994	898	4.78	10.6	366.1	2249	1008	4.83	10.79	399.1	2523	1124	4.89	10.98
14x14	374.6	2465	1124	5.44	11.94	409.6	2765	1253	5.50	12.13	445.6	3084	1391	5.56	12.32
16x16	415.1	2991	1376	6.10	13.27	453.1	3338	1528	6.16	13.46	492.1	3707	1689	6.22	13.65
18x18	455.6	3571	1655	6.77	14.61	496.6	3970	1832	6.83	14.80	538.6	4393	2018	6.89	14.99
20x20	496.1	4204	1961	7.43	15.94	540.1	4661	2164	7.49	16.13	585.1	5141	2378	7.55	16.33
22x22	536.6	4892	2294	8.1	17.28	583.6	5409	2526	8.16	17.47	631.6	5951	2768	8.21	17.66
24x24	577.1	5634	2654	8.76	18.61	627.1	6216	2916	8.82	18.80	678.1	6823	3190	8.88	18.99

COL. SIZE	h = 9 1/2 in., d = 8 1/4 in.					h = 10 in., d = 8 3/4 in.				
	A_c in.^{2}	J/c in.^{3}	c in.	c' in.	A_c in.^{2}	J/c in.^{3}	J/c' in.^{3}	c in.	c' in.	A_c in.^{2}
10x10	383.6	2275	992	4.29	9.83	415.6	2544	1104	4.35	10.02
12x12	433.1	2816	1248	4.95	11.17	468.1	3129	1380	5.01	11.36
14x14	482.6	3424	1537	5.62	12.51	520.6	3785	1691	5.67	12.70
16x16	532.1	4098	1858	6.28	13.85	573.1	4511	2037	6.34	14.04
18x18	581.6	4839	2213	6.94	15.18	625.6	5308	2418	7.00	15.37
20x20	631.1	5646	2601	7.61	16.52	678.1	6176	2834	7.67	16.71
22x22	680.6	6519	3021	8.27	17.85	730.6	7113	3284	8.33	18.04
24x24	730.1	7458	3474	8.94	19.19	783.1	8121	3770	9.00	19.38

Table 4-10 Properties of Critical Transfer Section—Corner Column

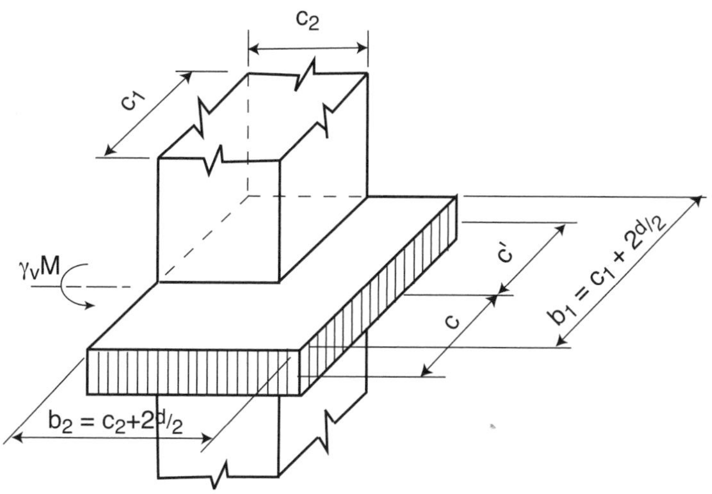

Concrete area of critical section:

$A_c = (2b_1 + b_2)d$

Modulus of critical section:

$\frac{J}{c} = [2b_1 d(b_1 + 2b_2) + d^3(2b_1 + b_2)/b_1]/6$

$\frac{J}{c'} = [2b_1^2 d(b_1 + 2b_2) + d^3(2b_1 + b_2)]/6(b_1 + b_2)$

where:

$c = b_1^2/(2b_1 + b_2)$

$c' = b_1(b_1 + b_2)/(2b_1 + b_2)$

COL. SIZE	h = 5 in., d = 3¾ in. A_c in.²	J/c in.³	J/c' in.³	c in.	c' in.	h = 5½ in., d = 4¼ in. A_c in.²	J/c in.³	J/c' in.³	c in.	c' in.	h = 6 in., d = 4¾ in. A_c in.²	J/c in.³	J/c' in.³	c in.	c' in.
10x10	89.1	458	153	2.97	8.91	103.1	546	182	3.03	9.09	117.6	642	214	3.09	9.28
12x12	104.1	619	206	3.47	10.41	120.1	732	244	3.53	10.59	136.6	854	285	3.59	10.78
14x14	119.1	805	268	3.97	11.91	137.1	946	315	4.03	12.09	155.6	1097	366	4.09	12.28
16x16	134.1	1016	339	4.47	13.41	154.1	1189	396	4.53	13.59	174.6	1372	457	4.59	13.78
18x18	149.1	1252	417	4.97	14.91	171.1	1460	487	5.03	15.09	193.6	1679	560	5.09	15.28
20x20	164.1	1513	504	5.47	16.41	188.1	1759	586	5.53	16.59	212.6	2017	672	5.59	16.78
22x22	179.1	1799	600	5.97	17.91	205.1	2087	696	6.03	18.09	231.6	2388	796	6.09	18.28
24x24	194.1	2110	703	6.47	19.41	222.1	2443	814	6.53	19.59	250.6	2789	930	6.59	19.78

COL. SIZE	h = 6½ in., d = 5¼ in. A_c in.²	J/c in.³	J/c' in.³	c in.	c' in.	h = 7 in., d = 5¾ in. A_c in.²	J/c in.³	J/c' in.³	c in.	c' in.	h = 7½ in., d = 6¼ in. A_c in.²	J/c in.³	J/c' in.³	c in.	c' in.
10x10	132.6	746	249	3.16	9.47	148.1	858	286	3.22	9.66	164.1	979	326	3.28	9.84
12x12	153.6	984	328	3.66	10.97	171.1	1124	375	3.72	11.16	189.1	1273	424	3.78	11.34
14x14	174.6	1257	419	4.16	12.47	194.1	1428	476	4.22	12.66	124.1	1609	536	4.28	12.84
16x16	195.6	1566	522	4.66	13.97	217.1	1770	590	4.72	14.16	239.1	1986	662	4.78	14.34
18x18	216.6	1909	636	5.16	15.47	240.1	2151	717	5.22	15.66	264.1	2406	802	5.28	15.84
20x20	237.6	2288	763	5.66	16.97	263.1	2571	857	5.72	17.16	289.1	2867	956	5.78	17.34
22x22	258.6	2701	900	6.16	18.47	286.1	3028	1009	6.22	18.66	314.1	3369	1123	6.28	18.84
24x24	279.6	3150	1050	6.66	19.97	309.1	3524	1175	6.72	20.16	339.1	3913	1304	6.78	20.34

Table 4-10 continued

COL. SIZE	h = 8 in., d = 6¾ in.					h = 8½ in., d = 7¼ in.					h = 9 in., d = 7¾ in.				
	A_c in.^{2}	J/c in.^{3}	J/c' in.^{3}	c in.	c' in.	A_c in.^{2}	J/c in.^{3}	J/c' in.^{3}	c in.	c' in.	A_c in.^{2}	J/c in.^{3}	J/c' in.^{3}	c in.	c' in.
10x10	180.6	1109	370	3.34	10.03	197.6	1249	416	3.41	10.22	215.1	1398	466	3.47	10.41
12x12	207.6	1432	477	3.84	11.53	226.6	1602	534	3.91	11.72	246.1	1783	594	3.97	11.91
14x14	234.6	1801	600	4.34	13.03	255.6	2004	668	4.41	13.22	277.1	2219	740	4.47	13.41
16x16	261.6	2214	738	4.84	14.53	284.6	2454	818	4.91	14.72	308.1	2706	902	4.97	14.91
18x18	288.6	2673	891	5.34	16.03	313.6	2952	984	5.41	16.22	339.1	3246	1082	5.47	16.41
20x20	315.6	3176	1059	5.84	17.53	342.6	3499	1166	5.91	17.72	370.1	3837	1279	5.97	17.91
22x22	342.6	3724	1241	6.34	19.03	371.6	4094	1365	6.41	19.22	401.1	4479	1493	6.47	19.41
24x24	369.6	4318	1439	6.84	20.53	400.6	4738	1579	6.91	20.72	432.1	5173	1724	6.97	20.91

COL. SIZE	h = 9½ in., d = 8¼ in.					h = 10 in., d = 8¾ in.				
	A_c in.^{2}	J/c in.^{3}	J/c' in.^{3}	c' in.	A_c in.^{2}	J/c in.^{3}	J/c' in.^{3}	c in.	c' in.	A_c in.^{2}
10x10	233.1	1559	520	3.53	10.59	251.6	1730	577	3.59	10.78
12x12	266.1	1975	658	4.03	12.09	286.6	2178	726	4.09	12.28
14x14	299.1	2446	815	4.53	13.59	321.6	2685	895	4.59	13.78
16x16	332.1	2972	991	5.03	15.09	356.6	3250	1083	5.09	15.28
18x18	365.1	3553	1184	5.53	16.59	391.6	3874	1291	5.59	16.78
20x20	398.1	4189	1396	6.03	18.09	426.6	4556	1519	6.09	18.28
22x22	431.1	4879	1626	6.53	19.59	461.6	5296	1765	6.59	19.78
24x24	464.1	5625	1875	7.03	21.09	496.6	6094	2031	7.09	21.28

References

4.1 *Notes on ACI 318-02,* Chapter 10: Deflections, EB702, Portland Cement Association, Skokie, Illinois, 2002.

4.2 "Aspects of Design of Reinforced Concrete Flat Plate Slab Systems," by S.K. Ghosh, *Analysis and Design of High-Rise Concrete Buildings,* SP-97, American Concrete Institute, Detroit, Michigan, 1985, pp. 139-157.

Chapter 5

Simplified Design for Columns

5.1 INTRODUCTION

Use of high-strength materials has had a significant effect on the design of concrete columns. Increased use of high-strength concretes has resulted in columns that are smaller in size and, therefore, are more slender. Consequently, in certain situations, slenderness effects must be considered, resulting in designs that are more complicated than when these effects may be neglected.

For buildings with adequate shearwalls, columns may be designed for gravity loads only. However, in some structures—especially low-rise buildings—it may not be desirable or economical to include shearwalls. In these situations, the columns must be designed to resist both gravity and lateral loads. In either case, it is important to be able to distinguish between a column that is slender and one that is not. A simplified design procedure is outlined in this chapter, which should be applicable to most columns. Design aids are given to assist the engineer in designing columns within the limitations stated.

5.2 DESIGN CONSIDERATIONS

5.2.1 Column Size

The total loads on columns are directly proportional to the bay sizes (i.e. tributary areas). Larger bay sizes mean more load to each column and, thus, larger column sizes. Bay size is often dictated by the architectural and functional requirements of the building. Large bay sizes may be required to achieve maximum unobstructed floor space. The floor system used may also dictate the column spacing. For example, the economical use of a flat plate floor system usually requires columns that are spaced closer than those supporting a pan joist floor system. Architecturally, larger column sizes can give the impression of solidity and strength, whereas smaller columns can express slender grace. Aside from architectural considerations, it is important that the columns satisfy all applicable strength requirements of the ACI Code, and at the same time, be economical. Minimum column size and concrete cover to reinforcement may be governed by fire-resistance criteria (see Chapter 10, Tables 10-2 and 10-6).

5.2.2 Column Constructability

Columns must be sized not only for adequate strength, but also for constructability. For proper concrete placement and consolidation, the engineer must select column size and reinforcement to ensure that the reinforcement is not congested. Bar lap splices and locations of bars in beams and slabs framing into the column must be considered. Columns designed with a smaller number of larger bars usually improve constructability.

5.2.3 Column Economics

Concrete is more cost effective than reinforcement for carrying compressive axial load; thus, it is more economical to use larger column sizes with lesser amounts of reinforcement. Also, columns with a smaller number of larger bars are usually more economical than columns with a larger number of smaller bars.

Reuse of column forms from story level to story level results in significant savings. It is economically unsound to vary column size to suit the load on each story level. It is much more economical to use the same column size for the entire building height, and to vary only the longitudinal reinforcement as required. In taller buildings, the concrete strength is usually varied along the building height as well.

5.3 PRELIMINARY COLUMN SIZING

It is necessary to select a preliminary column size for cost estimating and/or frame analysis. The initial selection can be very important when considering overall design time. In general, a preliminary column size should be determined using a low percentage of reinforcement; it is then possible to provide any additional reinforcement required for the final design (including applicable slenderness effects) without having to change the column size. Columns which have reinforcement ratios in the range of 1% to 2% will usually be the most economical.

The design chart presented in Fig. 5-1, based on ACI Eq. (10-2), can be used for nonslender tied square columns loaded at an eccentricity of no more than 0.1h, where h is the size of the column. Design axial load strengths $\phi P_{n(max)}$ for column sizes from 10 in. to 24 in. with reinforcement ratios between 1 and 8% are presented. For other columns sizes and shapes, and material strengths, similar design charts based on ACI Eq. (10-1) or (10-2) can be easily developed.

This design chart will provide quick estimates for a column size required to support a factored load P_u within the allowable limits of the reinforcement ratio (ACI 10.9). Using the total tributary factored load P_u for the lowest story of a multistory column stack, a column size should be selected with a low percentage of reinforcement. This will allow some leeway to increase the amount of steel for the final design, if required.

5.4 SIMPLIFIED DESIGN FOR COLUMNS

5.4.1 Simplified Design Charts

Numerous design aids and computer programs are available for determining the size and reinforcement of columns subjected to axial forces and/or moments. Tables, charts, and graphs provide design data for a wide variety of column sizes and shapes, reinforcement layouts, load eccentricities and other variables. These design aids eliminate the necessity for making complex and repetitious calculations to determine the strengths of columns, as preliminarily sized. The design aids presented in References 5.1 and 5.2 are widely used. In addition, extensive column load tables are available in the CRSI Handbook. Each publication presents the design data in a somewhat different format; however, the accompanying text in each reference readily explains the method of use.

PCA's computer program PCACOLUMN may also be used to design or investigate rectangular or circular column sections with any reinforcement layout or pattern.[5.4] The column section may be subjected to axial

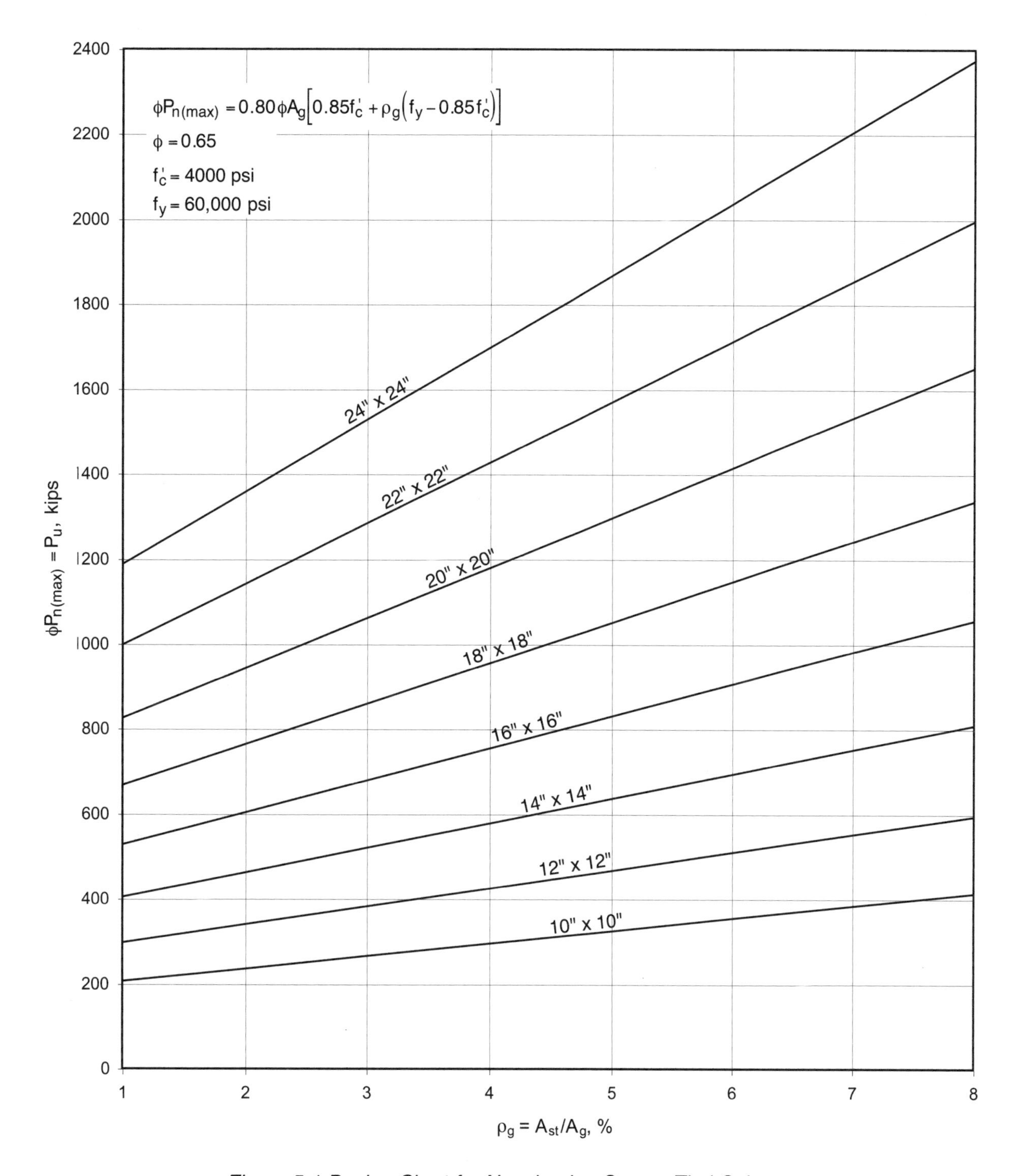

Figure 5-1 Design Chart for Nonslender, Square Tied Columns

compressive loads acting alone or in combination with uniaxial or biaxial bending. Slenderness effects may also be considered, if desired. PCACOLUMN will output all critical load values and the interaction diagram (or, moment contour) for any column section.

In general, columns must be designed for the combined effects of axial load and bending moment. As noted earlier, appreciable bending moments due to lateral loads may occur in the columns of buildings without shear-walls. To allow rapid selection of column size and longitudinal reinforcement for a factored axial load P_u and bending moment M_u, Figs. 5-16 through 5-23 are included at the end of this chapter. All design charts are based on f'_c = 4000 psi and f_y = 60,000 psi, and are valid for square, tied, nonslender columns with symmetrical bar arrangements as shown in Fig. 5-2. The number in parentheses next to the number of reinforcing bars is the reinforcement ratio, $\rho_g = A_{st}/A_g$, where A_{st} is the total area of the longitudinal bars and A_g is the gross area of the column section. A clear cover of 1.5 in. to the ties was used (ACI 7.7.1); also used was No. 3 ties with longitudinal bars No. 10 and smaller and No. 4 ties with No. 11 bars (ACI 7.10.5).

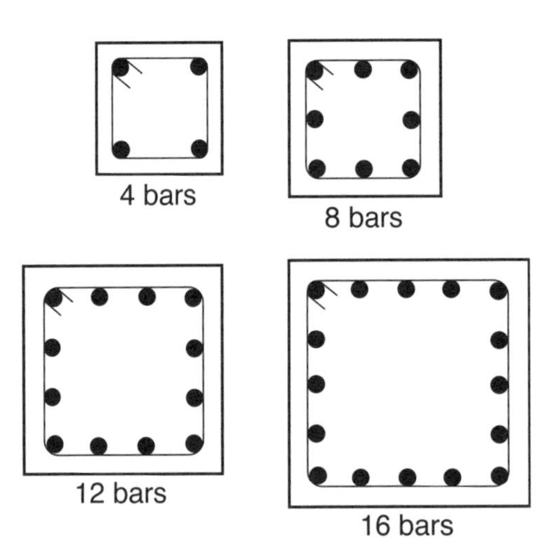

Figure 5-2 Bar Arrangements for Column Design Charts

For simplicity, each design curve is plotted with straight lines connecting a number of points corresponding to certain transition stages. In general, the transition stages are defined as follows (see Fig. 5-3):

Stage 1: Pure compression (no bending moment)

Stage 2: Stress in reinforcement closest to tension face = 0 ($f_s = 0$)

Stage 3: Stress in reinforcement closest to tension face = 0.5 f_y ($f_s = 0.5\ f_y$)

Stage 4: Balanced point; stress in reinforcement closest to tension face = f_y ($f_s = f_y$)

Stage 5: Pure bending (no axial load)

Note that Stages 2 and 3 are used to determine which type of lap splice is required for a given load combination (ACI 12.17). In particular, for load combinations falling within Zone 1, compression lap splices are allowed, since all of the bars are in compression. In Zone 2, either Class A (half or fewer of the bars spliced at one location) or Class B (more than one-half of the bars spliced at one location) tension lap splices must be used. Class B tension lap splices are required for load combinations falling within Zone 3.

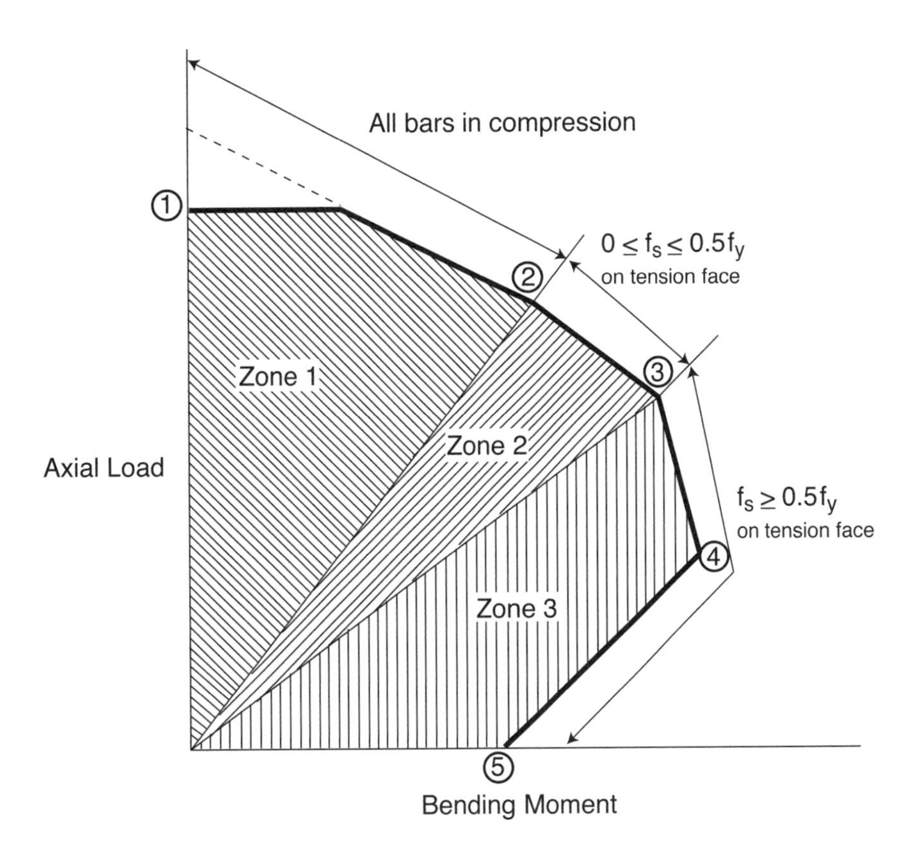

Figure 5-3 Transition Stages on Interaction Diagram

Simplified equations based on strain compatibility analysis can be derived to obtain the critical point on the design interaction diagram corresponding to each transition stage. The following equations are valid within the limitations stated above:

(1) Point 1 (see Fig. 5-1):

$$\phi P_{n(max)} = 0.80\phi[0.85\, f_c' (A_g - A_{st}) + f_y A_{st}] \qquad \text{ACI Eq. (10-2)}$$

$$= 0.80\phi A_g[0.85\, f_c' + \rho_g (f_y - 0.85\, f_c')]$$

where A_g = gross area of column, in.^{2}

A_{st} = total area of longitudinal reinforcing bars, in.^{2}

$\rho_g = A_{st}/A_g$

ϕ = strength reduction factor = 0.65

(2) Points 2-4 (see Fig. 5-4):

$$\phi P_n = \phi\, [C_1 h d_1 + 87 \sum_{i=1}^{n} A_{si}(1 - C_2 \frac{d_i}{d_1})] \qquad (5\text{-}1)$$

$$\phi M_n = \phi\,[0.5C_1hd_1(h - C_3d_1) + 87\sum_{i=1}^{n} A_{si}(1 - C_2\frac{d_i}{d_1})(\frac{h}{2} - d_i)]/12 \qquad (5\text{-}2)$$

where h = column dimension in the direction of bending (width or depth), in.

d_1 = distance from compression face to centroid of reinforcing steel in layer 1 (layer closest to tension face), in.

d_i = distance from compression face to centroid of reinforcing steel in layer i in.

A_{si} = total steel area in layer i, in.^{2}

n = total number of layers of reinforcement

ϕ = strength reduction factor = 0.65

C_1, C_2, C_3 = constants given in Table 5-1

Values of ϕP_n obtained from the above equations are in kips and ϕM_n are in ft-kips. To ensure that the stress in the reinforcing bars is less than or equal to fy, the quantity $(1\text{-}C_2d_i/d_1)$ must always be taken less than or equal to $60/87 = 0.69$.

(3) Point 5:

For columns with 2 or 3 layers of reinforcement:

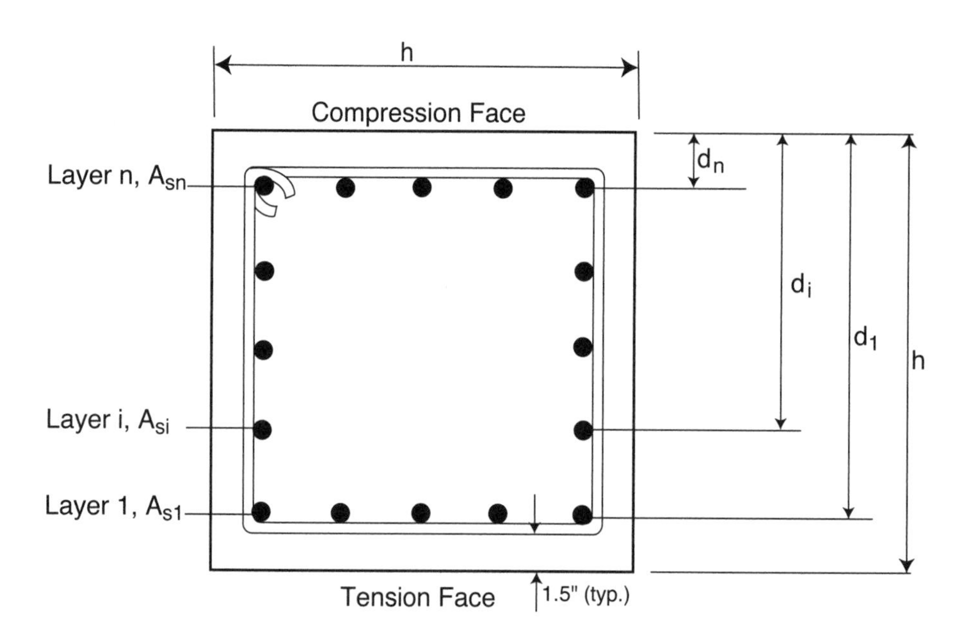

Figure 5-4 Notation for Eqs. (5-1) and (5-2)

Table 5-1 Constants for Points 2-4

Point No.	C_1	C_2	C_3
2	2.89	1.00	0.85
3	2.14	1.35	0.63
4	1.70	1.69	0.50

$$\phi M_n = 4A_{s1}d_1$$

For columns with 4 or 5 layers of reinforcement:

$$\phi M_n = 4(A_{s1} + A_{s2})(d_1 - \frac{s}{2})$$

where s = center to center spacing of the bars

In both equations, $\phi = 0.90$; also, A_{s1} and A_{s2} are in.^{2}, d_1 and s are in in., and ϕM_n is in ft-kips.

The simplified equations for Points 2-4 will produce values of ϕP_n and ϕM_n approximately 3% larger than the exact values (at most). The equations for Point 5 will produce conservative values of ϕM_n for the majority of cases. For columns subjected to small axial loads and large bending moments, a more precise investigation into the adequacy of the section should be made because of the approximate shape of the simplified interaction diagram in the tension-controlled region. However, for typical building columns, load combinations of this type are rarely encountered.

For a column with a larger cross-section than required for loads, a reduced effective area not less than one-half of the total area may be used to determine the minimum reinforcement and the design capacity (ACI 10.8.4), this provision must not be used in regions of high seismic risk.

Essentially this means that a column of sufficient size can be designed to carry the design loads, and concrete added around the designed section without having to increase the amount of longitudinal reinforcement to satisfy the minimum requirement in ACI 10.9.1. Thus, in these situations, the minimum steel percentage, based on actual gross cross-sectional area of column, may be taken less than 0.01, with a lower limit of 0.005 (the exact percentage will depend on the factored loads and the dimensions of the column). It is important to note that the additional concrete must not be considered as carrying any portion of the load, but must be considered when computing member stiffness (ACI R10.8.4).

Additional design charts for other column sizes and material strengths can obviously be developed. For rectangular or round columns, the graphs presented in Reference 5.2 may be used; these graphs are presented in a nondimensionalized format and cover an extensive range of column shapes and material strengths. Also, the CRSI Handbook[5.3] gives extensive design data for square, rectangular, and round column sections.

5.4.1.1 Example: Construction of Simplified Design Chart

To illustrate the simplified procedure for constructing column interaction diagrams, determine the points corresponding to the various transition stages for an 18 × 18 in. column reinforced with 8-No. 9 bars, as shown in Fig. 5-5.

(1) Point 1 (pure compression):

$$\rho_g = \frac{8.0}{18 \times 18} = 0.0247$$

$$\phi P_{n(max)} = (0.80 \times 0.65 \times 324)[(0.85 \times 4) + 0.0247(60-(0.85 \times 4))]$$

$$= 808 \text{ kips}$$

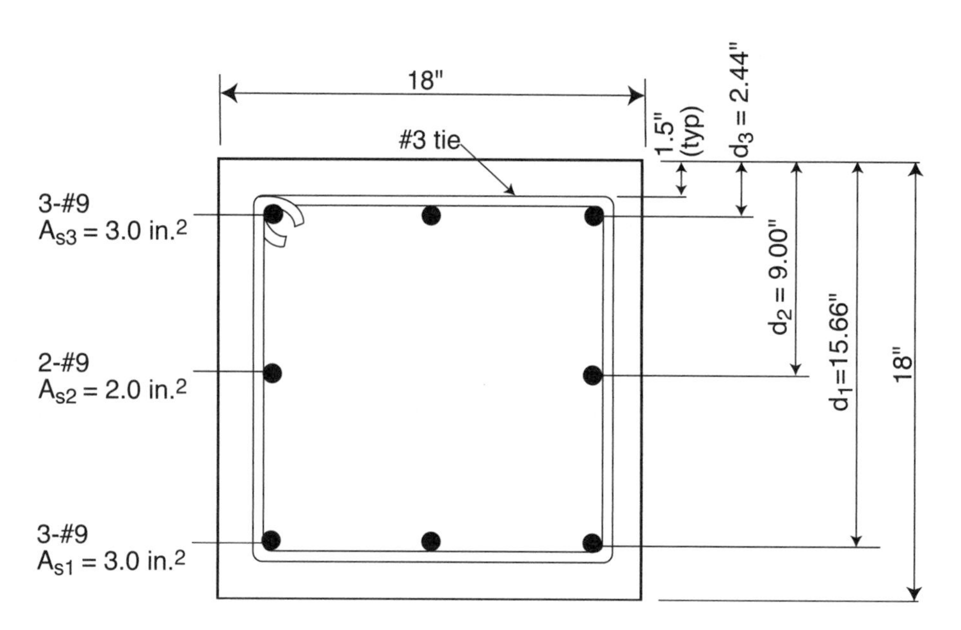

Figure 5-5 Column Cross-Section for Example Problem

(2) Point 2 ($f_{s1} = 0$):

Using Fig. 5-5 and Table 5-1:

Layer 1: $1 - C_2\dfrac{d_1}{d_1} = 1 - 1 = 0$

Layer 2: $1 - C_2\dfrac{d_2}{d_1} = 1 - 1 \times \left(\dfrac{9.00}{15.56}\right) = 0.42$

Layer 3: $1 - C_2\dfrac{d_3}{d_1} = 1 - 1 \times \left(\dfrac{2.44}{15.56}\right) = 0.84 > 0.69$

$1 - C_2d_3/d_1$ being greater than 0.69 in layer 3 means that the steel in layer 3 has yielded; therefore, use $1 - C_2d_3/d_1 = 0.69$.

$$\phi P_n = 0.65\,[C_1hd_1 + 87\sum_{i=1}^{3} A_{si}(1 - C_2\frac{d_i}{d_1})]$$

$$= 0.65\{(2.89 \times 18 \times 15.56) + 87[(3 \times 0) + (2 \times 0.42) + (3 \times 0.69)]\}$$

$$= 0.65\,(809.4 + 253.2)$$

$$= 690 \text{ kips}$$

$$\phi M_n = 0.65\,[0.5\,C_1hd_1(h - C_3d_1) + 87\sum_{i=1}^{3} A_{si}(1 - C_2\frac{d_i}{d_1})(\frac{h}{2} - d_i)]/12$$

$$= 0.65\{(0.5 \times 2.89 \times 18 \times 15.56)[18 - (0.85 \times 15.56)]$$

$$+ 87[(3.0 \times 0(9 - 15.56)) + (2.0 \times 0.42(9 - 9)) + (3.0 \times 0.69(9 - 2.44))]\}/12$$

$$= 0.65\,(1932.1 + 1181.4)/12$$

$$= 169 \text{ ft-kips}$$

(3) Point 3 ($f_{s1} = 0.5\ f_y$):

In this case, $C_1 = 2.14$, $C_2 = 1.35$, and $C_3 = 0.63$ (Table 5-1) (Table 5-1)

Layer 1: $1 - C_2 \frac{d_1}{d_1} = 1 - 1.35 = -0.35$

Layer 2: $1 - C_2 \frac{d_2}{d_1} = 1 - 1.35\left(\frac{9.00}{15.56}\right) = 0.22$

Layer 3: $1 - C_2 \frac{d_3}{d_1} = 1 - 1.35\left(\frac{2.44}{15.56}\right) = 0.79 > 0.69$ Use 0.69

$\phi P_n = 0.65\{(2.14 \times 18 \times 15.56) + 87[(3 - (-0.35)) + (2 \times 0.22) + (3 \times 0.69)]\}$

$= 0.65\ (599.4 + 127.0) = 472$ kips

$\phi M_n = 0.65\{(0.5 \times 2.14 \times 18 \times 15.56)[18 - (0.63 \times 15.56)]$

$+ 87[(3.0(-0.35) \times (9 - 15.56)) + 0 + (3.0 \times 0.69(9 - 2.44))]\}/12$

$= 0.65\ (2456.6 + 1780.7)/12 = 229$ ft-kips

(4) Point 4 ($f_{s1} = f_y$):

In this case, $C_1 = 1.70$, $C_2 = 1.69$, and $C_3 = 0.50$

Similar calculations yield the following:

$\phi P_n = 312$ kips

$\phi M_n = 260$ ft-kips

(5) Point 5 (pure bending):

For columns with 3 layers of reinforcement:

$\phi M_n = 4A_{s1}d_1 = 4 \text{ x } 3.0 \times 15.56 = 187$ ft-kips

Each of these points, connected by straight dotted lines, is shown in Fig. 5-6. The solid line represents the exact interaction diagram determined from PCACOLUMN. As can be seen, the simplified interaction diagram compares well with the one from PCACOLUMN except in the region where the axial load is small and the bending moment is large; there, the simplified diagram is conservative. However, as noted earlier, typical building columns will rarely have a load combination in this region. Note that PCACOLUMN also gives the portion of the interaction diagram for tensile axial loads (negative values of ϕP_n) and bending moments.

Simplified interaction diagrams for all of the other columns in Figs. 5-16 through 5-23 will compare just as well with the exact interaction diagrams; the largest discrepancies will occur in the region near pure bending only.

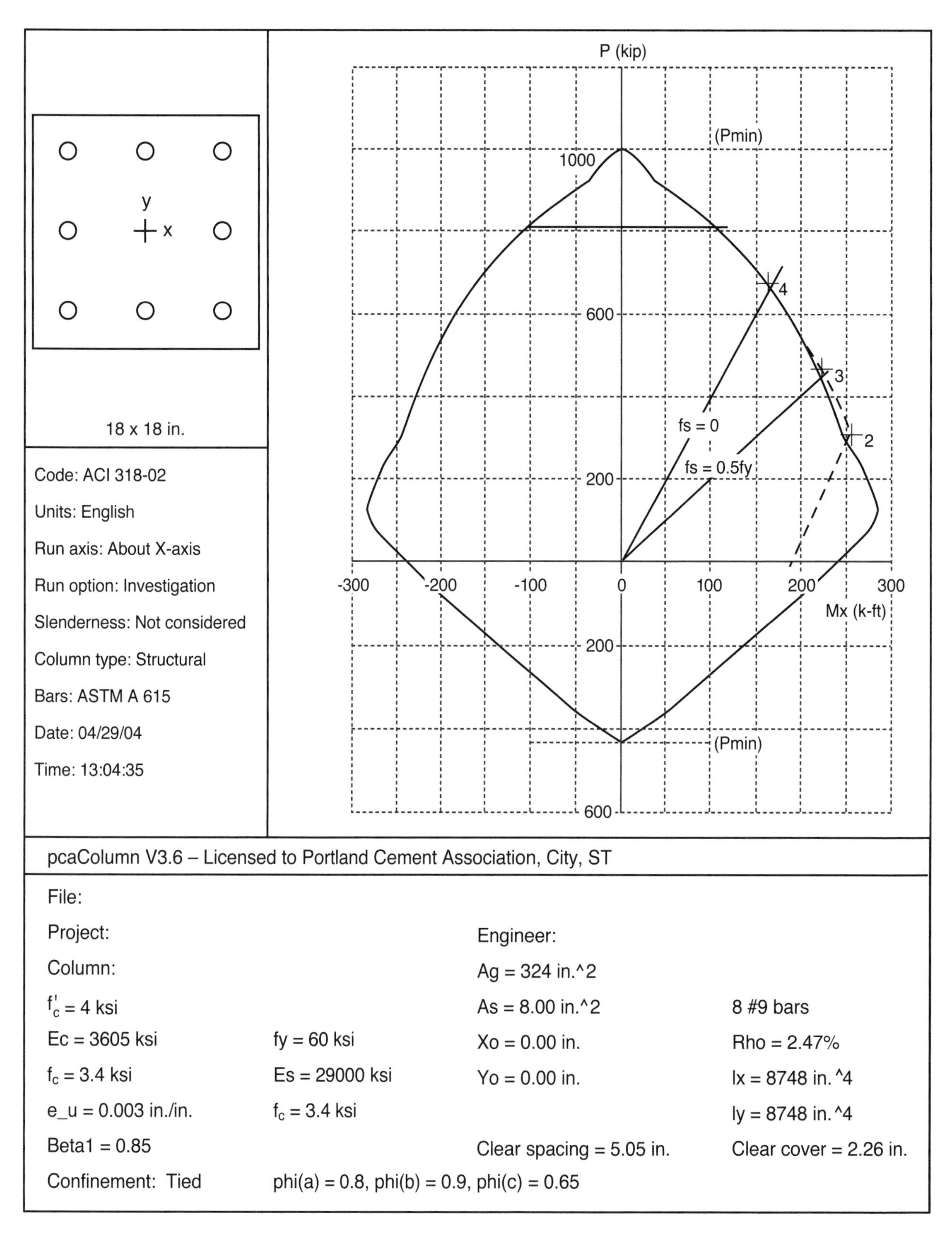

Figure 5-6 Comparison of Simplified and PCACOLUMN Interation Diagrams

5.4.2 Column Ties

The column tie spacing requirements of ACI 7.10.5 are summarized in Table 5-2. For No. 10 column bars and smaller, No. 3 or larger ties are required; for bars larger than No. 10, No. 4 or larger ties must be used. Maximum tie spacing shall not exceed the lesser of 1) 16 longitudinal bar diameters, 2) 48 tie bar diameters, and 3) the least column dimension

Table 5-2 Column Tie Spacing

Tie Size	Column Bars	Maximum Spacing* (in.)
#3	# 5	10
	# 6	12
	# 7	14
	# 8	16
	# 9	18
	#10	18
#4	#11	22

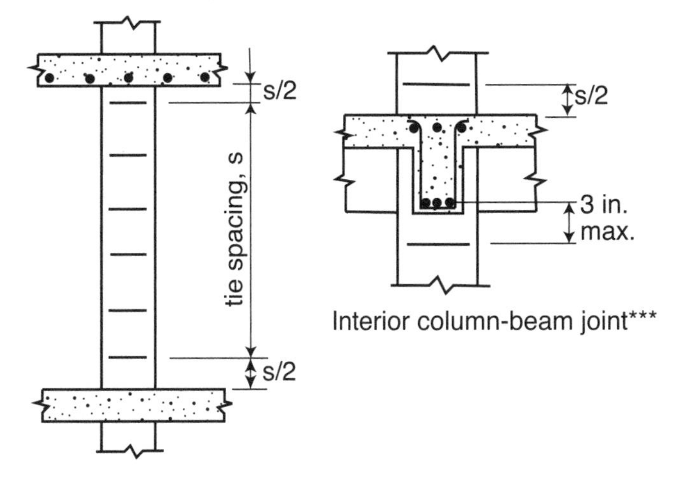

*Maximum spacing not to exceed least column dimension (ACI 7.10.5.2)
**Also valid for joints with beams on less than 4 sides of the column (ACI 7.10.5.4)
***Beams on all 4 sides of the column (ACI 7.10.5.5)

Suggested tie details to satisfy ACI 7.10.5.3 are shown in Fig. 5-7 for the 8, 12, and 16 column bar arrangements. In any square (or rectangular) bar arrangement, the four corner bars are enclosed by a single one-piece tie (ACI 7.10.5.3). The ends of the ties are anchored by a standard 90° or 135° hook (ACI 7.1.3). It is important to alternate the position of hooks in placing successive sets of ties. For easy field erection, the intermediate bars in the 8 and 16 bar arrangements can be supported by the separate crossties shown in Fig. 5-7. Again, it is important to alternate the position of the 90° hooked end at each successive tie location. The two-piece tie shown for the 12 bar arrangement should be lap spliced at least 1.3 times the tensile development length of the tie bar, ℓ_d, but not less than 12 in. To eliminate the supplementary ties for the 8, 12, and 16 bar arrangements, 2, 3, and 4 bar bundles at each corner may also be used; at least No. 4 ties are required in these cases (ACI 7.10.5.1).

Column ties must be located not more than one-half a tie spacing above top of footing or slab in any story, and not more than one-half a tie spacing below the lowest reinforcement in the slab (or drop panel) above (see ACI 7.10.5.4 and Table 5-2). Where beams frame into a column from four sides, ties may be terminated 3 in. below the lowest beam reinforcement (ACI 7.10.5.5). Note that extra ties are required within 6 in. from points of offset bends at column splices (see ACI 7.8.1 and Chapter 8).

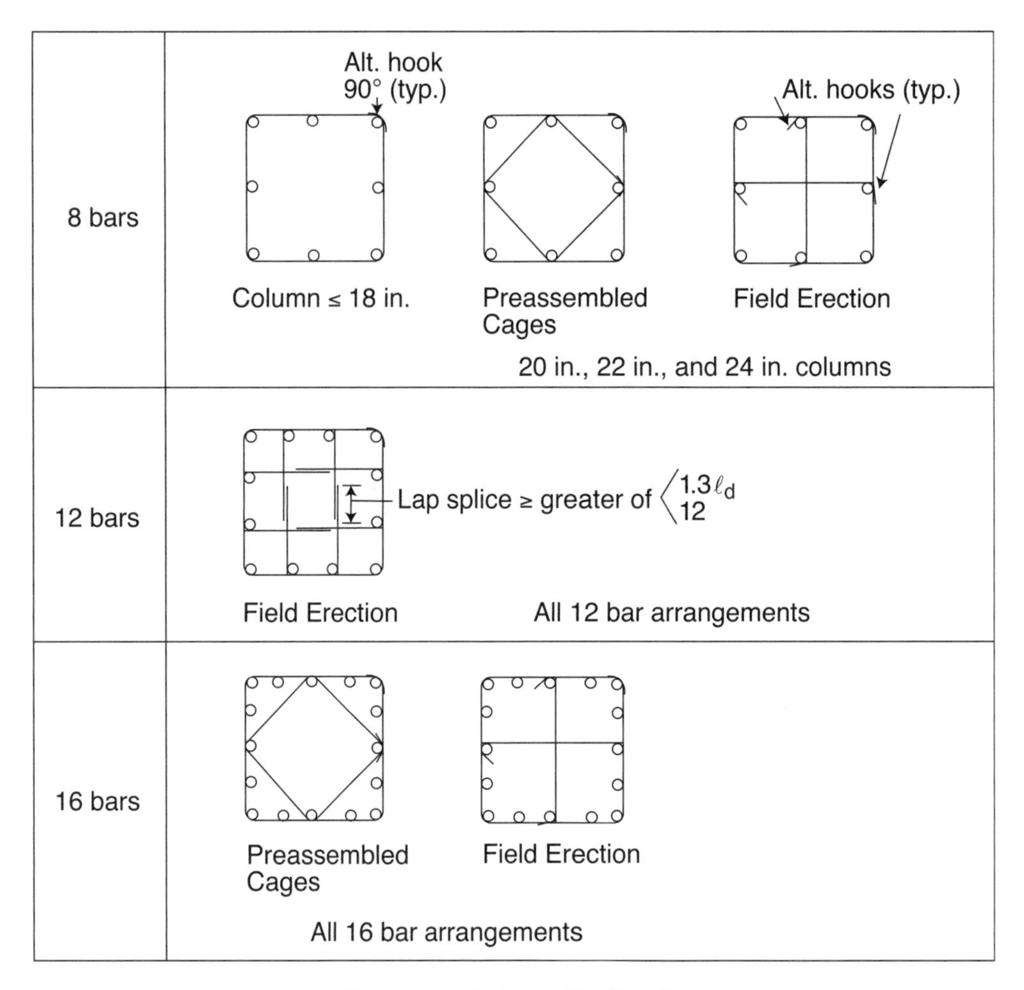

Figure 5-7 Column Tie Details

5.4.3 Biaxial Bending of Columns

Biaxial bending of a column occurs when the loading causes bending simultaneously about both principal axes. This problem is often encountered in the design of corner columns.

A general biaxial interaction surface is depicted in Fig. 5-8. To avoid the numerous mathematical complexities associated with the exact surface, several approximate techniques have been developed that relate the response of a column in biaxial bending to its uniaxial resistance about each principal axis (Reference 5.5 summarizes a number of these approximate methods). A conservative estimate of the nominal axial load strength can be obtained from the following (see ACI R10.3.6 and Fig. 5-9):

$$\frac{1}{\phi P_{ni}} = \frac{1}{\phi P_{nx}} + \frac{1}{\phi P_{ny}} - \frac{1}{\phi P_o}$$

where P_{ni} = nominal axial load strength for a column subjected to an axial load P_u at eccentricities e_x and e_y

P_{nx} = nominal axial load strength for a column subjected to an axial load P_u at eccentricity of e_x only ($e_y = 0$)

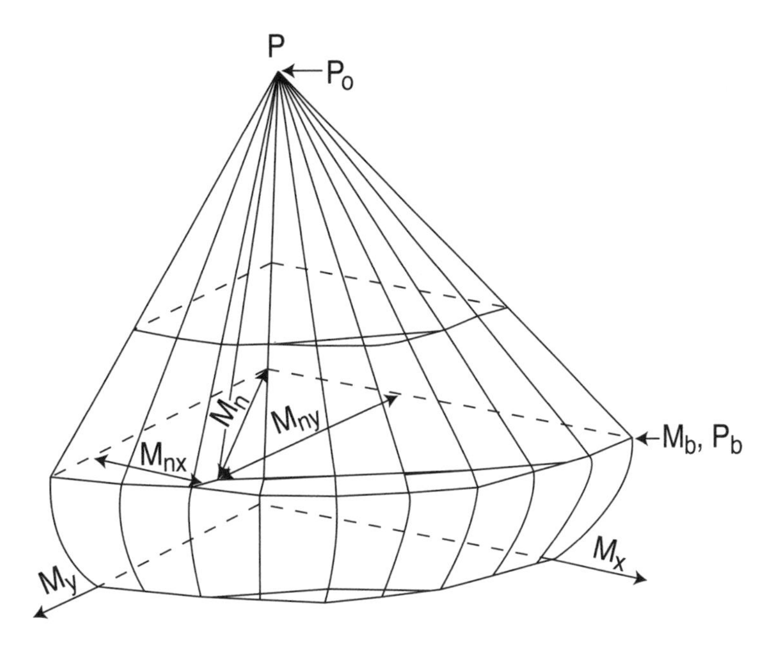

Figure 5-8 Biaxial Interaction Surface

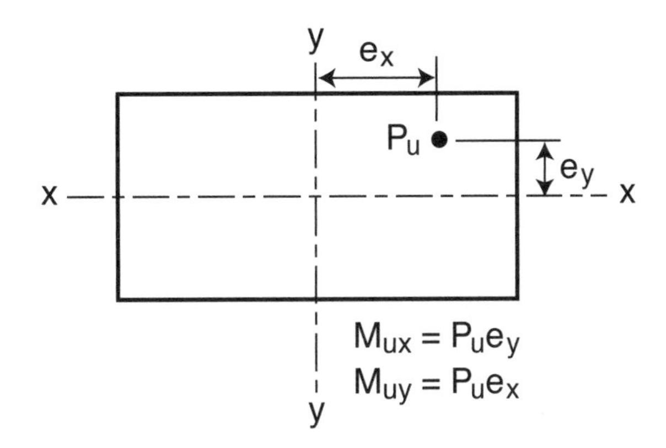

Figure 5-9 Notation for Biaxial Loading

P_{ny} = nominal axial load strength for a column subjected to an axial load P_u at eccentricity e_y only ($e_x = 0$)

P_o = nominal axial load strength for a column subjected to an axial load P_u at eccentricity of zero (i.e., $e_x = e_y = 0$)

$= 0.85\, f'_c\, (A_g - A_{st}) + f_y A_{st}$

The above equation can be rearranged into the following form:

$$\phi P_{ni} = \frac{1}{\dfrac{1}{\phi P_{nx}} + \dfrac{1}{\phi P_{ny}} - \dfrac{1}{\phi P_o}}$$

In design, $P_u < \phi P_{ni}$ where P_u is the factored axial load acting at eccentricities e_x and e_y. This method is most suitable when ϕP_{nx} and ϕP_{ny} are greater than the corresponding balanced axial loads; this is usually the case for typical building columns.

An iterative design process will be required when using this approximate equation for columns subjected to biaxial loading. A trial section can be obtained from Figs. 5-16 through 5-23 with the factored axial load P_u and the total factored moment taken as $M_u = M_{ux} + M_{uy}$ where $M_{ux} = P_u e_x$ and $M_{uy} = P_u e_y$. The expression for ϕP_{ni} can then be used to check if the section is adequate or not. Usually, only an adjustment in the amount of reinforcement will be required to obtain an adequate or more economical section.

5.4.3.1 Example: Simplified Design of a Column Subjected to Biaxial Loading

Determine the size and reinforcement for a corner column subjected to P_u = 360 kips, M_{ux} = 50 ft-kips, and M_{uy} = 25 ft-kips.

(1) Trial section

From Fig. 5-18 with P_u = 360 kips and M_u = 50 + 25 = 75 ft-kips, select a 14 × 14 in. column with 4-No.9 bars.

(2) Check the column using the approximate equation

For bending about the x-axis:

$$\phi P_{nx} = 455 \text{ kips for } M_{ux} = 50 \text{ ft-kips (see Fig. 5-18)}$$

For bending about the y-axis:

$$\phi P_{ny} = 464 \text{ kips for } M_{uy} = 25 \text{ ft-kips (see Fig. 5-18)}$$

$$\phi P_o = 0.65[0.85 \times 4(14^2 - 4.0) + (60 \times 4.0)] = 580 \text{ kips}$$

$$\phi P_{ni} = \frac{1}{\frac{1}{464} + \frac{1}{455} - \frac{1}{580}} = 380 \text{ kips} > 360 \text{ kips} \qquad \text{OK}$$

use a 14 × 14 in. column with 4-No.9 bars.

For comparison purposes, PCACOLUMN was used to check the adequacy of the 14 × 14 in. column with 4-No. 8 bars. Fig. 5-10 is the output from the program which is a plot of ϕM_{ny} versus ϕM_{nx} for $\phi P_n = P_u$ = 360 kips (i.e., a horizontal slice through the interaction surface at ϕP_n = 360 kips). Point 1 represents the position of the applied factored moments for this example. As can be seen from the figure, the section reinforced with 4-No. 8 bars is adequate to carry the applied load and moments. Fig. 5-11 is also output from the PCACOLUMN program; this vertical slice through the interaction surface also reveals the adequacy of the section. As expected, the approximate equation resulted in a more conservative amount of reinforcement (about 27% greater than the amount from PCACOLUMN).

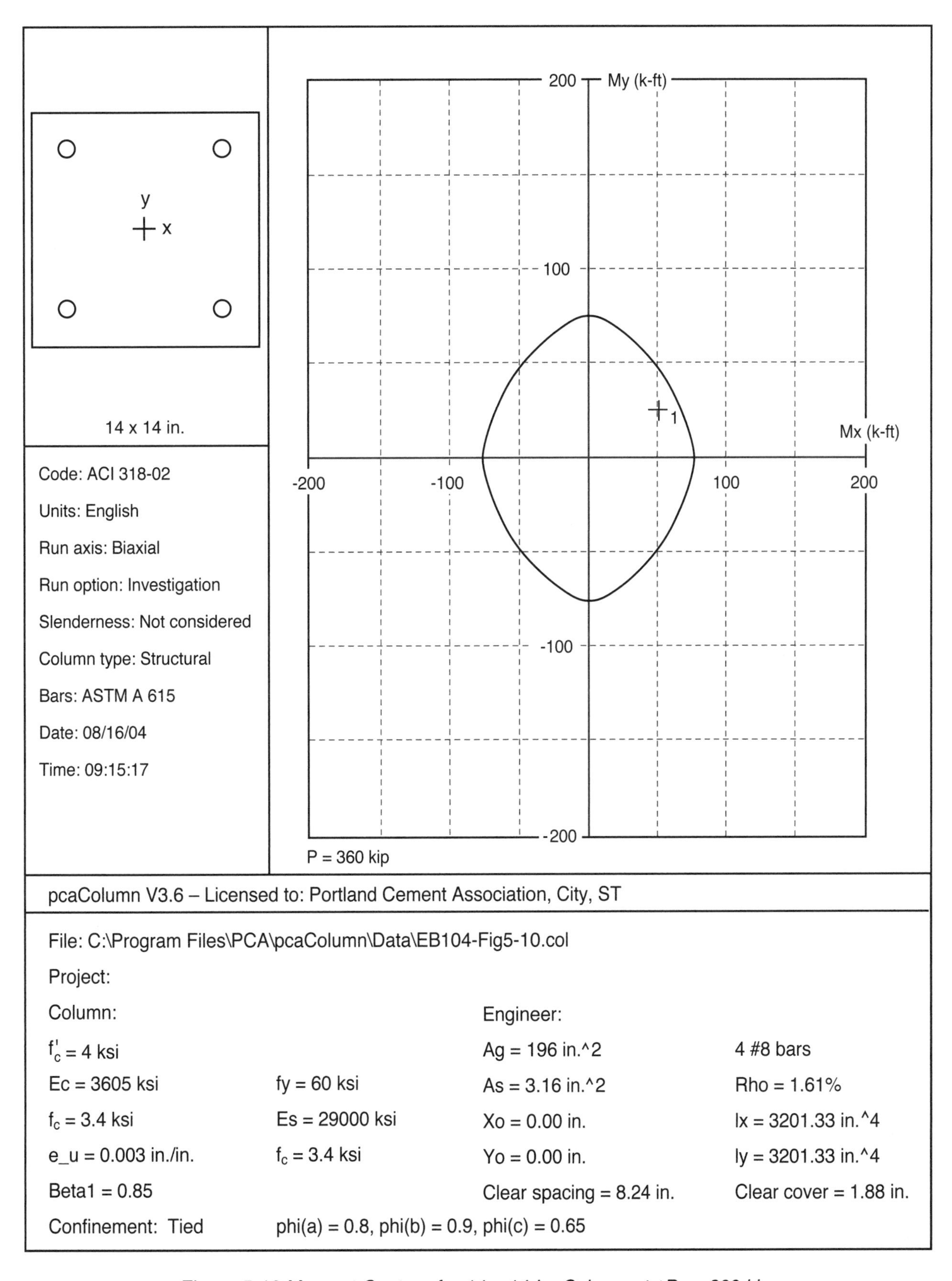

Figure 5-10 Moment Contour for 14 x 14 in. Column at ϕP_n = 360 kips

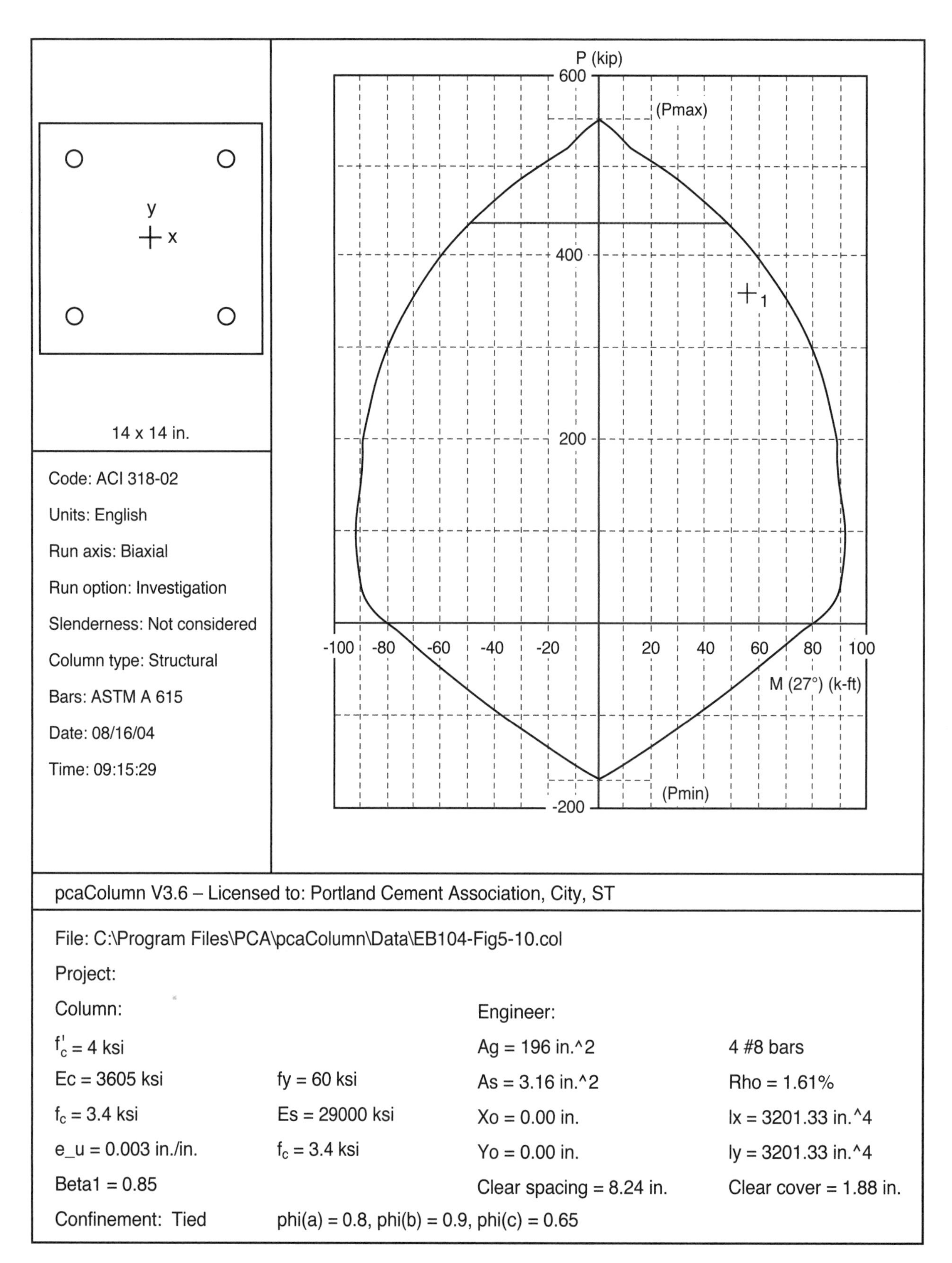

Figure 5-11 Interaction Diagram for 14 x 14 in. Column

5.5 COLUMN SLENDERNESS CONSIDERATIONS

5.5.1 Non-sway versus Sway Frames

When designing columns, it is important to establish whether or not the building frame is non-sway. A compression member may be assumed non-sway if located in a story in which the bracing elements (shear walls, shear trusses, or other types of lateral bracing) have a such substantial lateral stiffness, to resist lateral movement of the story that the resulting lateral deflection is not large enough to affect the column strength substantially (ACI R10.11.4). There is rarely a completely non-sway or a completely sway frame. Realistically, a column within a story can be considered non-sway when horizontal displacements of the story do not significantly affect the moments in the column. ACI 10.11.4 gives criteria that can be used to determine if column located within a story is non-sway or sway. What constitutes adequate non-sway structure must be left to the judgment of the engineer; in many cases, it is possible to ascertain by inspection whether or not a structure is non-sway.

5.5.2 Minimum Sizing for Design Simplicity

Another important aspect to consider when designing columns is whether slenderness effects must be included in the design (ACI 10.10). In general, design time can be greatly reduced if 1) the building frame is adequately braced by shearwalls (per ACI R10.11.4) and 2) the columns are sized so that effects of slenderness may be neglected. The criteria for the consideration of column slenderness, as prescribed in ACI 10.10, are summarized in Fig. 5-12. M_{2b} is the larger factored end moment and M_{1b} is the smaller end moment; both moments, determined from an elastic frame analysis, are due to loads that result in no appreciable side sway. The ratio M_{1b}/M_{2b} is positive if the column is bent in single curvature, negative if it is bent in double curvature. For non-sway columns, the effective length factor k = 1.0 (ACI 10.11.12.1).

In accordance with ACI 10.12.2, effects of slenderness may be neglected when non-sway columns are sized to satisfy the following:

$$\frac{\ell_u}{h} \leq 12$$

where ℓ_u is the clear height between floor members and h is the column size. The above equation is valid for columns that are bent in double curvature with approximately equal end moments. It can be used for the first story columns provided the degree of fixity at the foundation is large enough.* Table 5-3 gives the maximum clear height ℓ_u for a column size that would permit slenderness to be neglected.

For a sway column with a column-to-beam stiffness ratio $\psi = 1$ at both ends, the effects of slenderness may be neglected with ℓ_u/h is less than 5, assuming k = 1.3 (see the alignment chart, in ACI R 10.12).**

* *For a discussion of fixity of column bases, see PCI Design Handbook-Precast and Prestressed Concrete, 5th Ed., Precast/Prestressed Concrete Institute, Chicago, IL, 1999.*

** *The effective length factor k may be determined for a non-sway or sway frame using ACI R10.12 or using the simplified equations which are also given in ACI R10.12.*

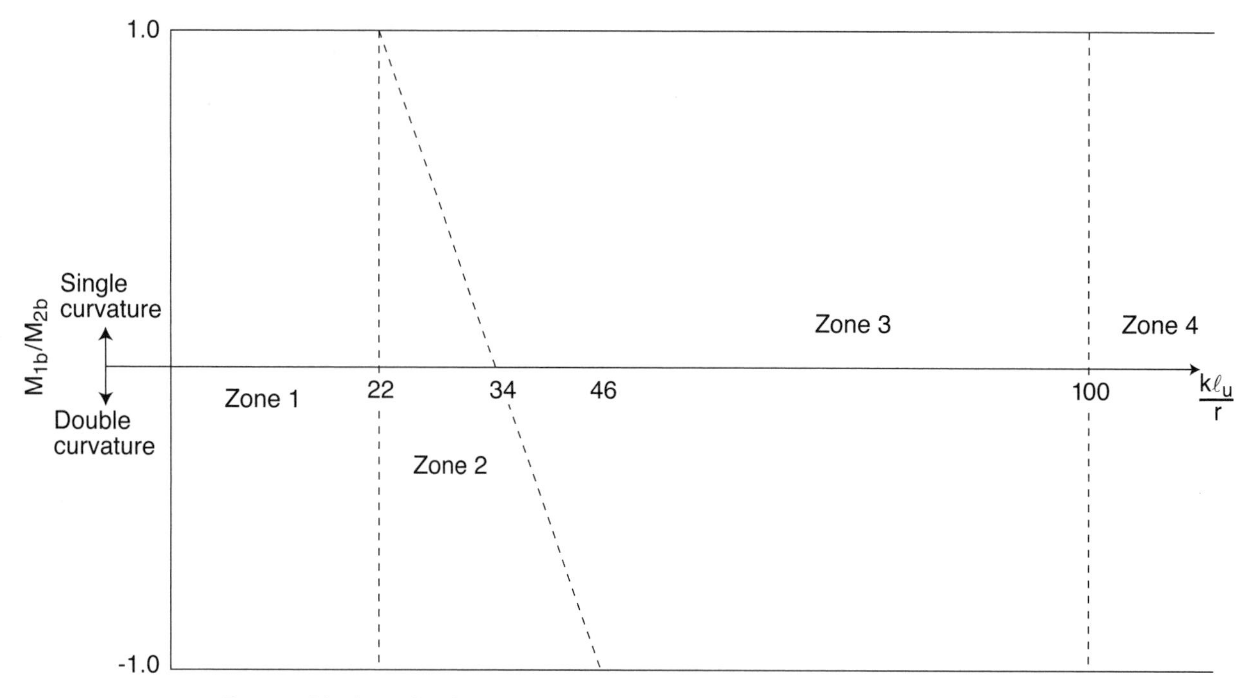

Zone 1: Neglect slenderness, braced and unbraced columns
Zone 2: Neglect slenderness, braced columns
Zone 3: Consider slenderness, moment magnifier method (ACI 10.11.5)
Zone 4: Consider slenderness. PΔ analysis (ACI 10.11.4.3)

Figure 5=12 Consideration of Column Slenderness

Table 5-3 Maximum Story Heights to Neglect Slenderness—Braced Columns

Column size h (in.)	Maximum clear height ℓ_u (ft)
10	10
12	12
14	14
16	16
18	18
20	20
22	22
24	24

If the beam stiffness is reduced to one-fifth of the column stiffness at each end, then k = 2.2; consequently, slenderness effects need not be considered as long as ℓ_u/h is less than 3. As can be seen from these two examples, beam stiffnesses at the top and the bottom of a column in a structure where sidesway is not prevented will have a significant influence on the degree of slenderness of the column.

Due to the complexities involved, the design of slender columns is not considered in this book. For a comprehensive discussion of this topic, the reader is referred to Chapter 11 of Reference 5.5. Also, PCACOLUMN can be used to design slender columns.[5.4]

5.6 PROCEDURE FOR SIMPLIFIED COLUMN DESIGN

The following procedure is suggested for design of a multistory column stack using the simplifications and column design charts presented in this chapter. For sway frames with non-slender columns, both gravity and wind loads must be considered in the design. Figs. 5-16 through 5-23 can be used to determine the required reinforcement. For non-sway frames with shearwalls resisting the lateral loads and the columns sized so that slenderness may be neglected, only gravity loads need to be considered; the reinforcement can be selected for Figs. 5-16 through 5-23 as well.

STEP (1) LOAD DATA

(a) Gravity Loads:

Determine factored loads P_u for each floor of the column stack being considered. Include a service dead load of 4 kips per floor for column weight. Determine column moments due to gravity loads. For interior columns supporting a two-way floor system, maximum column moments may be computed by ACI Eq. (13-4) (see Chapter 4, Section 4.5). Otherwise, a general analysis is required.

(b) Lateral Loads:

Determine axial loads and moments from the lateral loads for the column stack being considered.

STEP (2) LOAD COMBINATIONS

For gravity (dead + live) plus lateral loading, ACI 9.2 specifies five load combinations that need to be considered (Table 2-6).

STEP (3) COLUMN SIZE AND REINFORCEMENT

Determine an initial column size based on the factored axial load P_u in the first story using Fig. 5-1, and use this size for the full height of building. Note that the dimensions of the column may be preset by architectural (or other) requirements. Once a column size has been established, it should be determined if slenderness effects need to be considered (see Section 5.5). For columns with slenderness ratios larger than the limits given in ACI 10.10, it may be advantageous to increase the column size (if possible) so that slenderness effects may be neglected.

As noted earlier, for nonslender columns, Figs. 5-16 through 5-23 may be used to select the required amount of reinforcement for a given P_u and M_u. Ideally, a column with a reinforcement ratio in the range of 1% to 2% will result in maximum economy. Depending on the total number of stories, differences in story heights, and magnitudes of lateral loads, 4% to 6% reinforcement may be required in the first story columns. If the column bars are to be lap spliced, the percentage of reinforcement should usually not exceed 4% (ACI R10.9.1). For overall economy, the amount of reinforcement can be decreased at the upper levels of the building. In taller buildings, the concrete strength is usually varied along the building height as well, with the largest f'_c used in the lower level(s).

5.7 EXAMPLES: SIMPLIFIED DESIGN FOR COLUMNS

The following examples illustrate the simplified methods presented in this chapter.

5.7.1 Example: Design of an Interior Column Stack for Building #2 Alternate (1)—Slab and Column Framing Without Structural Walls (Sway Frame)

f'_c = 4000 psi (carbonate aggregate)

f_y = 60,000 psi

Required fire resistance rating = 2 hours

(1) LOAD DATA

Roof: LL = 20 psf Floors: LL = 50 psf
DL = 122 psf DL = 136 psf (8.5 in. slab)

Calculations for the first story interior column are as follows:

(a) Total factored load:*

Factored axial loads due to gravity are summarized in Table 5-4.

Table 5-4 Interior Column Gravity Load Summary for Building #2—Alternate (1)

Floor	Dead Load (psf)	Live Load (psf)	Tributary Area (sq ft)	Influence Area	RM	Reduced Live Load (psf)	Cumulative Dead Load (kips)	Cumulative Live Load (kips)	Cumulative Factored Load ACI-Eq. (9-2) (kips)
5th (roof)	122	20	480	—	1	20.0	63	9.6	80
4th	136	50	480	1920	0.59	29.5	132	23.8	186
3rd	136	50	480	3840	0.49	24.5	201	35.5	288
2nd	136	50	480	5760	0.45	22.5	270	46.3	388
1st	136	50	480	7680	0.42	21.0	340	56.4	487

(b) Factored moments:

gravity loads:

The moment due to dead load is small. Only moments due to live loads and wind loads will be considered.

* *Axial load from wind loads is zero (see Fig. 2-15).*

Live load moment $= 0.035\ w_{\ell}\ell_2\ell_n^2$

$= 0.035(0.05)(24)(18.83^2) = 14.9$ ft-kips ACI Eq. (13.4)

Portion of live load moment to first story column $= 14.9 \left[\dfrac{12}{12+15}\right] = 6.6$ ft-kip

(2) LOAD COMBINATIONS

For the 1st story column:

gravity loads: $P_u = 487$ kips ACI Eq. (9-2)

$M_u = 1.6(6.6) = 10.6$ ft-kips

gravity loads + wind loads: $P_u = 1.2(340) + 1.6\ (9.6) = 423$ kips ACI Eq. (9-3)

$M_u = 1.6(6.6) + 0.8(68.75) = 65.6$ ft-kips

or

$P_u = 1.2(340) + 0.5(56.4) = 436$ kips ACI Eq. (9-4)

$M_u = 0.50(6.6) + 1.6(68.75) = 113$ ft-kips

or

$P_u = 0.9(340) = 306$ kips ACI Eq. (9-6)

$M_u = 1.6(68.75) = 110$ ft-kips

Factored loads and moments, and load combinations, for the 2nd through 5th story columns are calculated in a similar manner, and are summarized in Table 5-5.

Table 5-5 Interior Column Load Summary for Building #2, Alternate (1)

Floor	ACI Eq. (9-2)		ACI Eq. (9-3)		ACI Eq. (9-4)		ACI Eq. (9-6)	
	P_u	M_u	P_u	M_u	P_u	M_u	P_u	M_u
5th (roof)	80	10	90	15	80	13	56	10
4th	186	12	174	25	170	34	119	30
3rd	288	12	257	34	259	54	181	50
2nd	388	13	340	44	348	73	243	68
1st	487	11	423	65	436	113	306	110

(3) COLUMN SIZE AND REINFORCEMENT

With P_u = 487 kips, try a 16 × 16 in. column with 1% reinforcement (see Fig. 5-1). Check for fire resistance: From Table 10-2, for a fire resistance rating of 2 hours, minimum column dimension = 10 in. < 16 in. O.K.

Determine if the columns are slender.

As noted above, a column in a sway frame is slender if $k\ell_u/r \geq 22$. In lieu of determining an "exact" value, estimate k to be 1.2 (a value of k less than 1.2 is usually not realistic for columns in a sway frame.

For a 1st story column:

$$\frac{k\ell_u}{r} = \frac{1.2[(15 \times 12) - 8.5]}{0.3(16)} = 43 > 22$$

For the 2nd through 5th story columns:

$$\frac{k\ell_u}{r} = \frac{1.2[(12 \times 12) - 8.5]}{0.3(16)} = 34 > 22$$

Therefore, slenderness must be considered for the entire column stack. To neglect slenderness effects, the size of the column h would have to be:

$$\frac{1.2[(15 \times 12) - 8.5]}{0.3h} < 22 \rightarrow h > 31.2 \text{ in.}$$

Obviously, this column would not be practical for a building of the size considered. Reference 5.4 or 5.5 can be used to determine the required reinforcement for the 16 × 16 in. column, including slenderness effects.

Figure 5.13 shows the results from PCACOL for an interior 1st story column, including slenderness effects. Thirty five percent of the gross moment of inertia of the slab column strip and seventy percent of the gross moment of inertia of the column section were used to account for the cracked cross section.* It was assumed that the column was fixed at the foundation; appropriate modifications can be made if this assumption is not true, based on the actual footing size and soil conditions. Points 1 to 4 correspond to the load combination given in ACI Eq. (9-2) through Eq. (9-4), point 5 is from ACI Eq. (9-6). As can be seen from the figure, 8-No.10 bars are required at the 1st floor. The amount of reinforcement can decrease at higher elevations in the column stack. Note that if the column is assumed to be hinged at the base, $k\ell_u/r$ is greater than 100, and a second-order frame analysis would be required.

Check for fire resistance: From Table 10-6, for a fire resistance rating of 4 hours or less, the required cover to the main longitudinal reinforcement = 1.5 in. < provided cover = 1.875 in. O.K.

---

* *The moments of inertia of the flexural and compression members are required in order to compute the effective length factor k of the column. ACI R10.12.1 recommends using a value of 0.35 Ig for flexural members (to account for the effect of cracking and reinforcement on relative stiffness) and 0.70Ig for compression members when computing the relative stiffness at each end of the compression member, where Ig is the gross moment of inertia of the section.*

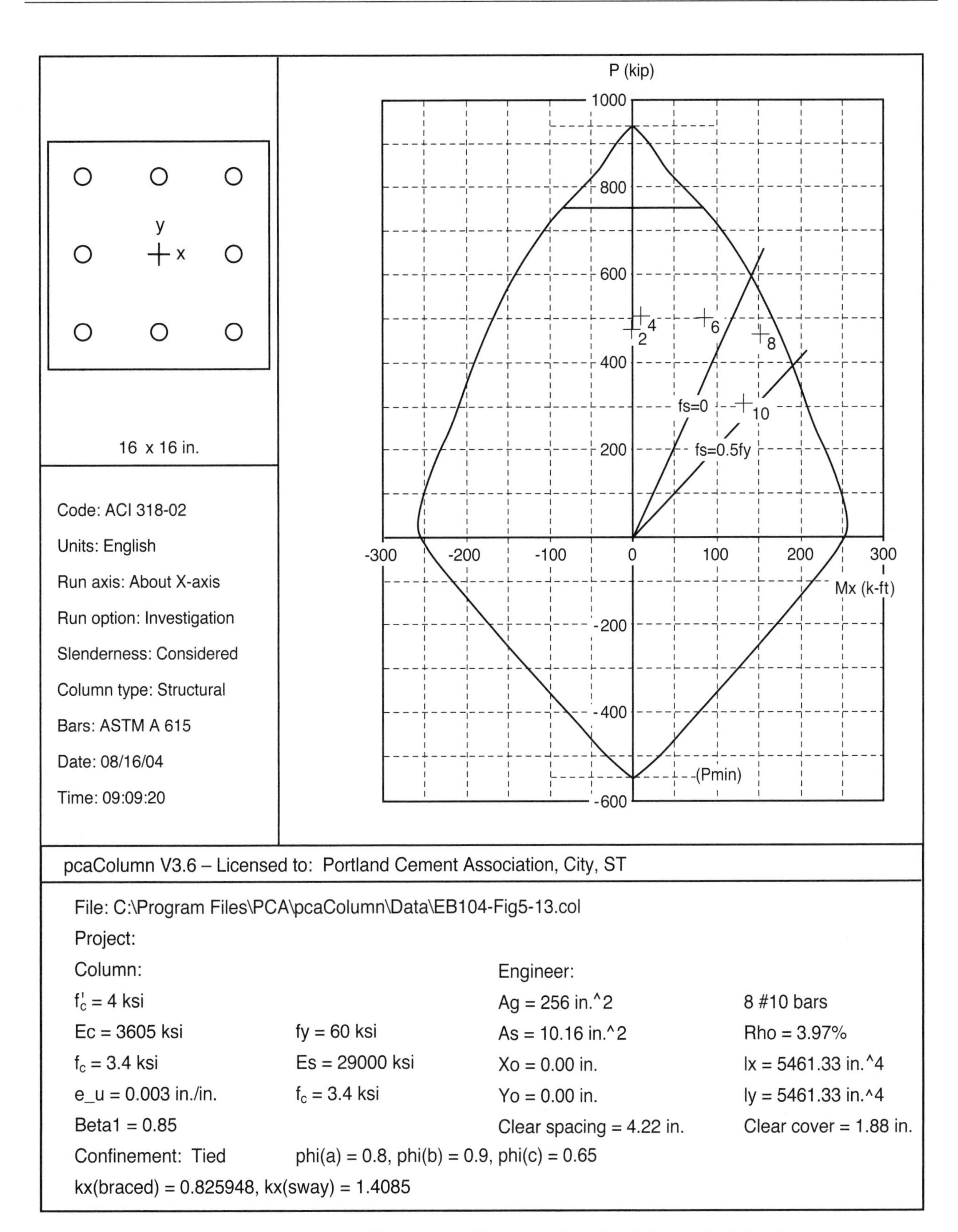

Figure 5-13 Interaction Diagram for First Story Interior Column, Building #2, Alternate (1), Including Slenderness

5.7.2 Example: Design of an Interior Column Stack for Building #2 Alternate (2) – Slab and Column Framing with Structural Walls (Non-sway Frame)

f'_c = 4000 psi (carbonate aggregate)

f_y = 60,000 psi

Required fire resistance rating = 2 hours

For the Alternate (2) framing, columns are designed for gravity loading only; the structural walls are designed to resist total wind loading.

(1) LOAD DATA

Roof: LL = 20 psf   Floors: LL = 50 psf
DL = 122 psf   DL = 142 psf (9 in. slab)

Calculations for the first story interior column are as follows:

(a) Total factored load (see Table 5-6):

Table 5-6 Interior Column Gravity Load Summary for Building #2, Alternate (2)

Floor	Dead Load (psf)	Live Load (psf)	Tributary Area (sq ft)	Influence Area (sq ft)	RM	Reduced Live Load (psf)	Cumulative Dead Load (kips)	Cumulative Live Load (kips)	Cumulative Factored Load ACI-Eq. (9-2) (kips)
5th (roof)	122	20	480	--	1	20	63	9.6	80
4th	142	50	480	1920	0.59	29.5	135	23.76	189
3rd	142	50	480	3840	0.49	24.5	207	35.52	295
2nd	142	50	480	5760	0.45	22.5	279	46.32	398
1st	142	50	480	7680	0.42	21	351	56.4	501

(b) Factored gravity load moment:

$$M_u = 0.035\, w_\ell \ell_2 \ell_n^2 = 0.035(1.6 \times 0.05)(24)(18.83^2) = 23.8 \text{ ft-kips}$$

portion of M_u to 1st story column = 24 (12/27) = 10.6 ft-kips

Similar calculations can be performed for the other floors.

(2) LOAD COMBINATIONS

The applicable load combination for each floor is summarized in Table 5-7. Note that only ACI Eq. (9-2) needs to be considered for columns in a non-sway frame.

Table 5-7 Interior Column Load Summary for Building #2, Alternate (2)

Floor	ACI Eq. (9-2)	
	P_u (kips)	M_u (ft-kips)
5th	80	9.5
4th	189	11.9
3rd	295	11.9
2nd	398	13.2
1st	501	10.6

(3) COLUMN SIZE AND REINFORCEMENT

With P_u = 501 kips, try a 16 × 16 in. column with 1% reinforcement (see Fig. 5-1).

Check for fire resistance: From Table 10-2, for a fire resistance rating of 2 hours, minimum column dimension = 10 in. < 16 in. O.K.

Determine if the columns are slender.

Using Table 5-3, for a 16 in. column, the maximum clear story height to neglect slenderness is 18.67 ft. Since the actual clear story heights are less than this value, slenderness need not be considered for the entire column stack.

- 1st story columns:

P_u = 501 kips, M_u = 10.6 ft-kips

From Fig. 5-19, use 4-No. 8 bars (ρ_g = 1.23%)

- 2nd through 5th story columns:

Using 4-No. 8 bars for the entire column stack would not be economical. ACI 10.8.4 may be used so that the amount of reinforcement at the upper levels may be decreased. The required area of steel at each floor can be obtained from the following:

$$\text{Required } A_{st} = (\text{area of 4-No. 8 bars})\left(\frac{P_u \text{ at floor level}}{546 \text{ kips}}\right)$$

where 546 kips is ϕP_n for the 16 × 16 in. column reinforced with 4-No. 8 bars. It is important to note that ρ_g should never be taken less than 0.5% (ACI 10.8.4). The required reinforcement for the column stack is summarized in Table 5-8.

Table 5-8 Reinforcement for Interior Column of Building #2, Alternat (2)

Floor	Required A_{st} (in^2)	Required ρ_g (%)	Reinforcement (ρ_g%)
5th (roof)	0.52	0.5	4-#6 (0.69)
4th	1.16	0.5	4-#6 (0.69)
3rd	1.76	0.69	4-#6 (0.69)
2nd	2	0.78	4-#7 (0.94)
1st	3.16	1.23	4-#8 (1.23)

Check for fire resistance: From Table 10-6, for a fire resistance rating of 4 hours or less, the required cover to the main longitudinal reinforcement = 1.5 in. < provided cover. O.K.

Column ties and spacing can be selected from Table 5-2.

5.7.3 Example: Design of an Edge Column Stack (E-W Column Line) for Building #1—3-story Pan Joist Construction (Sway Frame)

f'_c = 4000 psi (carbonate aggregate)

f_y = 60,000 psi

Required fire resistance rating = 1 hour (2 hours for columns supporting Alternate (2) floors).

(1) LOAD DATA

Roof: LL = 12 psf Floors: LL = 60 psf
DL = 105 psf DL = 130 psf

Calculations for the first story column are as follows:

(a) Total factored load (see Table 5-9):

Table 5-9 Edge Column Gravity Load Summary for Building #1

Floor	Dead Load (psf)	Live Load (psf)	Tributary Area (sq ft)	Influence Area	RM	Reduced Live Load (psf)	Cumulative Dead Load (kips)	Cumulative Live Load (kips)	Cumulative Factored Load ACI Eq. (9-2) (kips)
3rd (roof)	105	12	450	—	1	12	51	5.4	64
2nd	130	60	450	1800	0.6	36	114	21.6	165
1st	130	60	450	3600	0.5	30	176	35.1	262

(b) Factored moments in 1st story edge columns:

gravity loads: M_u = 327.6 ft-kips (see Section 3.8.3 – Step (2), M_u @ exterior columns) portion of M_u to 1st story column = 327.6/2 = 164 ft-kips

wind loads (see Fig. 2-13):

$P = 10.91$ kips

$M = 61.53$ ft-kips

(2) LOAD COMBINATIONS

For the 1st story column:

gravity loads:

$P_u = 268$ kips — ACI Eq. (9-2)

$M_u = 164$ ft-kips

gravity + wind loads:

$P_u = 1.2(176) + 0.8(10.91) = 220$ kips — ACI Eq. (9-3)

$M_u = 1.2(99.5) + 0.8(61.53) = 169$ ft-kips

or

$P_u = 1.2(176) + 0.50(35.1) + 1.6(10.91) = 247$ kips — ACI Eq. (9-4)

$M_u = 1.2(99.5) + 0.50(27.7) + 1.6(61.53) = 232$ ft-kips

or

$P_u = 0.9(176) + 1.6(10.91) = 176$ kips — ACI Eq. (9-6)

$M_u = 0.9(99.5) + 1.6(61.53) = 188$ ft-kips

Factored loads and moments, and load combinations, for the 2nd and 3rd story columns are calculated in a similar manner, and are summarized in Table 5-10.

Table 5-10 Edge Column Load Summary for Building #1

Floor	ACI Eq. (9-2)		ACI Eq. (9-3)		ACI Eq. (9-4)		ACI Eq. (9-6)	
	P	M	P_u	M_u	P_u	M_u	P_u	M_u
3rd	64	202	71	233	66	223	48	166
2nd	165	164	140	150	154	194	109	150
1st	262	164	220	169	247	232	176	188

(3) COLUMN SIZE AND REINFORCEMENT

For edge columns, initial selection of column size can be determined by referring directly to the column design charts and selecting an initial size based on required moment strength. For largest M_u = 233 kips, try a 16 × 16 in. column (see Fig. 5-19).

Check for fire resistance: From Table 10-2, for fire resistance ratings of 1 hour and 2 hours, minimum column dimensions of 8 in. and 10 in., respectively, are both less than 16 in. O.K.

Determine if the columns are slender.

Using k = 1.2, slenderness ratios for all columns:

$$\frac{k\ell_u}{r} = \frac{1.2\,[(13 \times 12) - 19.5]}{0.3\,(16)} = 34 > 22$$

Thus, all of the columns are slender. To neglect slenderness effects, the size of the column would have to be:

$$\frac{1.2\,[(13 \times 12) - 19.5]}{0.3h} < 22 \rightarrow h > 24.8 \text{ in.}$$

This column would probably not be practical for a building of the size considered.

Fig. 5-14 shows the results from PCACOLUMN for a first story edge column, including slenderness effects. Thirty five percent of the gross moment of inertia was used for the 36 × 19.5 in. column-line beam and seventy percent of the gross moment of inertia of the column cross section to account for cracking. The column was assumed fixed at the foundation. As can be seen from the figure, 8-No. 11 bars are required in this case.

Check for fire resistance: From Table 10-6, for fire resistance ratings of 4 hours or less, required cover to main longitudinal reinforcement is 1.5 in. < provided cover = 1.875 in. O.K.

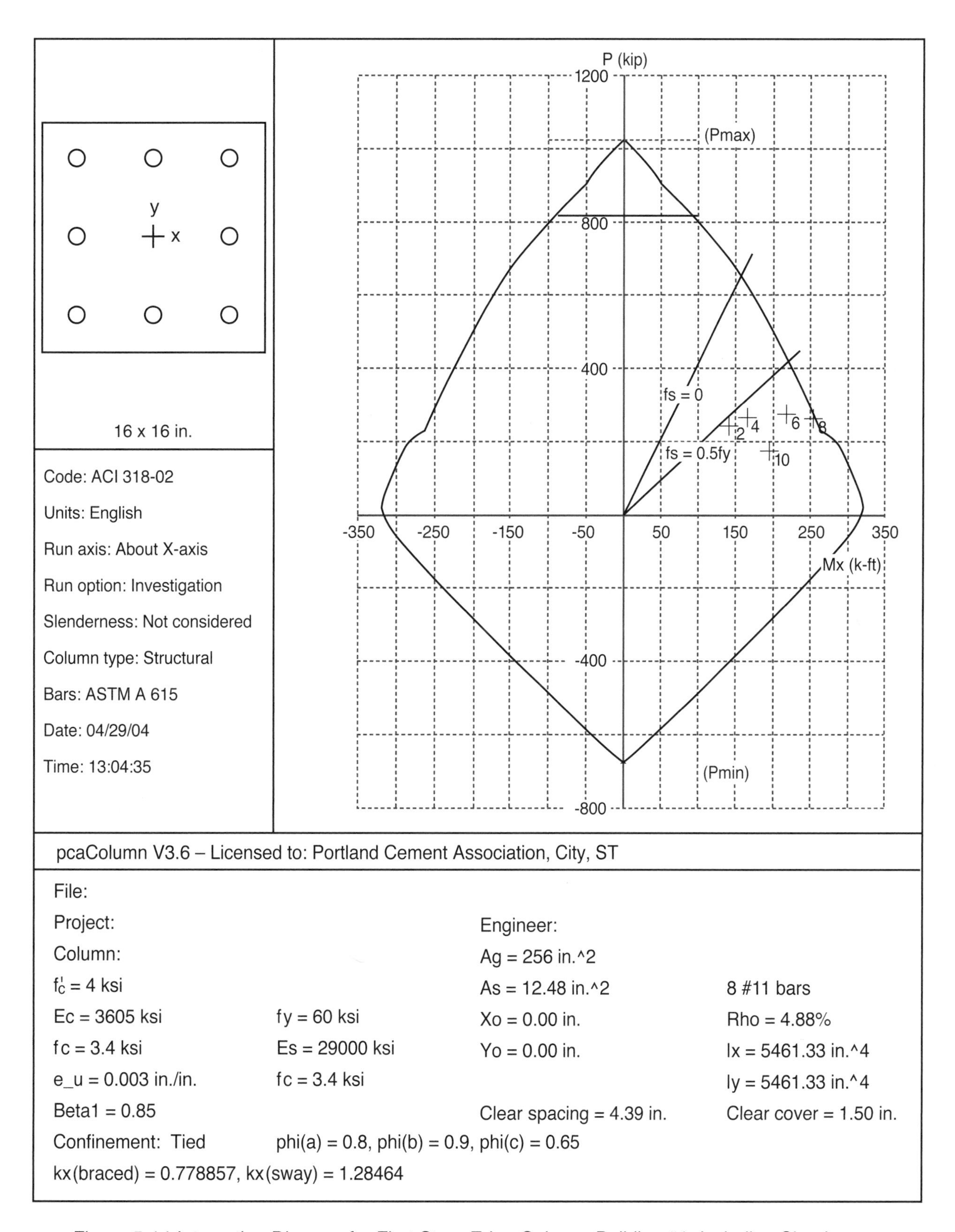

Figure 5-14 Interaction Diagram for First Story Edge Column, Building #1, Including Slenderness

5.8 COLUMN SHEAR STRENGTH

Columns in sway frames are required to resist the shear forces from lateral loads. For members subjected to axial compression, the concrete shear strength ϕV_c is given in ACI Eq. (11-4). Fig. 5-15 can be used to obtain this quantity for the square column sizes shown. The largest bar size from the corresponding column design charts of Figs. 5-16 through 5-23 were used to compute ϕV_c (for example, for a 16 × 16 in. column, the largest bar size in Fig. 5-19 is No. 11).

ACI Eq. (9-6) should be used to check column shear strength:

$$U = 0.9D + 1.6W$$

$$N_u = P_u = 0.9D$$

$$V_u = 1.6W$$

If V_u is greater than ϕV_c, spacing of column ties can be reduced to provide additional shear strength ϕV_s. Using the three standard spacings given in Chapter 3, Section 3.6, the values of ϕV_s given in Table 5-11 may be used to increase column shear strength.

Table 5-11 Shear Strength Provided by Column Ties

Tie Spacing	V_s - #3 ties*	V_s - #4 ties*
d/2	19 kips	35 kips
d/3	29 kips	54 kips
d/4	40 kips	71 kips

*2 legs, Grade 60 bars

For low-rise buildings, column shear strength ϕV_c will usually be more than adequate to resist the shear forces from wind loads.

5.8.1 Example: Design for Column Shear Strength

Check shear strength for the 1st floor interior columns of Building No. 2, Alternate (1) – slab and column framing without structural walls. For wind in the N-S direction, V = 9.17 kips (see Fig. 2-15).

$$N_u = P_u = 306 \text{ kips (see Example 5.7.1)}$$

$$V_u = 1.6(9.17) = 14.67 \text{ kips}$$

From Fig. 5-15, for a 16 × 16 in. column with N_u = 306 kips:

$$\phi V_c \cong 32 \text{ kips} > 16.42 \text{ kips} \quad \text{O.K.}$$

Column shear strength is adequate. With No. 10 column bars, use No. 3 column ties at 16 in. on center (least column dimension governs; see Table 5-2).

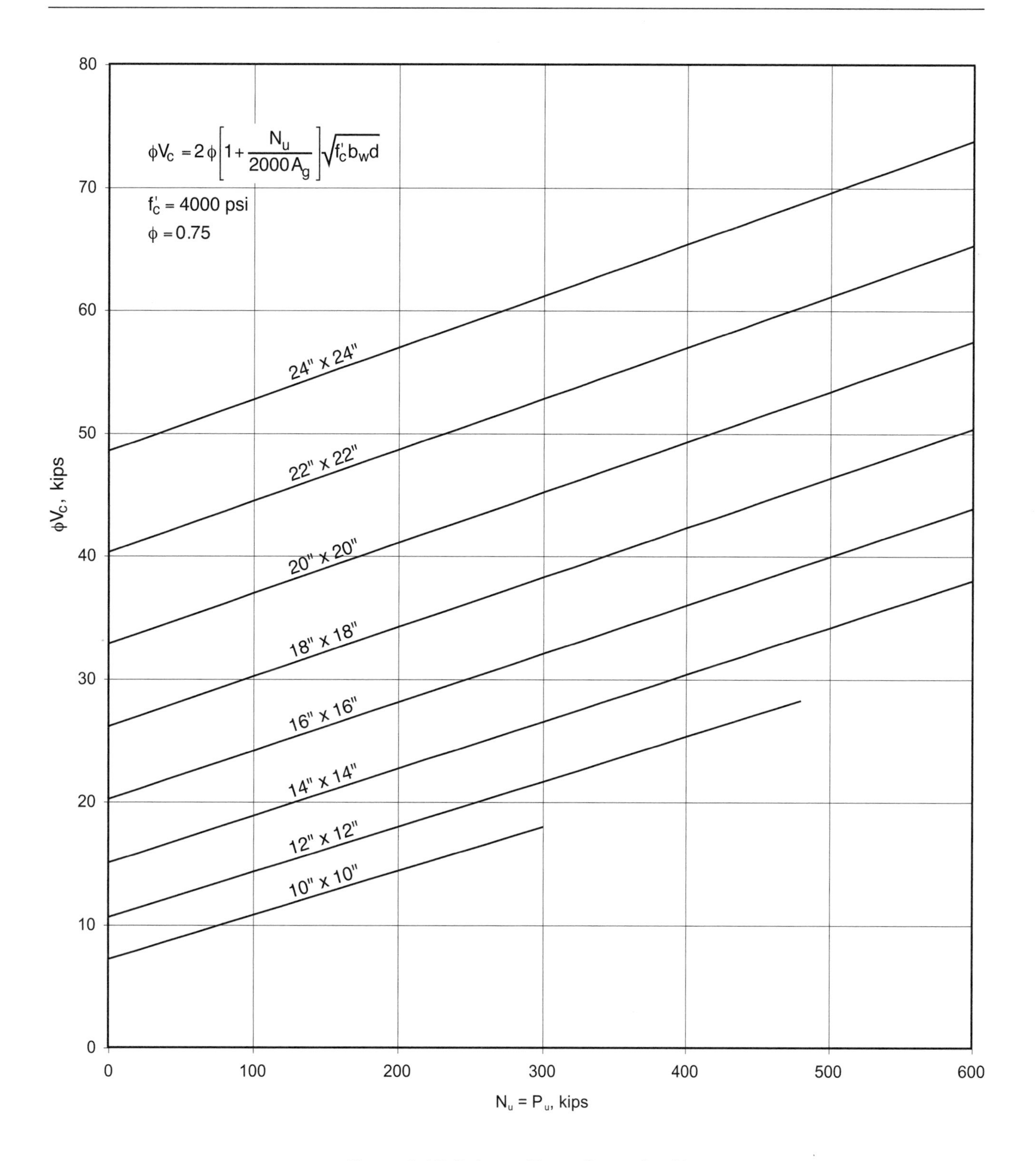

Figure 5-15 Column Shear Strength, ϕV_c

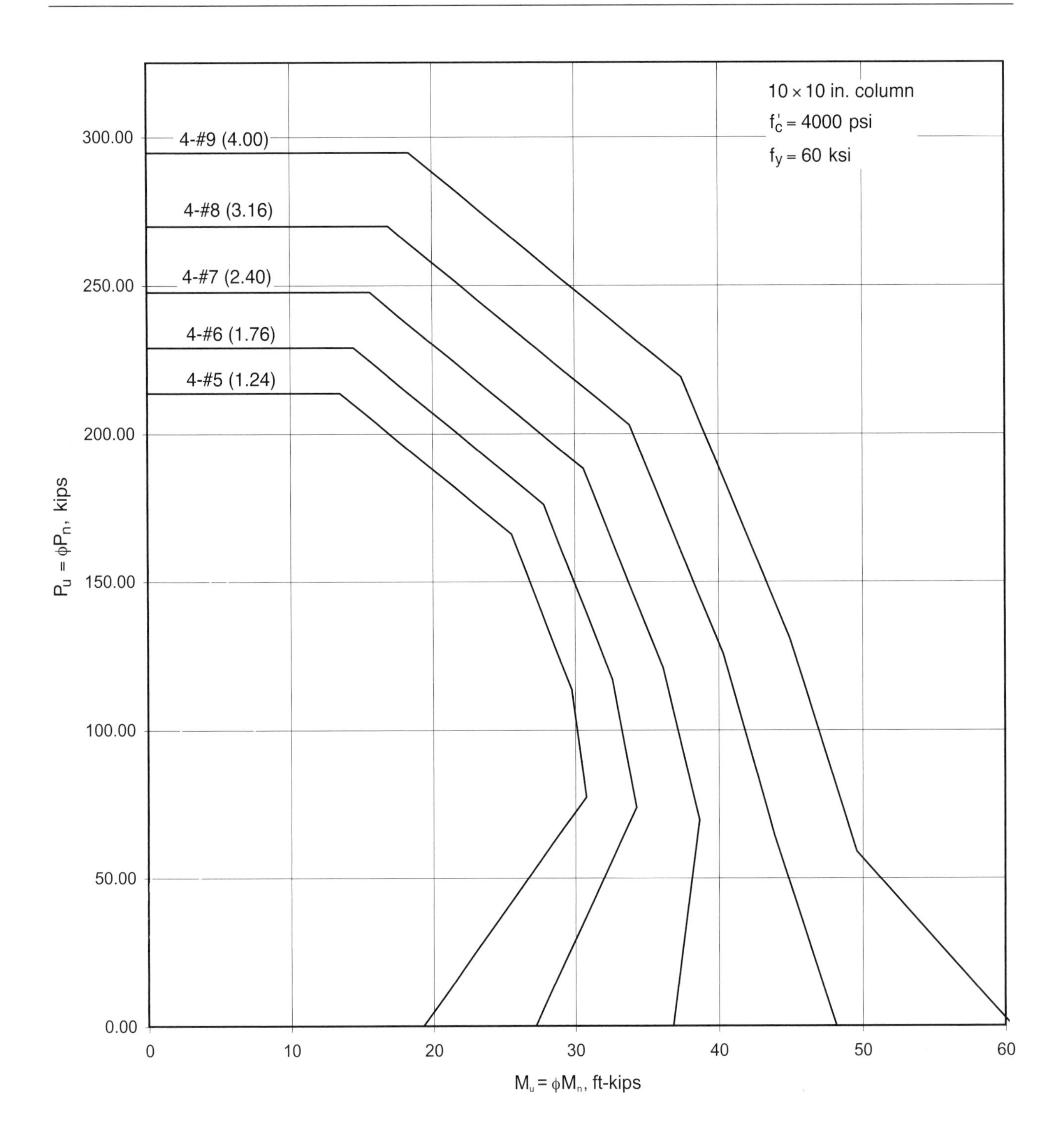

Figure 5-16 – 10 x 10 in. Column Design Chart

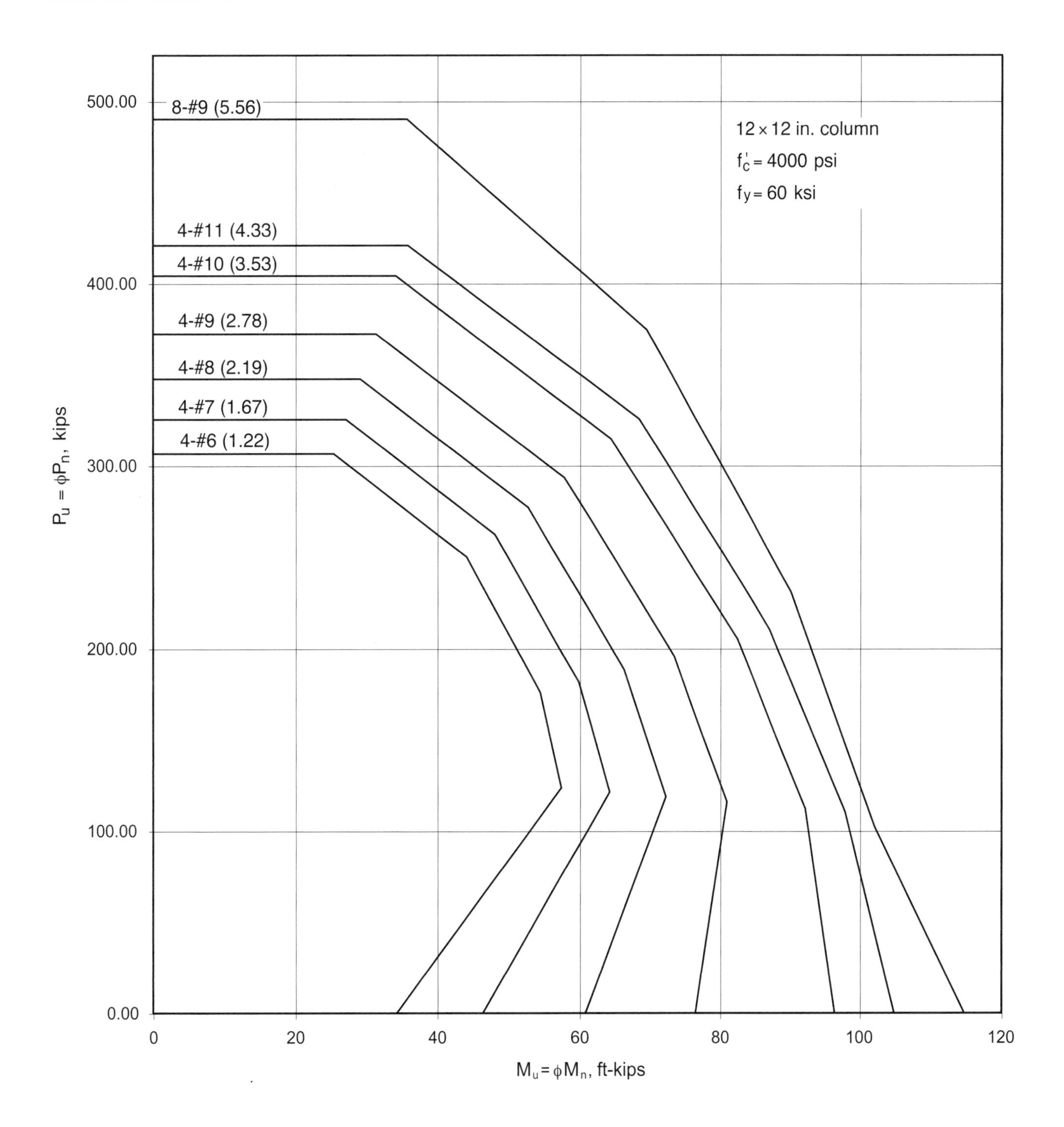

Figure 5-17 – 12 x 12 in. Column Design Chart

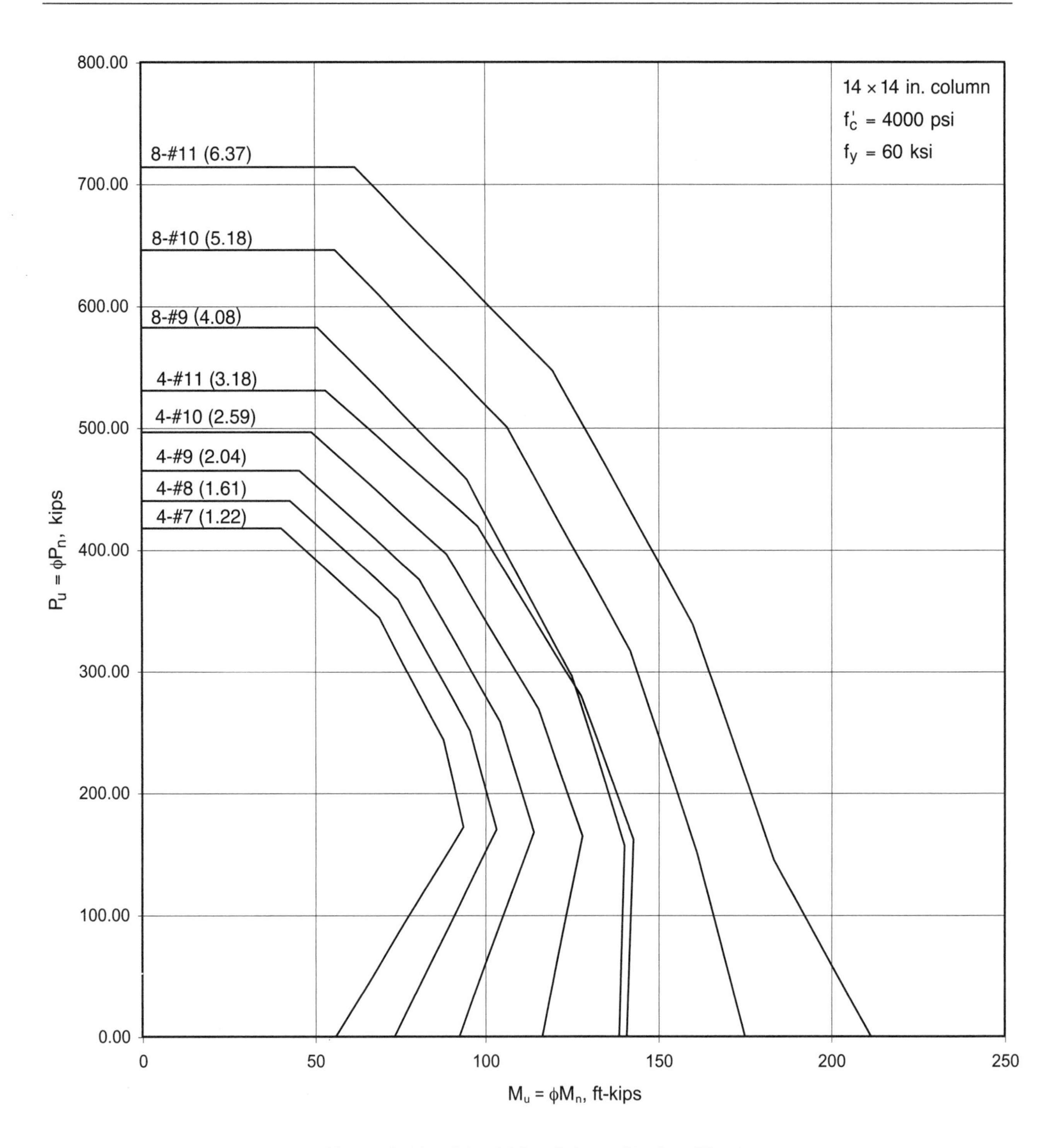

Figure 5-18 – 14 x 14 in. Column Design Chart

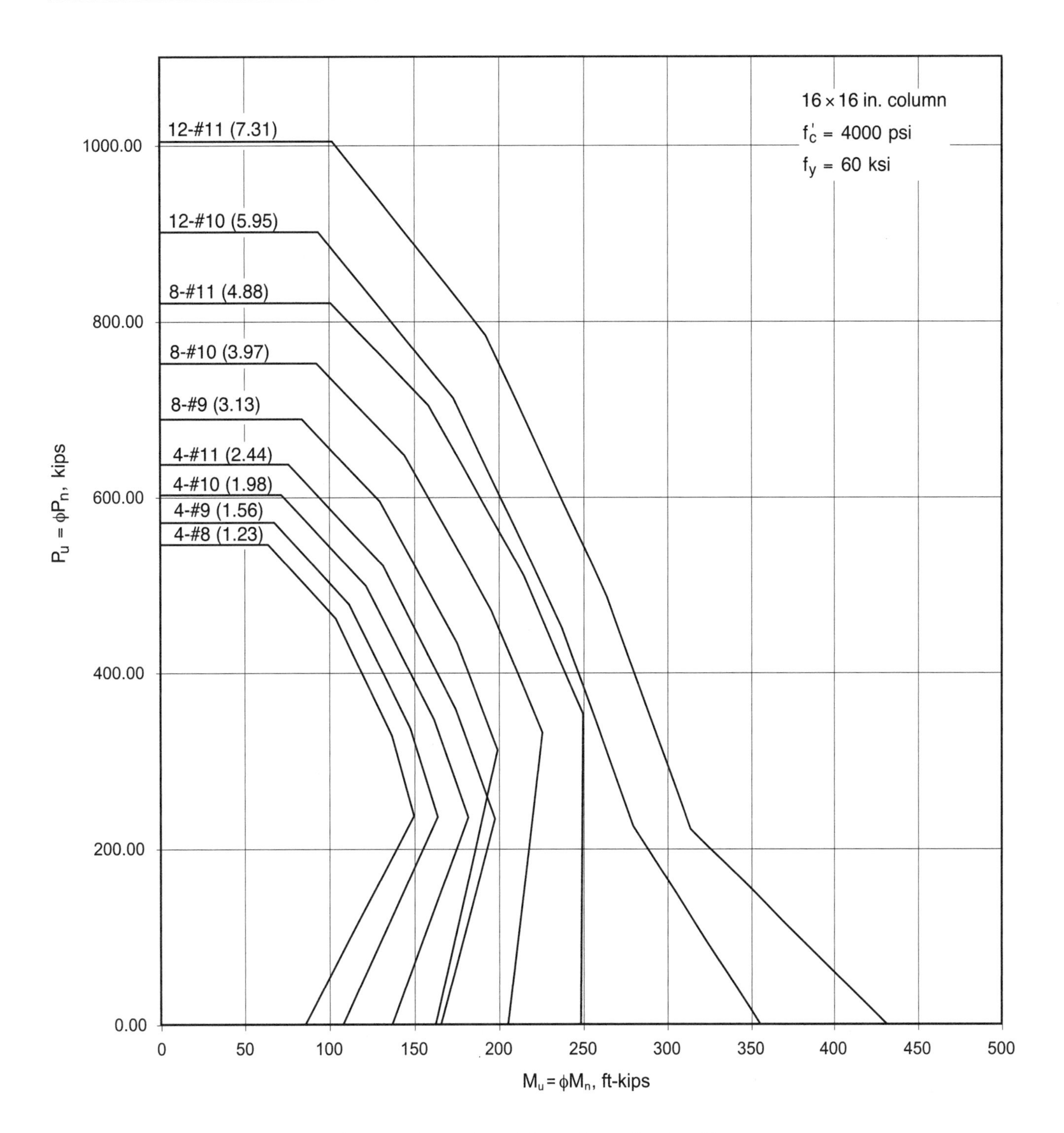

Figure 5-19 – 16 x 16 in. Column Design Chart

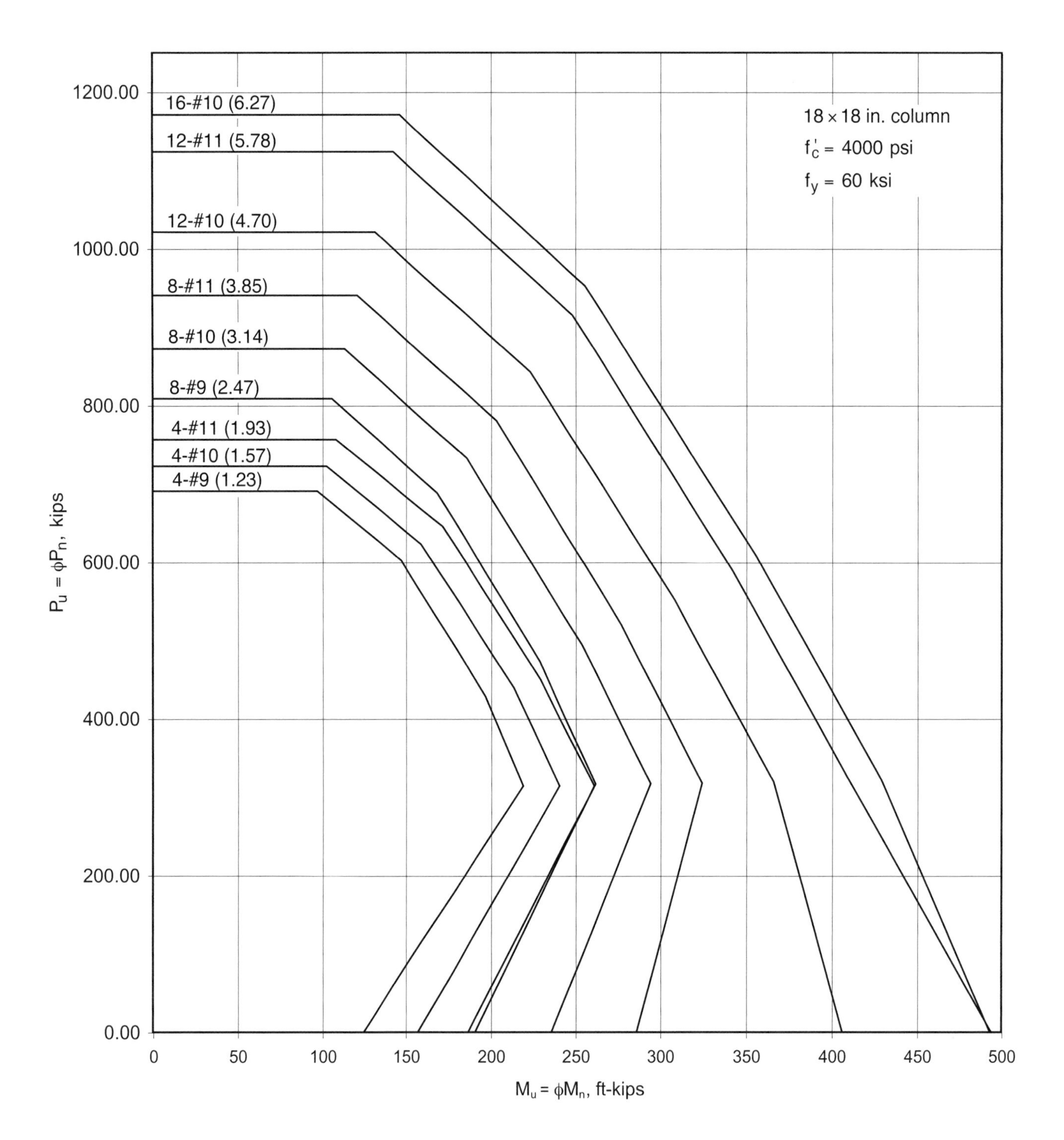

Figure 5-20 – 18 x 18 in. Column Design Chart

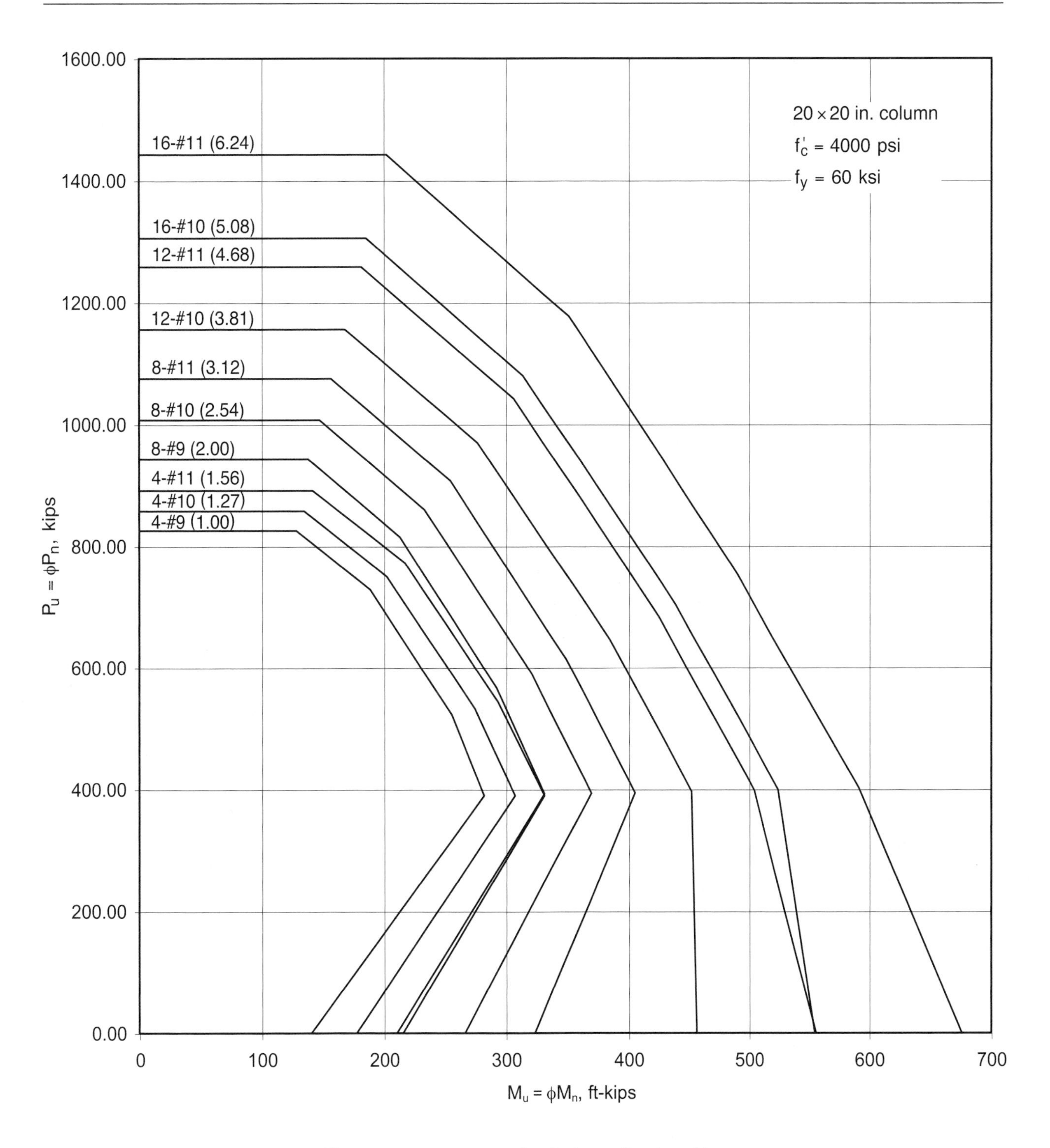

Figure 5-21 – 20 x 20 in. Column Design Chart

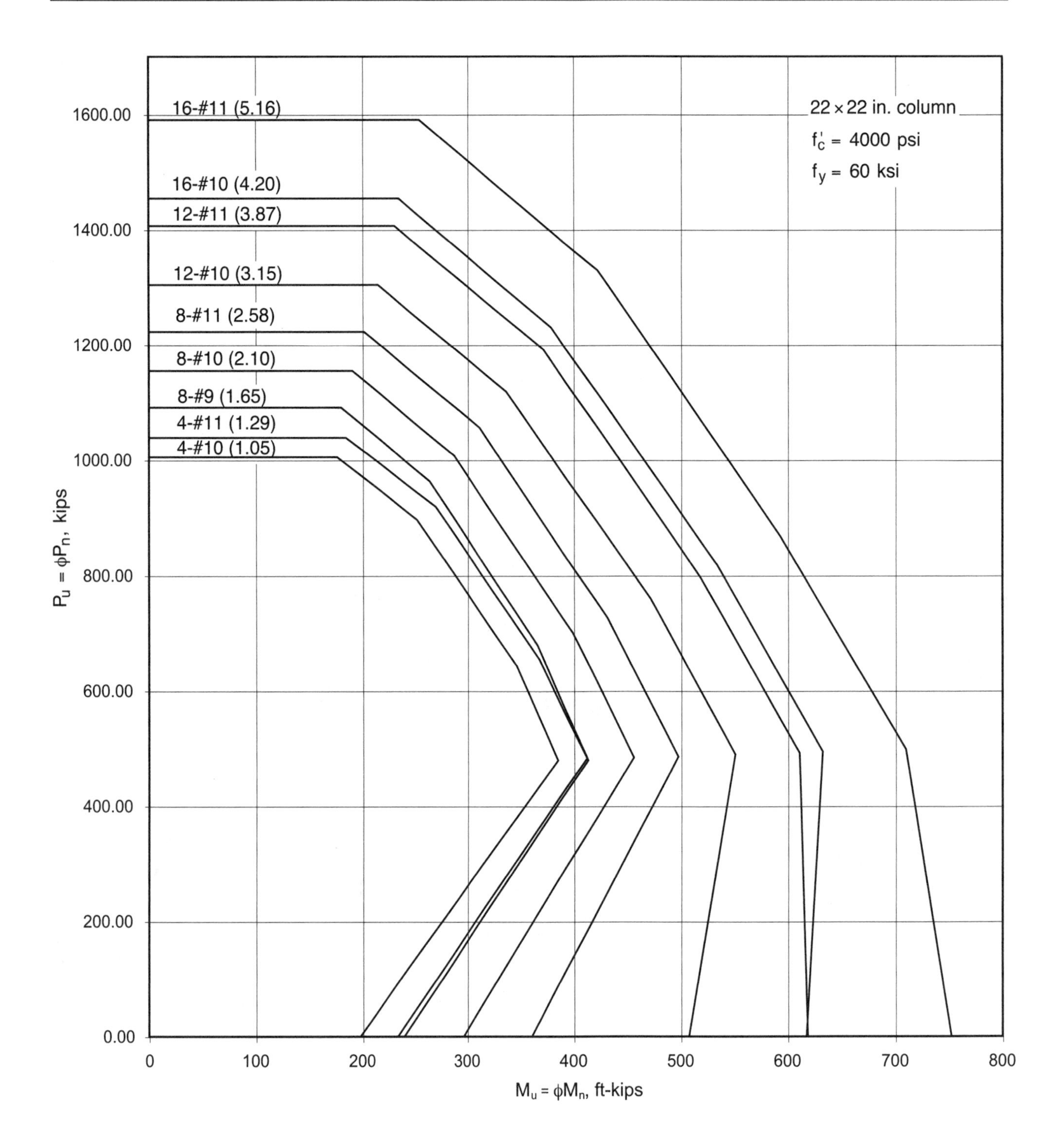

Figure 5-22 – 22 x 22 in. Column Design Chart

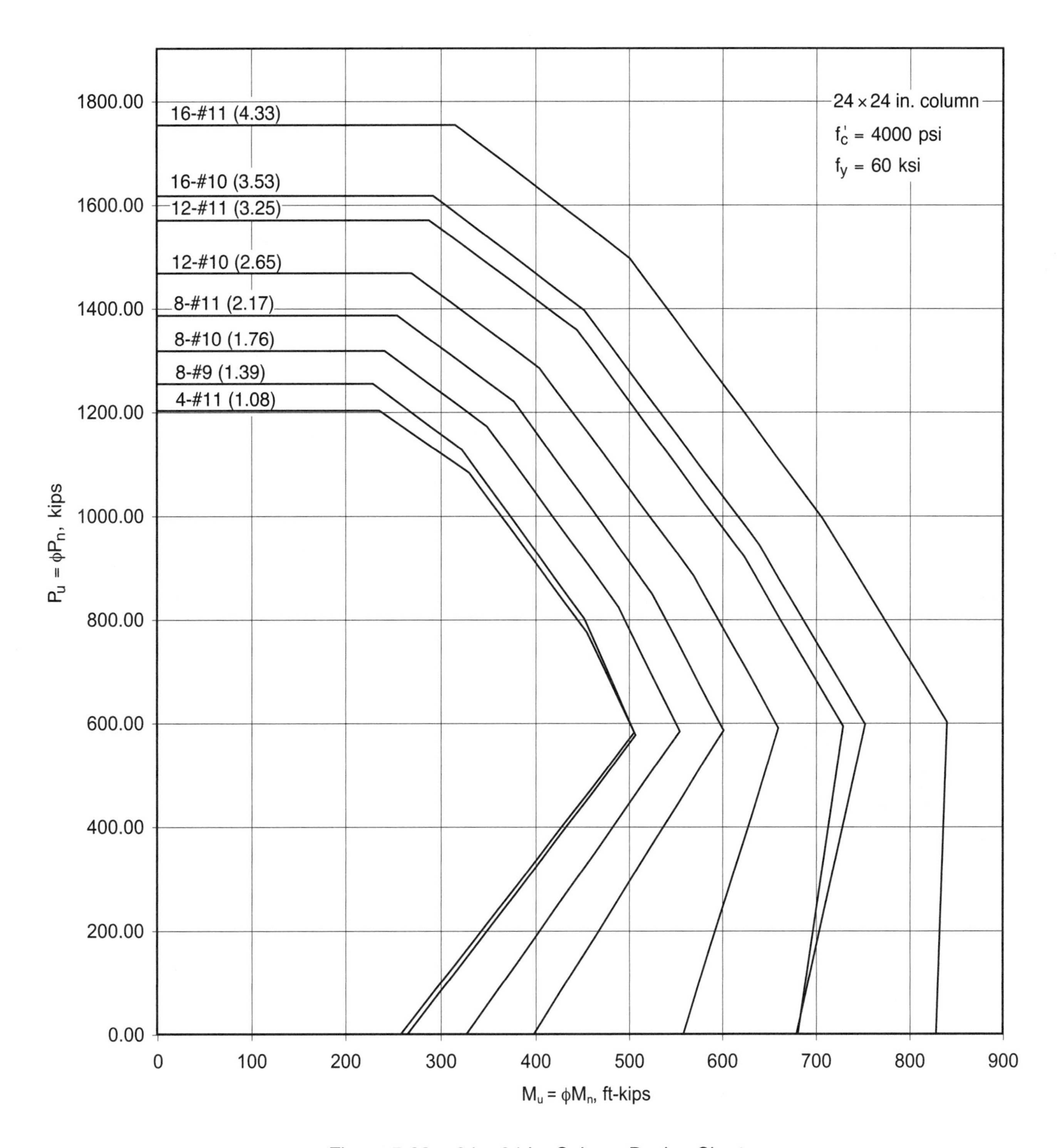

Figure 5-23 – 24 x 24 in. Column Design Chart

References

5.1 *Strength Design of Reinforced Concrete Columns,* Portland Cement Association, Skokie, Illinois, EB009.02D, 1977, 49 pp.

5.2 *Design Handbook in Accordance with the Strength Design Method of ACI 318-95: Vol. 2–Columns,* SP17A(90).CT93, American Concrete Institute, Detroit, Michigan, 1997, 222 pp.

5.3 *CRSI Handbook,* 9th Edition, Concrete Reinforcing Steel Institute, Schaumburg, Illinois, 2002.

5.4 *pcaColumn, Design and investigation of reinforced concrete column sections,* pcaStructure Point, Portland Cement Association, Skokie, Illinois, 2003.

5.5 *Notes on ACI 318-02,* 8th Edition, EB070, Portland Cement Association, Skokie, Illinois 2002.

Chapter 6

Simplified Design for Structural Walls

6.1 INTRODUCTION

For buildings in the low to moderate height range, frame action alone is usually sufficient to provide adequate resistance to lateral loads. Whether directly considered or not, nonstructural walls and partitions can also add to the total rigidity of a building and provide reserve capacity against lateral loads.

Structural walls or shearwalls are extremely important members in high-rise buildings. If unaided by walls, high-rise frames often could not be efficiently designed to satisfy strength requirements or to be within acceptable lateral drift limits. Since frame buildings depend primarily on the rigidity of member connections (slab-column or beam-column) for their resistance to lateral loads, they tend to be uneconomical beyond a certain height range (11-14 stories in regions of high to moderate seismicity, 15-20 stories elsewhere). To improve overall economy, structural walls are usually required in taller buildings.

If structural walls are to be incorporated into the framing system, a tentative decision needs to be made at the conceptual design stage concerning their location in plan. Most multi-story buildings are constructed with a central core area. The core usually contains, among other things, elevator hoistways, plumbing and HVAC shafts, and possibly exit stairs. In addition, there may be other exit stairs at one or more locations remote from the core area. All of these involve openings in floors, which are generally required by building codes to be enclosed with walls having a fire resistance rating of one hour or two hours, depending on the number of stories connected. In general, it is possible to use such walls for structural purposes.

If at all possible, the structural walls should be located within the plan of the building so that the center of rigidity of the walls coincides with the line of action of the resultant wind loads or center of mass for seismic design (see Chapter 11). This will prevent torsional effects on the structure. Since concrete floor systems act as rigid horizontal diaphragms, they distribute the lateral loads to the vertical framing elements in proportion to their rigidities. The structural walls significantly stiffen the structure and reduce the amount of lateral drift. This is especially true when shearwalls are used with a flat plate floor system.

6.2 FRAME-WALL INTERACTION

The analysis and design of the structural system for a building frame of moderate height can be simplified if the structural walls are sized to carry the entire lateral load. Members of the frame (columns and beams or slabs) can then be proportioned to resist the gravity loads only. Neglecting frame-wall interaction for buildings of moderate size and height will result in reasonable member sizes and overall costs. When the walls stiffness is much higher than the stiffness of the columns in a given direction within a story, the frame takes only a small portion of the lateral loads. Thus, for low-rise buildings, neglecting the contribution of frame action in resisting lateral loads and assigning the total lateral load resistance to walls is an entirely reasonable assumption. In contrast, frame-wall interaction must be considered for high-rise structures where the walls have a significant effect on the frame: in the upper stories, the frame must resist more than 100% of the story shears caused by the wind loads. Thus, neglecting frame-wall interaction would not be conservative at these levels. Clearly, a more economical high-rise structure will be obtained when frame-wall interaction is considered.

With adequate wall bracing, the frame can be considered non-sway for column design. Slenderness effects can usually be neglected, except for very slender columns. Consideration of slenderness effects for sway and non-sway columns is discussed in Chapter 5, Section 5.5.

6.3 WALL SIZING FOR LATERAL BRACING

The size of openings required for stairwells and elevators will usually dictate minimum wall plan layouts. From a practical standpoint, a minimum thickness of 6 in. will be required for a wall with a single layer of reinforcement, and 10 in. for a wall with a double layer. While fire resistance requirements will seldom govern wall thickness, the building code requirements should not be overlooked. See Chapter 10 for design considerations for fire resistance. The above requirements will, in most cases, provide stiff enough walls so that the frame can be considered non-sway.

The designer has to distinguish between sway and non-sway frames. This can be done by inspection by comparing the total lateral stiffness of the columns in a story to that of the bracing elements. A compression member may be considered non-sway if it is located in a story in which the bracing elements (shearwalls) have such substantial lateral stiffness to resist the lateral deflection of the story that any resulting deflection is not large enough to affect the column strength substantially. The ACI 318-89 contained a simple criterion to establish whether structural walls provide sufficient lateral bracing to qualify the frame as non-sway. The shear walls must have a total stiffness at least six times the sum of the stiffness of all columns in a given direction within the story:

$$I_{(walls)} \geq 6I_{(columns)}$$

The above criterion can be used to size the structural walls within the range of structures covered in this publication so that the frame can be considered non-sway.

6.3.1 Example: Wall Sizing for Non-Sway Condition

Using the approximate criteria, size the structural walls for Alternate (2) of Building #2 (5 story flat plate)*. In general, both the N-S and E-W directions must be considered. The E-W direction will be considered in this example since the moment of inertia of the walls will be less in this direction. The plan of Building #2 is shown in Fig. 6-1.

Required fire resistance rating of exit stair enclosure walls = 2 hours

For interior columns: $I = (^1/_{12})(16^4) = 5461$ in.^{4}
For edge columns: $I = (^1/_{12})(12^4)$ 1728 in.^{4}
$I_{(columns)} = 8(5461) + 12(1728) = 64{,}424$ in.^{4}
$6I_{(columns)} = 386{,}544$ in.^{4}

* *The 5-story flat plate frame of Building #2 is certainly within the lower height range for structural wall consideration. Both architectural and economic considerations need to be evaluated to effectively conclude if structural walls need to be included in low-to-moderate height buildings.*

Try an 8 in. wall thickness. To accommodate openings required for stairwells, provide 8 ft flanges as shown in Fig. 6-2.

From Table 10-1, for a fire resistance rating of 2 hours, required wall thickness = 4.6 in. ≤ 8 in. O.K.

E-W direction
$A_g = (248 \times 8) + (88 \times 8 \times 2) = 1984 + 1408 = 3392$ in.^{2}
$\bar{x} = [(1984 \times 4) + (1408 \times 52)]/3392 = 23.9$ in.
$I_y = [(248 \times 8^3/12) + (1984 \times 19.9^2)] + [2(8 \times 88^3/12) + (1408 \times 28.1^2)] = 2{,}816{,}665$ in.^{4}
For two walls: $I_{(walls)} = 2(2{,}816{,}665) = 5{,}663{,}330$ in.^{4} >> 386,544 in.^{4}

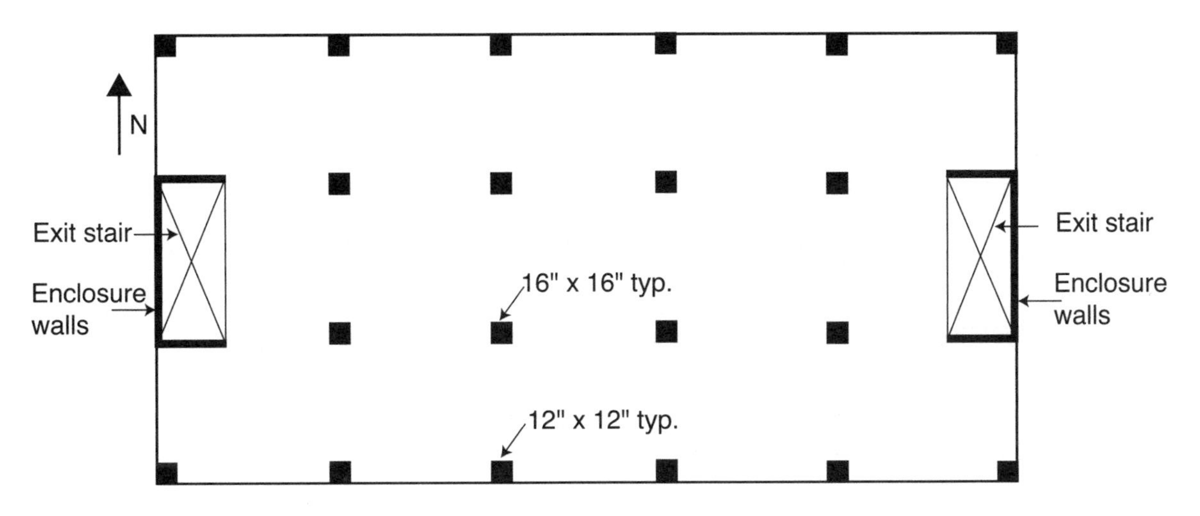

Figure 6-1 Plan of Building #2, Alternate (2)

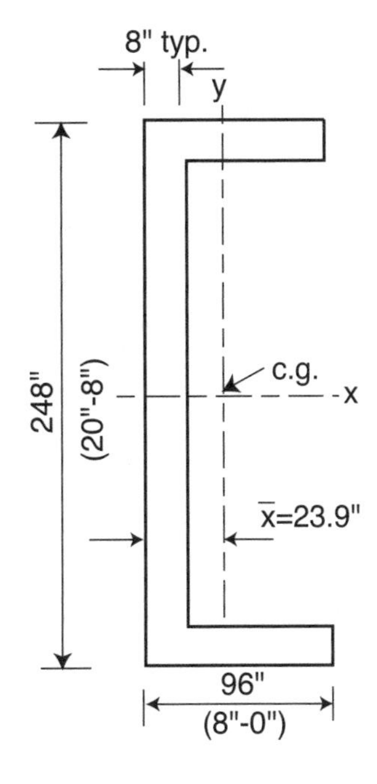

Figure 6-2 Plan View of Shearwall

Therefore, the frame can be considered non-sway for column design. Since the wall segments in the E-W direction provide most of the stiffness in this direction, the 8 ft length provided for the stairwell enclosure is more than adequate.

6.4 DESIGN FOR SHEAR

Design for horizontal shear forces (in the plane of the wall) can be critical for structural walls with small height-to-length ratios (i.e., walls in low-rise buildings). Special provisions for walls are given in ACI 11.10. In addition to shear, the flexural strength of the wall must also be considered (see Section 6.5).

Walls with minimum amounts of vertical and horizontal reinforcement are usually the most economical. If much more than the minimum amount of reinforcement is required to resist the factored shear forces, a change in wall size (length or thickness) should be considered. The amounts of vertical and horizontal reinforcement required for shear depends on the magnitude of the factored shear force, V_u. Table 6-1 summarizes the amounts of vertical and horizontal reinforcement required for shear for structural walls.

Table 6-1 Shear Reinforcement for Structural Walls

V_u	Horizontal Shear Reinforcement	Vertical Shear Reinforcement
$V_u \le \phi V_c / 2$	$\rho_h = 0.0020$ for #5 and smaller $\rho_h = 0.0025$ for other bars	$\rho_v = 0.0012$ for #5 and smaller $\rho_v = 0.0015$ for other bars
$\phi V_c / 2 < V_u \le \phi V_c$	$\rho_h = 0.0025$	$\rho_v = 0.0025$
$V_u > \phi V_c$	$\phi V_s = \phi A_v f_y d/s_2$ $\phi V_s + \phi V_c = V_u$ $\phi V_s + \phi V_c \le 10\phi\sqrt{f'_c}h(0.8\ell_w)$	$\rho_v = 0.0025 + 0.5[2.5 - h_w / \ell_w)](\rho_h - 0.0025)$

(1) When the factored shear force is less than or equal to one-half the shear strength provided by concrete ($V_u > \phi V_c/2$), minimum wall reinforcement according to ACI 14.3 must be provided. For walls subjected to axial compressive forces, ϕV_c may be taken as $\phi 2\sqrt{f'_c}hd$ where h is the thickness of the wall, $d = 0.8\ell_w$ (ACI 11.10.4), and ℓ_w is the length of the wall (ACI 11.10.5). Suggested vertical and horizontal reinforcement for this situation is given in Table 6-2.

(2) When the design shear force is more than one-half the shear strength provided by concrete ($V_u > \phi V_c/2$), minimum shear reinforcement according to ACI 11.10.9 must be provided. Suggested reinforcement (both vertical and horizontal) for this situation is given in Table 6-3.

(3) When the design shear force exceeds the concrete shear strength ($V_u > \phi V_c$), horizontal shear reinforcement must be provided according to ACI Eq. (11-31). Note that the vertical and horizontal reinforcement must not be less than that given in Table 6-3.

Table 6-2 Minimum Wall Reinforcement ($V_u \leq \phi V_c/2$)

Wall Thickness h (in.)	Vertical		Horizontal	
	Minimum A_s[a] ($in.^2/ft$)	Suggested Reinforcement	Minimum A_s[b] ($in.^2/ft$)	Suggested Reinforcement
6	0.09	#3 @ 15	0.14	#4 @ 16
8	0.12	#3 @ 11	0.19	#4 @ 12
10	0.14	#4 @ 16	0.24	#5 @ 15
12	0.17	#3 @ 15[c]	0.29	#4 @ 16[c]

[a]Minimum A_s/ft of wall = 0.0012(12)h = 0.0144h for #5 bars and less (ACI 14.3.2)
[b]Minimum A_s/ft of wall = 0.0020(12)h = 0.0240h for #5 bars and less (ACI 14.3.3)
[c]Two layers of reinforcement are required (ACI 14.3.4)

Table 6-3 Minimum Wall Reinforcement ($\phi V_c/2 \leq V_u \leq \phi V_c$)

Wall Thickness h (in.)	Vertical and Horizontal	
	Minimum A_s[a] ($in.^2/ft$)	Suggested Reinforcement
6	0.18	#4 @ 13
8	0.24	#4 @ 10
10	0.30	#5 @ 12
12	0.36	#4 @ 13[b]

[a]Minimum A_s/ft of wall = 0.0025(12)h = 0.03h (ACI 11.10.9)
[b]Two layers of reinforcement are required (ACI 14.3.4)

Using the same approach as in Section 3.6 for beams, design for required horizontal shear reinforcement in walls when $V_u > \phi V_c$ can be simplified by obtaining specific values for the design shear strength ϕV_s provided by the horizontal reinforcement. As noted above, ACI Eq. (11-31) must be used to obtain ϕV_s:

$$\phi V_s = \phi \frac{A_v f_y d}{s_2}$$

where A_v is the total area of the horizontal shear reinforcement within a distance s_2, $\phi = 0.75$. $f_y = 60{,}000$ psi, and $d = 0.8\ell_w$ (ACI 11.10.4). For a wall reinforced with No. 4 bars at 12 in. in a single layer, ϕV_s becomes:

$$\phi V_s = 0.75 \times 0.20 \times 60 \times (0.8 \times 12\ell_w)/12 = 7.2\ell_w \text{ kips}$$

where ℓ_w is the horizontal length of wall in feet.

Table 6-4 gives values of ϕV_s per foot length of wall based on various horizontal bar sizes and spacings.

Table 6-4 Shear Strength ϕV_s Provided by Horizontal Shear Reinforcement*

Bars Spacing S_2 (in.)	ϕV_s (kips/ft length of wall)			
	#3	#4	#5	#6
6	7.9	14.4	22.3	31.7
7	6.8	12.3	19.1	27.2
8	5.9	10.8	16.7	23.8
9	5.3	9.6	14.9	21.1
10	4.8	8.6	13.4	19.0
11	4.3	7.9	12.2	17.3
12	4.0	7.2	11.2	15.8
13	3.7	6.6	10.3	14.6
14	3.4	6.2	9.6	13.6
15	3.2	5.8	8.9	12.7
16	3.0	5.4	8.4	11.9
17	2.8	5.1	7.9	11.2
18	2.6	4.8	7.4	10.6

* Values of ϕV_s are for walls with a single layer of reinforcement.

Tabulated values should be doubled for walls with two layers

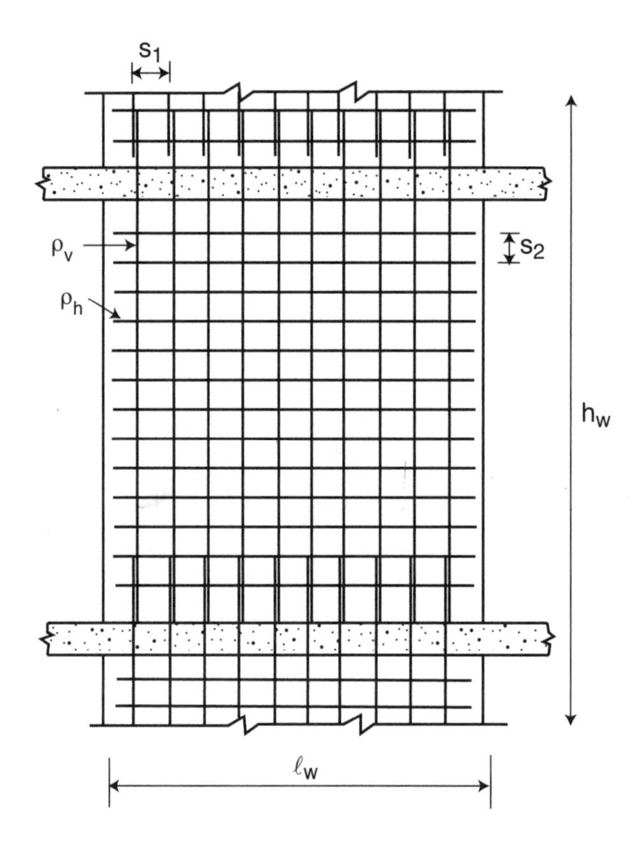

Table 6-5 gives values of $\phi V_c = \phi 2\sqrt{f'_c}h(0.8\ell_w)$ and limiting values of $\phi V_n = \phi V_c + \phi V_s = \phi 10\sqrt{f'_c}h(0.8\ell_w)$, both expressed in kips per foot length of wall.

Table 6-5 Design Values of ϕV_c and Maximum Allowable ϕV_n

Wall Thickness h (in.)	V_c (kips/ft length of wall)	Max. V_n (kips/ft length of wall)
6	5.5	27.3
8	7.3	36.4
10	9.1	45.5
12	10.9	54.6

The required amount of vertical shear reinforcement is given by ACI Eq. (11-32):

$$\rho_v = 0.0025 + 0.5(2.5 - h_w/\ell_w)(\rho_h - 0.0025)$$

where h_w = total height of wall
$\rho_v = A_{vn}/s_1h$
$\rho_h = A_{vh}/s_2h$

When the wall height-to-length ratio h_w/ℓ_w is less than 0.5, the amount of vertical reinforcement is equal to the amount of horizontal reinforcement (ACI 11.10.9.4).

6.4.1 Example 1: Design for Shear

To illustrate the simplified methods described above, determine the required shear reinforcement for the wall shown in Fig. 6-3. The service shear force from the wind loading is 160 kips. Assume total height of wall from base to top is 20 ft.

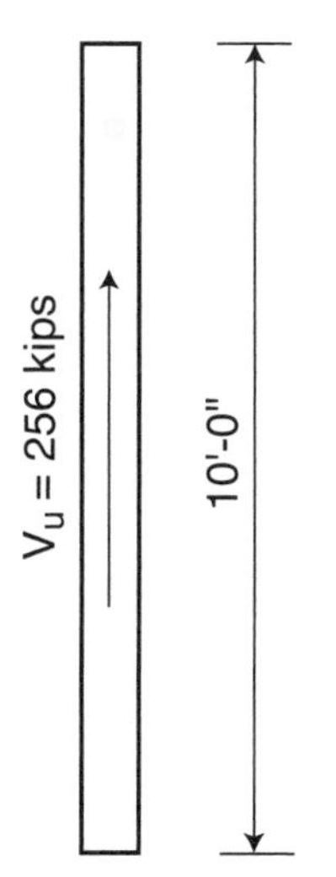

Figure 6-3 Plan View of Shearwall

(1) Determine factored shear force. Use ACI Eq. (9-4) for wind loads only

$$V_u = 1.6(160) = 256 \text{ kips}$$

(2) Determine ϕV_c and maximum allowable ϕV_n

From Table 6-5 $\quad \phi V_c = 7.3 \times 10 = 73$ kips
$\phi V_n = 36.4 \times 10 = 364$ kips

Wall cross section is adequate ($V_u <$ maximum ϕV_n); however, shear reinforcement is determined from ACI Eq. (11-31) must be provided ($V_u > \phi V_c$).

(3) Determine required horizontal shear reinforcement

$$\phi V_s = V_u - \phi V_c = 256 - 73 = 183 \text{ kips}$$
$$\phi V_s = 183/10 = 18.3 \text{ kips/ft length of wall}$$

Select horizontal bars from Table 6-4

For No. 5 @ 7 in., $\phi V_s = 19.1$ kips/ft > 18.3 kips/ft O.K.

$$s_{max} = 18 \text{ in.} > 7 \text{ in. O.K.}$$

Use No. 5 @ 7 in. horizontal reinforcement

Note: The use of minimum shear reinforcement No. 4 @ 10 in (Table 6-3) for an 8 in. wall thickness is not adequate:

$\phi V_s = 8.6$ kips/ft only (Table 6-4).

(4) Determine required vertical shear reinforcement

$$\begin{aligned} \rho_v &= 0.0025 + 0.5(2.5 - h_w/\ell_w)(\rho_h - 0.0025) \\ &= 0.0025 + 0.5(2.5 - 2)(0.0055 - 0.0025) \\ &= 0.0033 \end{aligned}$$

where $h_w/\ell_w = 20/10 = 2$

$$\rho_h = A_{vh}/s_2 h = 0.31/(8 \times 7) = 0.0055$$

Required $A_{vn}/s_1 = \rho_v h = 0.0033 \times 8 = 0.026$ in.^{2}/in.

For No. 5 bars: $s_1 = 0.31/0.026 = 11.9$ in. < 18 in. O.K.

Use No. 5 @ 12 in. vertical reinforcement.

6.4.2 Example 2: Design for Shear

For Alternate (2) of Building #2 (5-story flat plate), select shear reinforcement for the two shearwalls. Assume that the total wind forces are resisted by the walls, with slab-column framing resisting gravity loads only.

(1) E-W direction

Total shear force at base of building (see Chapter 2, Section 2.2.1.1):

$$V = 6.9 + 13.4 + 12.9 + 12.2 + 12.6 = 58 \text{ kips}$$

For each shearwall, V = 58/2 = 29 kips

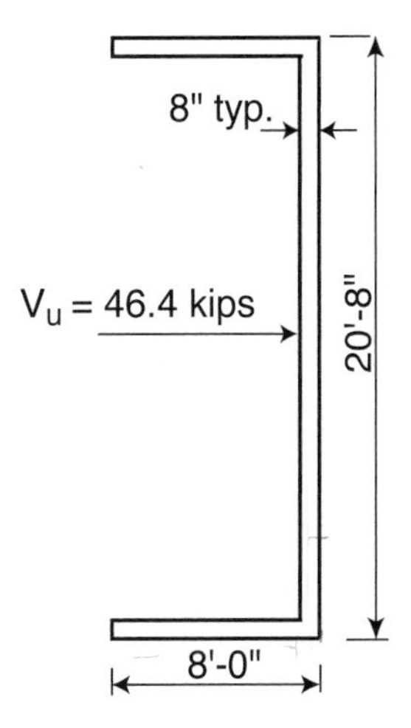

Factored shear force (use ACI Eq. (9-4) for wind load only):

$$V_u = 1.6(29) = 46.4 \text{ kips}$$

For the E-W direction, assume that the shear force is resisted by the two 8 ft flange segments only. For each segment:

$$\phi V_c = 7.3 \times 8 = 58.4 \text{ kips} \qquad \text{(see Table 6-5)}$$

Since V_u for each 8 ft segment = 46.4/2 = 23.2 kips which is less than $\phi V_c/2$ = 58.4/2 = 29.2 kips, provide minimum wall reinforcement from Table 6-2 For 8 in. wall, use No. 4 @ 12 in. horizontal reinforcement and No. 3 @ 11 in. vertical reinforcement.

(2) N-S direction

Total shear force at base of building (see Chapter 2, Section 2.2.1.1):

$$V = 16.2 + 31.6 + 30.6 + 29.2 + 30.7 = 138.3 \text{ kips}$$

For each shearwall, V = 138.3/2 = 69.2 kips

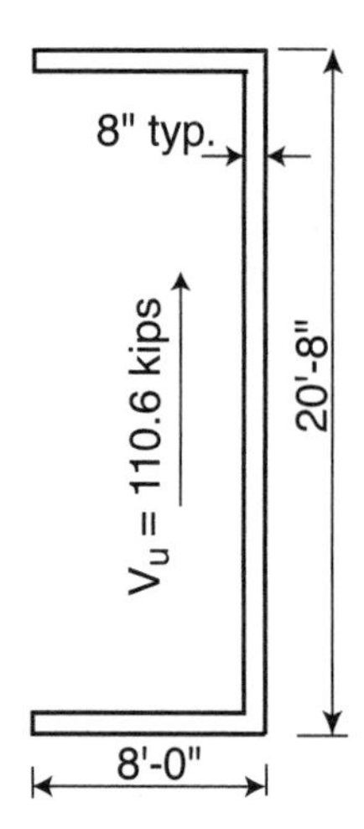

Factored shear force:

$$V_u = 1.6(69.2) = 110.6 \text{ kips}$$

For the N-S direction, assume that the shear force is resisted by the 20 ft-8 in. web segment only. From Table 6-5

$$\phi V_c = 7.3 \times 20.67 = 150.9 \text{ kips}$$

Since $\phi V_c/2 = 75.5$ kips $< V_u = 110.6$ kips $< \phi V_c = 150.9$ kips, provide minimum shear reinforcement from Table 6-3 For 8 in. wall, use No. 4 @ 10 in. horizontal as well as vertical reinforcement.

(3) Check shear strength in 2nd story in the N-S direction

$$V_u = 1.6(16.2 + 31.6 + 30.6 + 29.2)/2 = 86.1 \text{ kips}$$

The minimum shear reinforcement given in Table 6-3 s still required in the 2nd story since $\phi V_c/2 = 75.5$ kips $< V_u = 86.1$ kips $< \phi V_c = 150.9$ kips. For the 3rd story and above, the minimum wall reinforcement given in Table 6-2 can be used for all wall segments (V_u @ 3rd story = 62.7 kips $< \phi V_c/2 = 75.5$ kips). For horizontal reinforcement, use No. 4 @ 12 in., and for vertical reinforcement, use No. 3 @ 11 in.

(4) Summary of Reinforcement

Vertical bars:	Use No. 4 @ 10 in. for 1st and 2nd stories* No. 3 @ 10 in. for 3rd through 5th stories**
Horizontal bars:	Use No. 4 @ 10 in. for 1st and 2nd stories No. 4 @ 12 in. for 3rd through 5th stories

* *For moment strength, No. 6 @ 10 in. are required in the 8 ft. wall segments within the first story (see Example 6.5.1).*

** *Spacing of vertical bars reduced from 11 in. to 10 in. so that the bars in the 3rd story can be spliced with the bars in the 2nd story.*

6.5 DESIGN FOR FLEXURE

For buildings of moderate height, walls with uniform cross-sections and uniformly distributed vertical and horizontal reinforcement are usually the most economical. Concentration of reinforcement at the extreme ends of a wall (or wall segment) is usually not required except in high and moderate seismic zones. Uniform distribution of the vertical wall reinforcement required for shear will usually provide adequate moment strength as well. Minimum amounts of reinforcement will usually be sufficient for both shear and moment requirements.

In general, walls that are subjected to axial load or combined flexure and axial load need to be designed as compression members according to the provisions given in ACI Chapter 10 (also see Chapter 5)*. For rectangular shearwalls containing uniformly distributed vertical reinforcement and subjected to an axial load smaller than that producing balanced failure, the following approximate equation can be used to determine the nominal moment capacity of the wall[6.1] (see Fig. 6-4):

$$\phi M_n = \phi\left[0.5A_{st}f_y\ell_w\left(1+\frac{P_u}{A_{st}f_y}\right)\left(1-\frac{c}{\ell_w}\right)\right]$$

where A_{st} = total area of vertical reinforcement, in.^{2}

ℓ_w = horizontal length of wall, in.

P_u = factored axial compressive load, kips

f_y = yield strength of reinforcement = 60 ksi

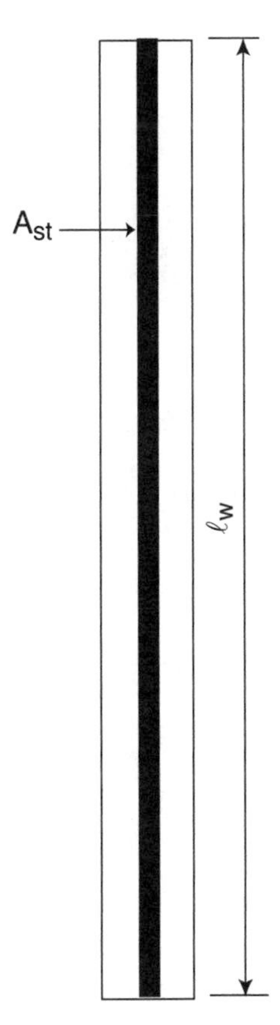

Figure 6-4 Plan View of Shearwall for Approximate Nominal Moment Capacity

* *In particular, ACI 10.2, 10.10, 10.11, 10.12, and 10.15 are applicable for walls.*

$$\frac{c}{\ell_w} = \frac{\omega + \alpha}{2\omega + 0.85\beta_1}, \text{where } \beta_1 = 0.85 \text{ for } f_c' = 4000 \text{ psi}$$

$$\omega = \left(\frac{A_{st}}{\ell_w h}\right)\frac{f_y}{f_c'}$$

$$\alpha = \frac{P_u}{\ell_w h f_c'}$$

h = thickness of wall, in.

ϕ = 0.90 (strength primarily controlled by flexure with low axial load)

Note that this equation should apply in a majority of cases since the wall axial loads are usually small.

6.5.1 Example: Design for Flexure

For Alternate (2) of Building #2 (5-story flat plate), determine the required amount of moment reinforcement for the two shearwalls. Assume that the 8 ft wall segments resist the wind moments in the E-W direction and the 20 ft-8 in. wall segments resist the wind moments in the N-S direction.

Roof: DL = 122 psf
Floors: DL = 142 psf

(1) Factored loads and load combinations

When evaluating moment strength, the load combination given in ACI Eq. (9-6) will govern.

U = 0.9D + 1.6W

(a) Dead load at first floor level:

Tributary floor area = 12 × 40 = 480 sq ft/story

Wall dead load = (0.150 × 3392)/144 = 3.53 kips/ft of wall height (see Sect. 6.3.1)

P_u = 0.9[(0.122 × 480) + (0.142 × 480 × 4) + (3.53 × 63)] = 498 kips

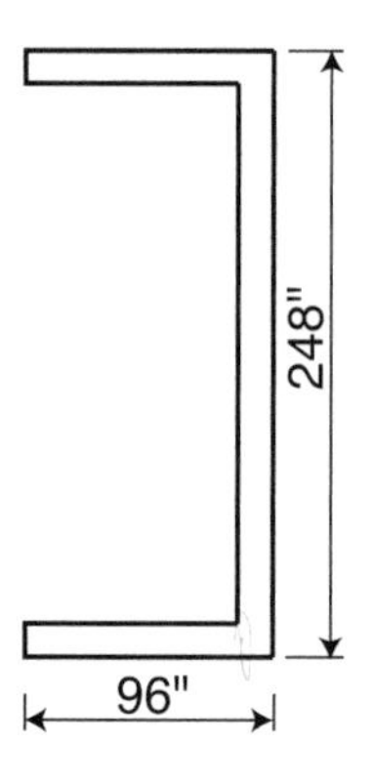

Proportion total P_u between wall segments:

2-8 ft segments:	2 × 96 = 192 in.	192/440 = 0.44
1-20 ft-8 in. segment:	248 in.	248/440 = 0.56

For 2-8 ft segments: P_u = 0.44(498) = 219 kips
1-20 ft-8 in. segment P_u = 0.56(498) = 279 kips

(b) Wind moments at first floor level:

From wind load analysis (see Chapter 2, Section 2.2.1.1):

E-W direction

$$M_u = 1.6\,[(6.9 \times 63) + (13.4 \times 51) + (12.9 \times 39) + (12.2 \times 27) + (12.6 \times 15)]/2$$
$$= 1712 \text{ ft-kips/shearwall}$$

N-S direction:

$$M_u = 1.6\,[(16.2 \times 63) + (31.6 \times 51) + (30.6 \times 39) + (29.2 \times 27) + (30.7 \times 15)]/2$$
$$= 4060 \text{ ft-kips/shearwall}$$

(c) Values of P_u and M_u for the 2nd and 3rd floor levels are obtained in a similar manner:

For 2nd floor level:

2-8 ft segments: $P_u = 171$ kips
1-20 ft-8 in. segment: $P_u = 218$ kips

E-W direction: $M_u = 1016$ ft-kips/shearwall
N-S direction: $M_u = 2400$ ft-kips/shearwall

For 3rd floor level:

2-8 ft segment: $P_u = 128$ kips
1-20 ft-8 in. segment: $P_u = 162$ kips

E-W direction: $M_u = 580$ ft-kips/shearwall
N-S direction: $M_u = 1367$ ft-kips/shearwall

(2) Design for Flexure in E-W direction

Initially check moment strength based on the required vertical shear reinforcement No. 4 @ 10 in. (see Example 6.4.2).

(a) For 2-8 ft wall segments at first floor level:

$P_u = 219$ kips

$M_u = 1712$ ft-kips

$\ell_w = 96$ in.

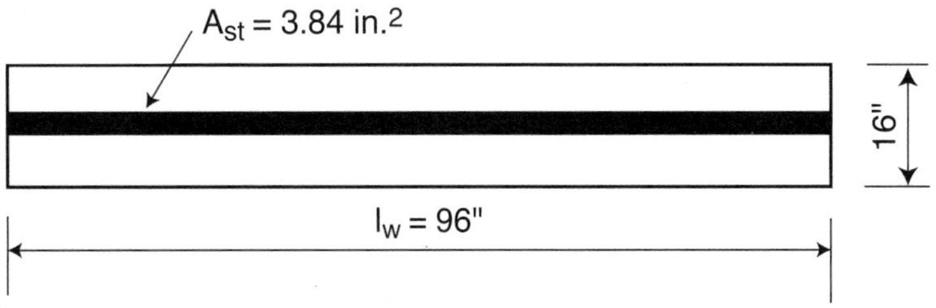

$A_{st} = 3.84$ in.^{2}

16"

$l_w = 96"$

For No. 4 @ 10 in. (2 wall segments):

$$A_{st} = 2 \times 0.24 \times 8 = 3.84 \text{ in.}^2$$

$$\omega = \left(\frac{A_{st}}{\ell_w h}\right)\frac{f_y}{f_c'} = \left(\frac{3.84}{96 \times 16}\right)\frac{60}{4} = 0.038$$

$$\alpha = \frac{P_u}{\ell_w h f_c'} = \frac{219}{96 \times 16 \times 4} = 0.036$$

$$\frac{c}{\ell_w} = \frac{\omega + \alpha}{2\omega + (0.85 \times 0.85)} = \frac{0.038 + 0.036}{2(0.038) + 0.72} = 0.092$$

$$M_n = 0.5A_{st}f_y\ell_w\left(1 + \frac{P_u}{A_{st}f_y}\right)\left(1 - \frac{c}{\ell_w}\right)$$

$$= 0.5 \times 3.84 \times 60 \times 96\left(1 + \frac{219}{3.84 \times 60}\right)(1 - 0.093)/12 = 1633 \text{ ft-kips}$$

$$\phi M_n = 0.9(1633) = 1469 \text{ ft-kips} < M_u = 1712 \text{ ft-kips} \qquad \text{N.G.}$$

No. 4 @ 10 in. is not adequate for moment strength in the E-W direction at the first story level.

Try No. 5 @ 10 in.:

$$A_{st} = 2 \times 0.39 \times 8 = 6.2 \text{ in.}^2$$

$$\omega = \left(\frac{6.2}{96 \times 16}\right)\frac{60}{4} = 0.061$$

$$\frac{c}{\ell_w} = \frac{0.061 + 0.036}{2(0.061) + 0.72} = 0.114$$

$$M_n = 0.5 \times 6.2 \times 60 \times 96\left(1 + \frac{219}{6.2 \times 60}\right)(1 - 0.114)/12 = 2094 \text{ ft-kips}$$

$$\phi M_n = 0.9(2094) = 1885 \text{ ft-kips} > M_u = 1712 \text{ ft-kips} \qquad \text{O.K.}$$

(b) For 2-8 ft wall segments at 2nd floor level:

P_u = 171 kips
M_u = 1016 ft-kips

Check No. 4 @ 10 in.:

A_{st} = 3.84 in.^{2}
ω = 0.038

$$\alpha = \frac{171}{96 \times 16 \times 4} = 0.028$$

$$\frac{c}{\ell_w} = \frac{0.038 + 0.028}{2(0.038) + 0.72} = 0.082$$

$$M_n = 0.5 \times 3.84 \times 60 \times 96\left(1 + \frac{171}{3.84 \times 60}\right)(1 - 0.082)/12 = 1474 \text{ ft-kips}$$

$$\phi M_n = 0.9(1474) = 1327 \text{ ft-kips} > M_u = 1016 \text{ ft-kips} \qquad \text{O.K.}$$

No. 4 @ 10 in. (required shear reinforcement) is adequate for moment strength above the first floor.

(c) For 2-8 ft wall segments at 3rd floor level:

P_u = 128 kips
M_u = 580 ft-kips

Check No. 3 @ 10 in. (required shear reinforcement above 2nd floor):

$$A_{st} = 2 \times 0.13 \times 8 = 2.08 \text{ in.}^2$$

$$\omega = \left(\frac{2.08}{96 \times 16}\right)\frac{60}{4} = 0.020$$

$$\alpha = \frac{128}{96 \times 16 \times 4} = 0.021$$

$$\frac{c}{\ell_w} = \frac{0.020 + 0.021}{2(0.020) + 0.72} = 0.054$$

$$M_n = 0.5 \times 2.08 \times 60 \times 96\left(1 + \frac{128}{2.08 \times 60}\right)(1 - 0.054)/12 = 957 \text{ ft-kips}$$

$$\phi M_n = 0.9(957) = 861 \text{ ft-kips} > M_u = 580 \text{ ft-kips} \qquad \text{O.K.}$$

No. 3 @ 10 in. (required shear reinforcement) is adequate for moment strength above the 2nd floor.

(3) Design for flexure in N-S direction

Initially check moment strength for required vertical shear reinforcement No. 4 @ 10 in. (see Example 6.4.2)

(a) For 1-20 ft-8 in. wall segment at first floor level:

P_u = 279 kips
M_u = 4060 ft-kips
ℓ_w = 248 in.
h = 8 in.

A_{st} = 4.96 in.^{2}

l_w = 248"

8"

For No. 4 @ 10 in.:

$$A_{st} = 0.24 \times 20.67 = 4.96 \text{ in.}^2$$

$$\omega = \left(\frac{4.96}{248 \times 8}\right)\frac{60}{4} = 0.038$$

$$\alpha = \frac{2.79}{248 \times 8 \times 4} = 0.035$$

$$\frac{c}{\ell_w} = \frac{0.038 + 0.035}{2(0.038) + 0.72} = 0.091$$

$$M_n = 0.5 \times 4.96 \times 60 \times 248\left(1 + \frac{279}{4.96 \times 60}\right)(1 - 0.091)/12 = 5415 \text{ ft-kips}$$

$$\phi M_n = 0.9(5415) = 4874 \text{ ft-kips} > M_u = 4060 \text{ ft-kips} \qquad \text{O.K.}$$

(b) For one-20 ft-8 in. wall segment at 3rd floor level:

P_u = 162 kips
M_u = 1367 ft-kips

Check No. 3 @ 10 in. (required shear reinforcement above 2nd floor):

$$A_{st} = 0.13 \times 20.67 = 2.69 \text{ in.}^2$$

$$\omega = \left(\frac{4.96}{248 \times 8}\right)\frac{60}{4} = 0.020$$

$$\alpha = \frac{162}{248 \times 8 \times 4} = 0.020$$

$$\frac{c}{\ell_w} = \frac{0.020 + 0.020}{2(0.020) + 0.72} = 0.053$$

$$M_n = 0.5 \times 2.69 \times 60 \times 2.48\left(1 + \frac{162}{2.69 \times 60}\right)(1 - 0.053)/12 = 3163 \text{ ft-kips}$$

$$\phi M_n = 0.9(3163) = 2847 \text{ ft-kips} > M_u = 1367 \text{ ft-kips} \quad \text{O.K.}$$

The required shear reinforcement for the 20 ft-8 in. wall segments is adequate for moment strength for full height of building.

(4) Summary

Required shear reinforcement determined in Example 6.4.2 can be used for the flexural reinforcement except for the 8 ft wall segments within the 1st floor where No. 5 @ 10 in. are required (see Fig. 6-5).

For comparison purposes, the shearwall was input into PCACOLUMN, using the add-on module PCA-COLUMN+ which enables the user to investigate any irregularly shaped reinforced concrete column.[6.2] For the reinforcement shown in Fig. 6-5 at the 1st story level, the shearwall was analyzed for the combined factored axial load (due to the dead loads) and moments (due to the wind loads) about each principal axis. The results are shown for the x and y axes in Figs. 6-6 and 6-7, respectively. As expected, the load combination point (represented by point 1 in the figures) is in the lower region of the interaction diagram, with the applied axial load well below the balanced point. Since PCACOLUMN uses the entire cross-section when computing the moment capacity (and not only certain segments as was done in the steps above), the results based on the reinforcement from the approximate analysis will be conservative.

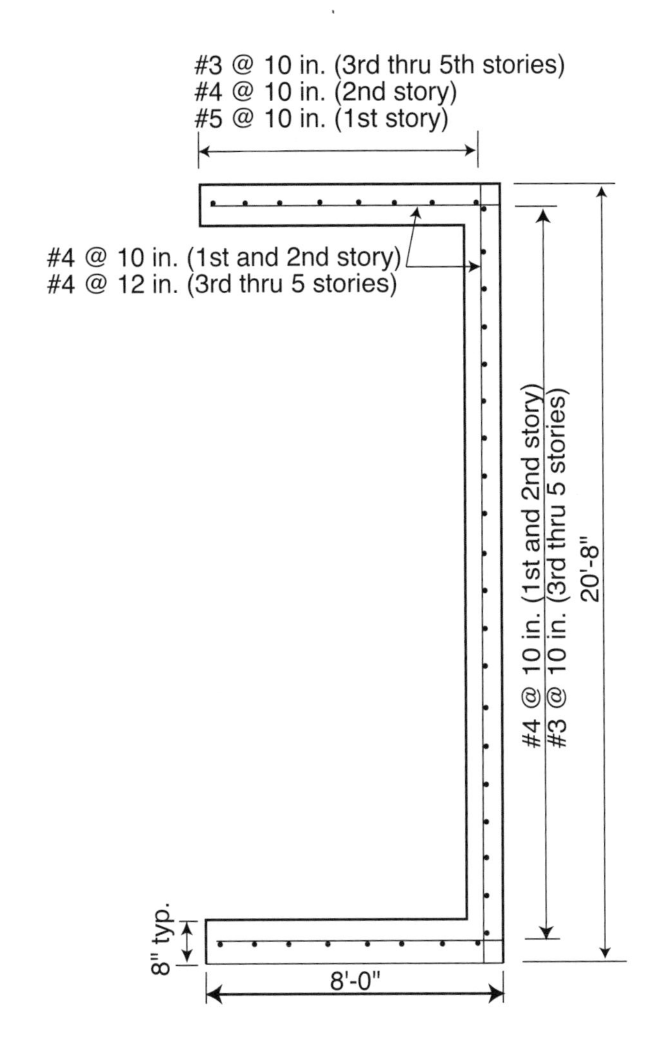

Figure 6-5 Required Reinforcement for Shearwall in Building #2

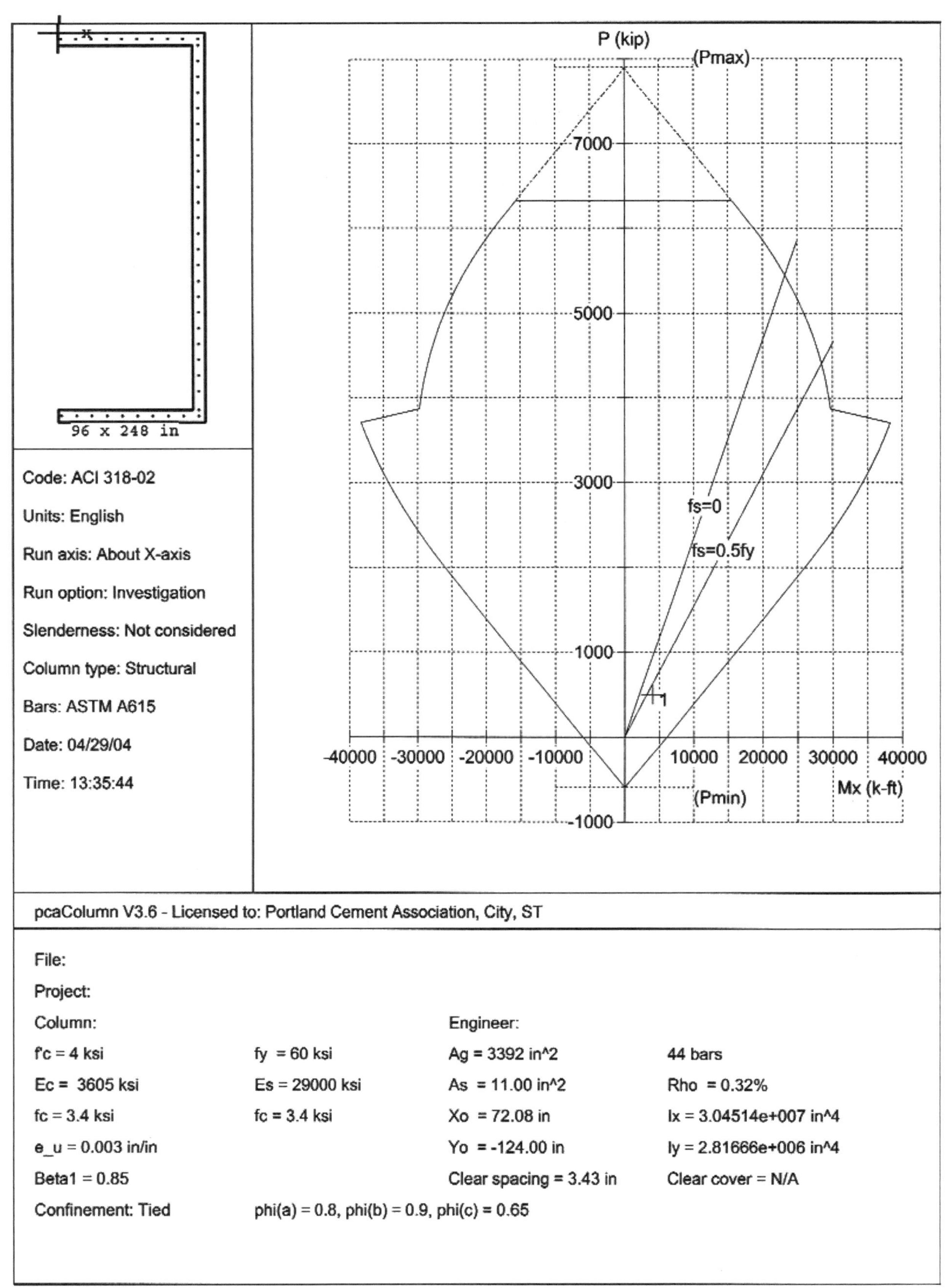

Figure 6-6 Interaction Diagram for Shearwall Bending About the X-axis

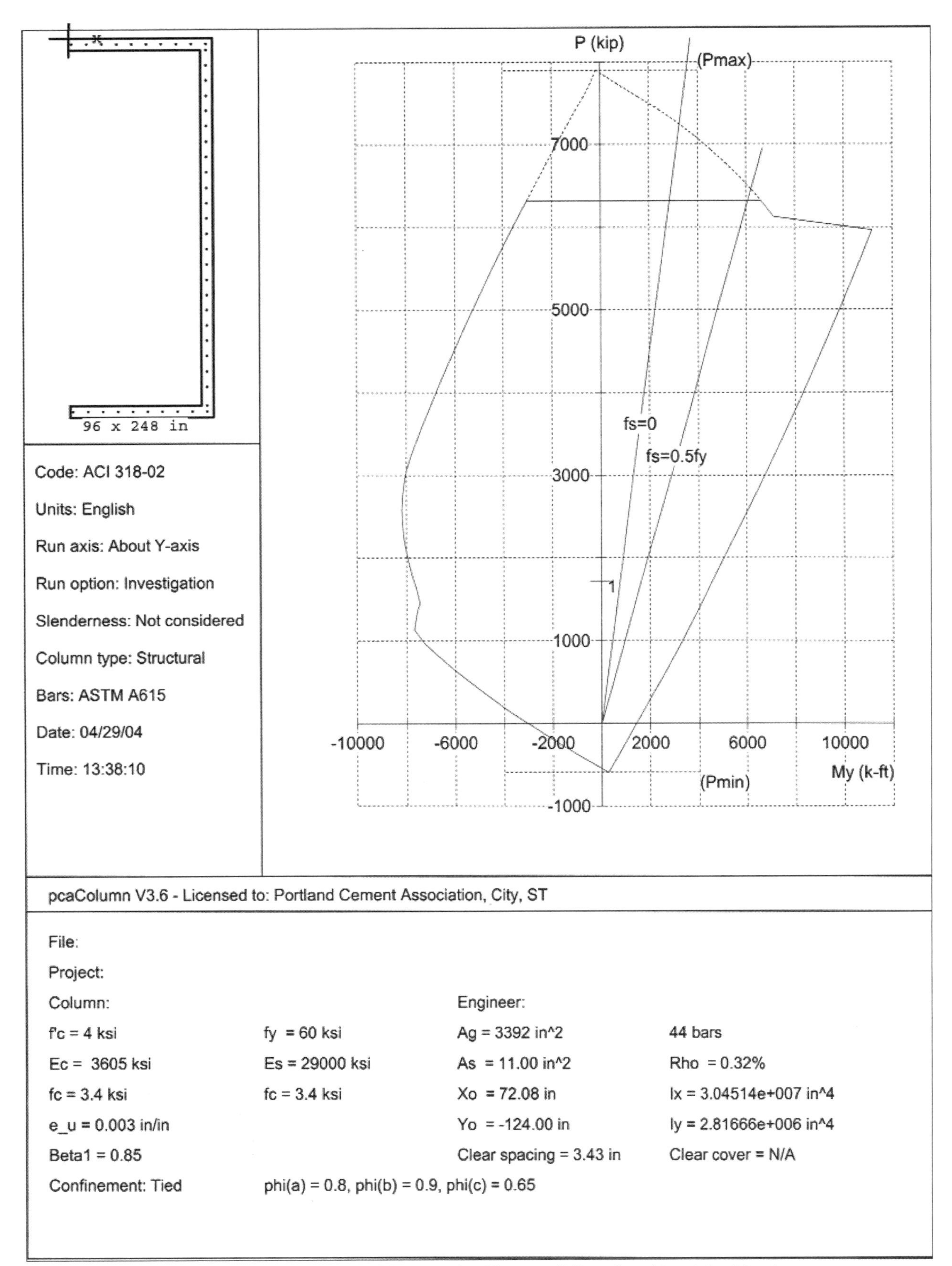

Figure 6-7 Interaction Diagram for Shearwall Bending About the Y-axis

References

6.1 Cardenas, A.E., Hanson, J.M., Corley, W.G., Hognestad, E., "Design Provisions for Shearwalls", *Journal of the American Concrete Institute*, Vol. 70, No. 3, March 1973, pp. 221-230.

6.2 *pcaColumn, Design and investigation of reinforced concrete column sections,* pcaStructure Point, Portland Cement Association, Skokie, Illinois, 2003.

Chapter 7
Simplified Design for Footings

7.1 INTRODUCTION

A simplified method for design of spread footings is presented that can be used to obtain required footing thickness with a one-step design equation based on minimum footing reinforcement. Also included are simplified methods for shear, footing dowels, and horizontal load transfer at the base of a column. A simplified one-step thickness design equation for plain concrete footings is also given. The discussion will be limited to the use of individual square footings supporting square (or circular) columns and subject to uniform soil pressure. The design methods presented are intended to address the usual design conditions for footings of low-to-moderate height buildings. Footings that are subjected to uplift or overturning are beyond the scope of the simplified method.

A concrete strength of f'_c = 3000 psi is the most common and economical choice for footings. Higher strength concrete can be used where footing depth or weight must be minimized, but savings in concrete volume do not usually offset the higher unit price of such concrete. In certain situations, data are presented for both 3000 and 4000 psi concrete strengths. Also, all of the design equations and data are based on Grade 60 bars, which are the standard grade recommended for overall economy.

7.2 PLAIN VERSUS REINFORCED FOOTINGS

Reinforced footings are often used in smaller buildings without considering plain footings. Many factors need to be considered when comparing the two alternatives, the most important being economic considerations. Among the other factors are soil type, job-site conditions, and building size (loads to be transferred). The choice between using reinforcement or not involves a trade-off between the amounts of concrete and steel. The current market prices of concrete and reinforcement are important decision-making parameters. If plain footings can save considerable construction time, then the cost of the extra concrete may be justified. Also, local building codes should be consulted to determine if plain concrete footings are allowed in certain situations. For a given project, both plain and reinforced footings can be quickly proportioned by the simplified methods in this chapters and an overall cost comparison made (including both material and construction costs). For the same loading conditions, the thickness of a plain footing will be about twice that of a reinforced footing with minimum reinforcement (see Section 7.8).

7.3 SOIL PRESSURE

Soil pressures are usually obtained from a geotechnical engineer or set by local building codes. In cities where experience and tests have established the allowable (safe) bearing pressures of various soils, local building codes may be consulted to determine the bearing capacities to be used in design. In the absence of such information or for conditions where the nature of the soils is unknown, borings or load tests should be made. For larger buildings, borings or load tests should always be made.

In general, the base area of the footing is determined using unfactored loads and allowable soil pressures (ACI 15.2.2), while the footing thickness and reinforcement are obtained using factored loads (ACI 15.2.1). Since column design is based on factored loads, it may be expedient to use a composite load factor, C, where U=C (D+L) and increase the allowable soil pressure q_a by the same factor. This allows the use of the factored loads for the total footing design.

For ordinary buildings the factored load combination used most often is:

$$U = 1.2D + 1.6L \qquad \text{ACI Eq. (9-2)}$$

Table 7-1 gives composite load factors for various types of floor systems and building occupancies. A composite load factor may be taken directly from the table and interpolated, if necessary for any unusual L/D ratio. Alternately a single conservative value of C = 1.5 could be used. It is usually convenient to increase the allowable soil pressure q_a by the composite load factor and use factored loads for the total footing design.

Table 7-1 Composite Load Factors For Different Floor Systems

Floor System	D (psf)	Use	L (psf)	L/D	C	Approx C	% Difference
9" Flat Plate	120	roof	30	0.25	1.28	1.5	14.7
3' Waffle (25' span)	120	resid.	40	0.33	1.30	1.5	13.3
5' Waffle (50' span)	175	office	70	0.40	1.31	1.5	12.4
5" Flat Plate	70	roof	35	0.50	1.33	1.5	11.1
5" Flat Plate	60	resid.	40	0.67	1.36	1.5	9.3
5" Waffle (25 span)	100	office	100	1.00	1.40	1.5	6.7
5" Waffle (30 span)	173	indust.	250	1.45	1.44	1.5	4.2
3" Waffle (20 span)	125	indust.	250	2.00	1.47	1.5	2.2
Joist (15' x 15')	100	indust.	100	1.00	1.40	1.5	6.7
5' Waffle (25' span)	133	library	400	3.01	1.50	1.5	0.0

7.4 SURCHARGE

In cases where the top of the footing is appreciably below grade (for example, below the frost line) allowances need to be made for the weight of soil on top of the footing. In general, an allowance of 100 pcf is adequate for soil surcharge; unless wet packed conditions exist that warrant a higher value (say 130 pcf). Total surcharge (or overburden) above base of footing can include the loads from a slab on grade, the soil surcharge, and the footing weight.

7.5 ONE-STEP THICKNESS DESIGN FOR REINFORCED FOOTINGS

A simplified footing thickness equation can be derived for individual footings with minimum reinforcement using the strength design data developed in Reference 7.1. The following derivation is valid for $f'_c = 3000$ psi, $f_y = 60{,}000$ psi, and a minimum reinforcement ratio of 0.0018 (ACI 10.5.4).

$$\text{Set } \rho = 0.0018 \times 1.11 = 0.002\,^*$$

$$R_n = \rho f_y\left(1 - \frac{0.5\rho f_y}{0.85 f'_c}\right)$$

$$= 0.002 \times 60{,}000\left(1 - \frac{0.5 \times 0.002 \times 60{,}000}{0.85 \times 3000}\right)$$

$$= 117.2 \text{ psi}$$

For a 1ft wide design strip:

$$d^2_{regd} = \frac{M_u}{\phi R_n} = \frac{M_u \times 1000}{0.9 \times 117.2} = 9.48 M_u$$

where M_u is in ft-kips.
Referring to Fig. 7-1, the factored moment M_u at the face of the column (or wall) is (ACI 15.4.2):

$$M_u = q_u\left(\frac{c^2}{2}\right) = \frac{P_u}{A_f}\left(\frac{c^2}{2}\right)$$

where c is the largest footing projection from face of column (or wall). Substituting M_u into the equation for d^2_{reqd} results in the following:

$$d^2_{reqd} = 4.74 q_u c^2 = 4.74 \frac{P_u c^2}{A_f}$$

$$d_{reqd} = 2.2\sqrt{q_u c^2} = 2.2c\sqrt{\frac{P_u}{A_f}}$$

The above equation is in mixed units: P_u is in kips, c is in feet, A_f is in square feet, and d is in inches.

The one-step thickness equation derived above is applicable for both square and rectangular footings (using largest value of c) and wall footings. Since f_y has a larger influence on d than does f'_c, the simplified equation can be used for other concrete strengths without a substantial loss in accuracy. As shown in Fig. 7-1, this derivation assumes uniform soil pressure at the bottom of the footing; for footings subject to axial load plus moment, an equivalent uniform soil pressure can be used.

* *The minimum value of ρ is multiplied by 1.11 to account for the ratio of effective depth d to overall thickness h, assumed as $d/h \cong 0.9$.*

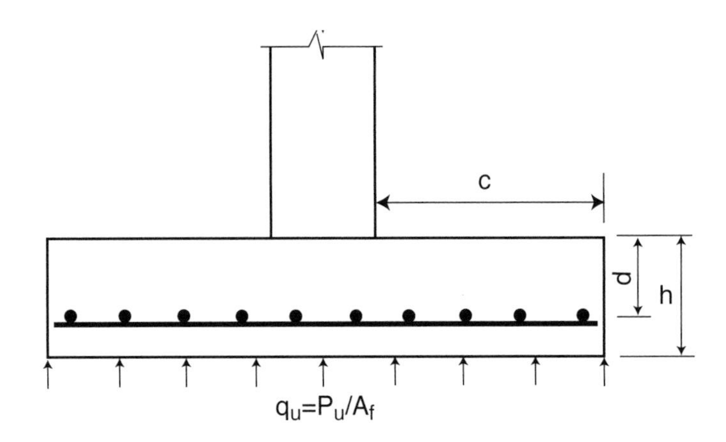

Figure 7-1 Reinforced Footing

According to ACI 11.12, the shear strength of footings in the vicinity of the column must be checked for both one-way (wide-beam) action and two-way action. Fig. 7-2 illustrates the tributary areas and critical sections for a square column supported by a square footing.

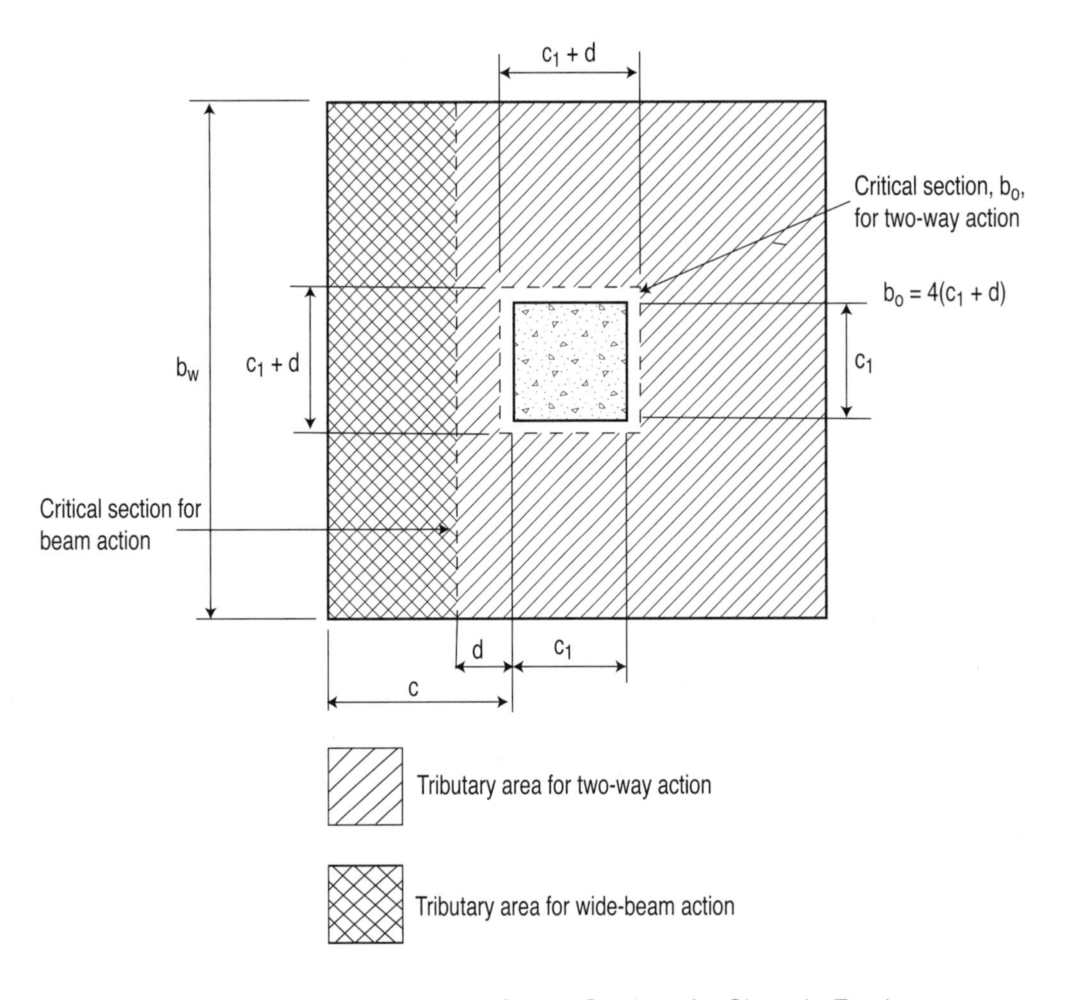

Figure 7-2 Tributary Areas and Critical Sections for Shear in Footings

For wide-beam action:

$$V_u \le 2\phi\sqrt{f_c'}b_w d$$

where b_w is the width of the footing, and V_u is the factored shear on the critical section (at a distance d from the face of the column). In general,

$$V_u = q_u b_w (c - d)$$

The minimum depth d can be obtained from the following equation:

$$\frac{d}{c} = \frac{q_u}{q_u + 2\phi\sqrt{f_c'}} \quad \text{where } q_u \text{ in psi}$$

This equation is shown graphically in Fig. 7-3 for $f_c' = 3000$ psi.

For a footing supporting a square column, the two-way shear strength will be the lesser of the values of V_c obtained from ACI Eqs. (11-34) and (11-35). Eq. (11-34) will rarely govern since the aspect ratio b_o/d will usually be considerably less than the limiting value to reduce the shear strength below $4\sqrt{f_c'}b_o d$.* Therefore, for two-way action,

$$V_u \le 4\phi\sqrt{f_c'}b_o d$$

where, for a square column, the perimeter of the critical section is $b_o = 4(c_1 + d)$. The factored shear V_u on the critical section (at d/2 from the face of the column) can be expressed as:

$$V_u = q_u\left[A_f - (c_1 + d)^2\right]$$

$$\left(\frac{q_u}{4} + \phi v_c\right)d^2 + \left(\frac{q_u}{2} + \phi v_c\right)c_1 d - \frac{q_u}{4}(A_f - A_c) = 0$$

where A_c = area of the column = c_1^2 and $v_c = 4\sqrt{f_c'}$. Fig. 7-4 can be used to determine d for footings with $f_c' = 3000$ psi: given q_u and A_f/A_c, the minimum value of d/c_1 can be read from the vertical axis.

Square footings, that are designed based on minimum flexural reinforcement will rarely encounter any one-way or two-way shear problems when supporting square columns. For other footing and column shapes, shear strength will more likely control the footing thickness. In any case, it is important to ensure that shear strength of the footing is not exceeded.

7.5.1 Procedure for Simplified Footing Design

(1) Determine base area of footing A_f from service loads (unfactored loads) and allowable (safe) soil pressure q_a determined for the site soil conditions and in accordance with the local building code.

---

* *For square interior columns, Eq. (11-37) will govern where $c/d_1 \le 0.25$.*

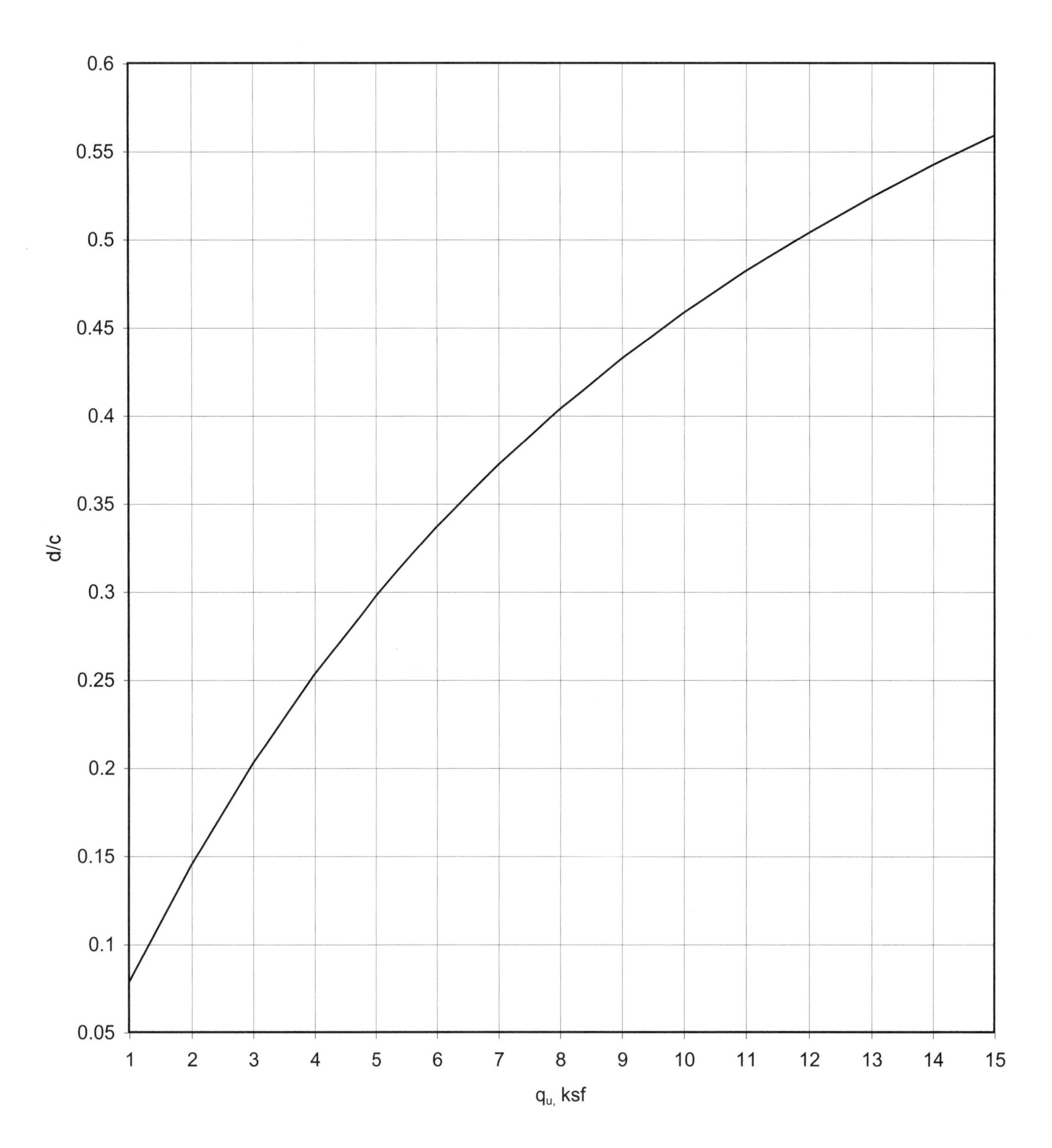

Figure 7-3 Minimum d for Wide-Beam Action

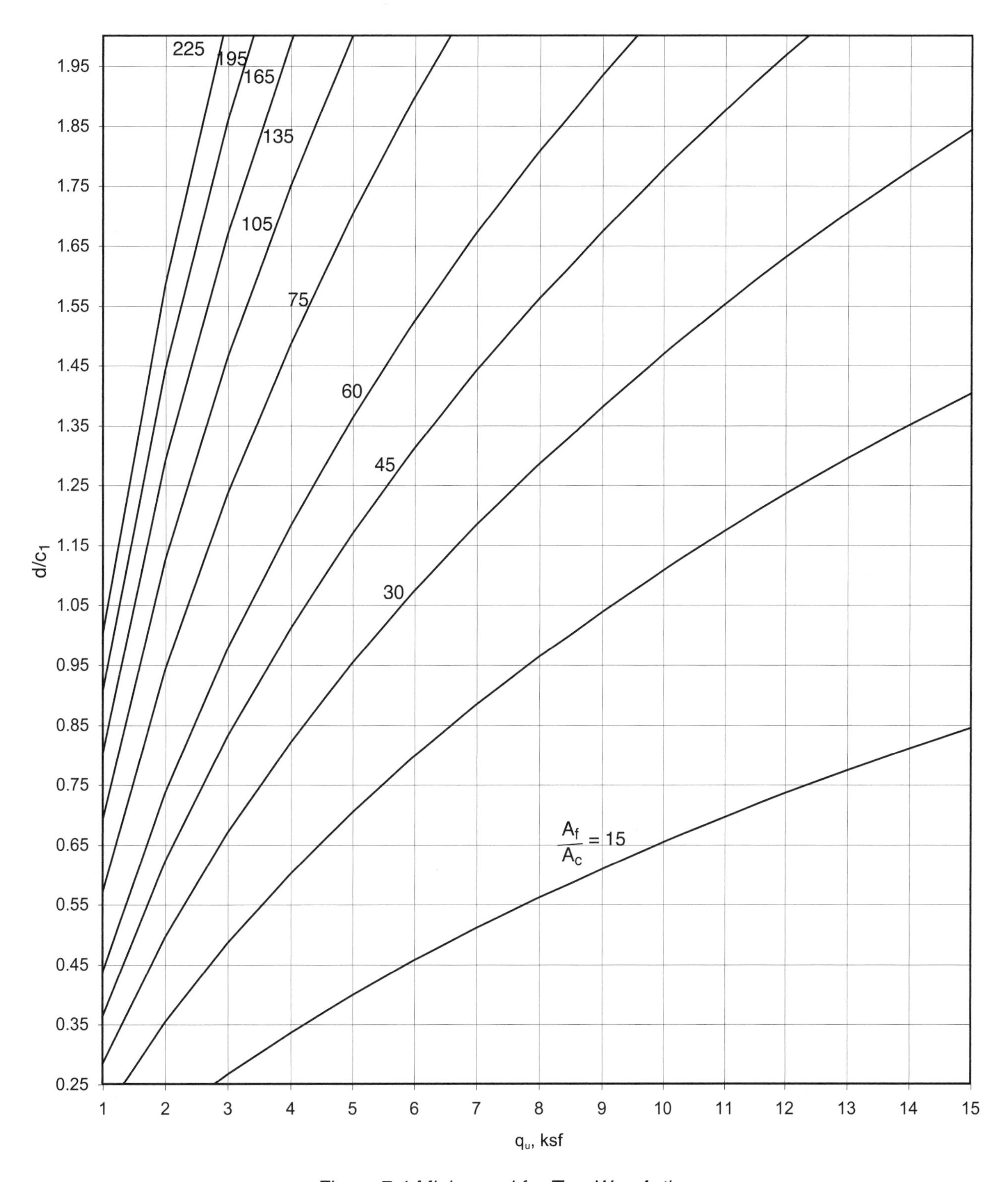

Figure 7-4 Minimum d for Two-Way Action

$$A_f = \frac{D + L + W + \text{surcharge (if any)}}{q_a}$$

Using a composite load factor of 1.5, the above equation can be rewritten as

$$A_f = \frac{P_u}{1.5q_a}$$

The following equations for P_u usually govern (for live load ≤ 100 lb/ft^2):

$$P_u = 1.2D + 1.6L + 0.5L_r \qquad \text{ACI Eq. (9-2)}$$
$$P_u = 1.2D + 1.6L_r + 0.5L \qquad \text{ACI Eq. (9-3)}$$
$$P_u = 1.2D \pm 1.6W + 0.5L + 0.5L_r \qquad \text{ACI Eq. (9-4)}$$

(2) Determine required footing thickness h from one-step thickness equation:

$$h = d + 4 \text{ in.}^*$$

$$h = 2.2c\sqrt{\frac{P_u}{A_f}} + 4 \text{ in.} \geq 10 \text{ in.}$$

where P_u = factored column load, kips
A_f = base area of footing, sq ft
c = greatest distance from face of column to edge of footing, ft **
h = overall thickness of footing, in.

(3) Determine minimum d for wide-beam action and two-way action from Figs. 7-3 and 7-4, respectively. Use the larger d obtained from the two figures, and compare it to the one obtained in step (2). In general, the value of d determined in step (2) will govern. Note that it is permissible to treat circular columns as square columns with the same cross-sectional area (ACI 15.3).

(4) Determine reinforcement:

$$A_s = 0.0018 \text{ bh}$$

A_s per foot width of footing:

$$A_s = 0.022h \text{ (in.}^2\text{/ft)}$$

Select bar size and spacing from Table 3-7. Note that the maximum bar spacing is 18 in. (ACI 7.6.5). Also, the provisions in ACI 10.6.4, which cover the maximum bar spacing for crack control, do not apply to footings.

---

* *3 in. cover (ACI 7.7.1) + 1 bar diameter (≅1 in.) = 4 in.*

** *For circular columns, c = distance from the face of an imaginary square column with the same area (ACI 15.3) to edge of footing.*

The size and spacing of the reinforcement must be chosen so that the bars can become fully developed. The bars must extend at least a distance ℓ_d from each face of the column, where ℓ_d is the tension development length of the bars (ACI 15.6). In every situation, the following conditions must be satisfied (see Fig. 7-5):

$$L > 2\,\ell_d + c_1 + 6 \text{ in.}$$

where L is the width of the footing and c_1 is the width of the column.

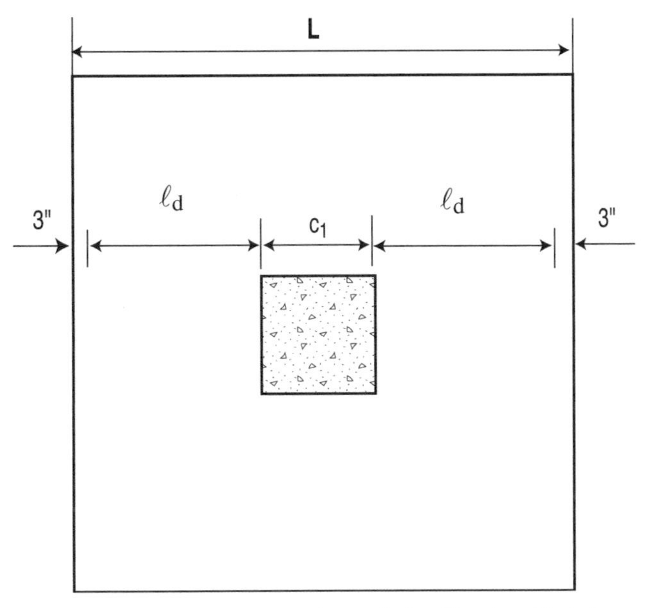

Figure 7-5 Avaialble Development Length for Footing Reinforcement

All of the spacing and cover criteria depicted in Fig. 7-6 are usually satisfied in typical situations; therefore, ℓ_d can becomputed from the following (ACI 12.2):

For No. 6 bars and smaller:

$$\ell_d = \left(\frac{f_y \alpha \beta \lambda}{25\sqrt{f'_c}} \right)$$

For No. 7 bars and larger:

$$\ell_d = \left(\frac{f_y \alpha \beta \lambda}{20\sqrt{f'_c}} \right)$$

Where $\alpha\beta\lambda$ are factor depend on the bar location, coating and concrete type (light weight or normal weight). For normal weight concrete reinforced with uncoated bars the multiplier $\alpha\beta\lambda$ can be taken as 1.0.

Values of ℓ_d for $f'_c = 3000$ psi and $f'_c = 4000$ psi are given in Table 7-2. In cases where the spacing and/or cover are less than those given in Fig. 7-6, a more detailed analysis using the appropriate modification factors in ACI 12.2 must be performed to obtain ℓ_d.

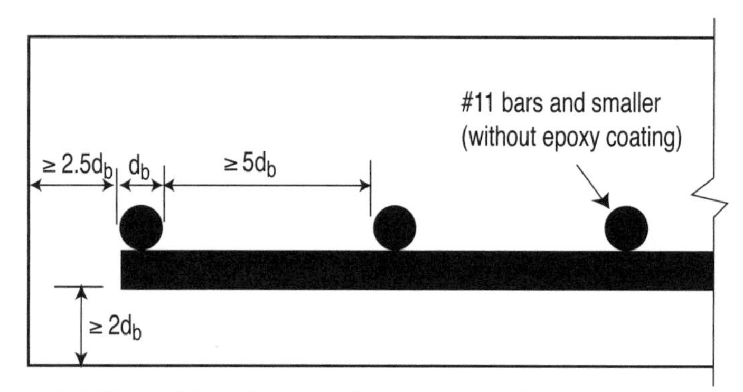

Figure 7-6 Typical Spacing and Cover of Reinforcement in Footings

Table 7-2 Minimum Development Length ℓ_d for Flexural Reinforcement in Footings (Grade 60)

Bar Size	Development length ℓ_d (in.)	
	f'_c = 3000 psi	f'_c = 4000 psi
# 4	22	19
# 5	27	24
# 6	33	28
# 7	48	42
# 8	55	47
# 9	62	54
#10	70	60
#11	77	67

7.6 FOOTING DOWELS

7.6.1 Vertical Force Transfer at Base of Column

The following discussion addresses footing dowels designed to transfer compression forces only. Tensile forces created by moments, uplift, or other causes must be transferred to the footings entirely by reinforcement (ACI 15.8.1.2).

Compression forces must be transferred by bearing on concrete and by reinforcement (if required). Bearing strength must be adequate for both column concrete and footing concrete. For the usual case of a footing with a total area considerably larger than the column area, bearing on column concrete will always govern until f'_c of the column concrete exceeds twice that of the footing concrete (ACI 10.17.1). For concrete strength f'_c = 4000 psi, the allowable bearing force ϕP_{nb} (in kips) on the column concrete is equal to $\phi P_{nb} = 2.21\ A_g$, where A_g is the gross area of the column in square inches. Values of ϕP_{nb} are listed in Table 7-3 for the column sizes given in Chapter 5.

When the factored column load P_u exceeds the concrete bearing capacity ϕP_{nb} the excess compression must be transferred to the footing by reinforcement (extended column bars or dowels; see ACI 15.8.2). Total area of reinforcement across the interface cannot be less than 0.5% of the column cross-sectional area (see Table 7-3). For the case when dowel bars are used, it is recommended that at least 4 dowels (one in each corner of the column) be provided.

Table 7-3 Bearing Capacity and Minimum Area of Reinforcement Across Interface

Column Size in.	ϕP_{nb} (kips)	Min. area of reinforcement (in^2)
10x10	221	0.5
12x12	318	0.72
14x14	433	0.98
16x16	566	1.28
18x18	716	1.62
20x20	884	2
22x22	1070	2.42
24x24	1273	2.88

Figure 7-7 shows the minimum dowel embedment lengths into the footing and column. The dowels must extend into the footing a compression development length of $\ell_{db} = 0.02\, d_b f_y / \sqrt{f'_c}$, but not less than $0.0003 d_b f_y$, where d_b is the diameter of the dowel bar (ACI 12.3.2)*. Table 7-4 gives the minimum values of ℓ_{db} for concrete with f'_c = 3000 psi and f'_c = 4000 psi**. The dowel bars are usually extended down to the level of the flexural steel of the footing and hooked 90°as shown in Fig. 7-7. The hooks are tied to the flexural steel to hold the dowels in place. It is important to note that the bent portions of the dowels cannot be considered effective for developing the bars in compression. In general, the following condition must be satisfied when hooked dowels are used:

$$h > \ell_{db} + r + d_{bd} + 2d_{bf} + 3 \text{ in.}$$

where r = minimum radius of dowel bar bend (ACI Table 7.2), in.
d_{bd} = diameter of dowel, in.
d_{bf} = diameter of flexural steel, in.

Table 7-4 Minimum Compression Development Length for Grade 60 Bars

Bar Size	Development length (in.)	
	f'_c = 3000 psi	f'_c = 4000 psi
# 4	11	10
# 5	14	12
# 6	17	15
# 7	20	17
# 8	22	19
# 9	25	22
#10	28	25
#11	31	27

*Tabulated values may be reduced by applicable modification factor in ACI 12.3.3

For the straight dowels, the minimum footing thickness h must be ℓ_{db} + 3 in.

* *The compression development length may be reduced by the applicable factor given in ACI 12.3.3.*

** *ℓ_{db} can conservatively be taken as 22 d_b for all concrete with f'_c > 3000 psi.*

In certain cases, the thickness of the footing must be increased in order to accommodate the dowels. If this is not possible, a greater number of smaller dowels can be used.

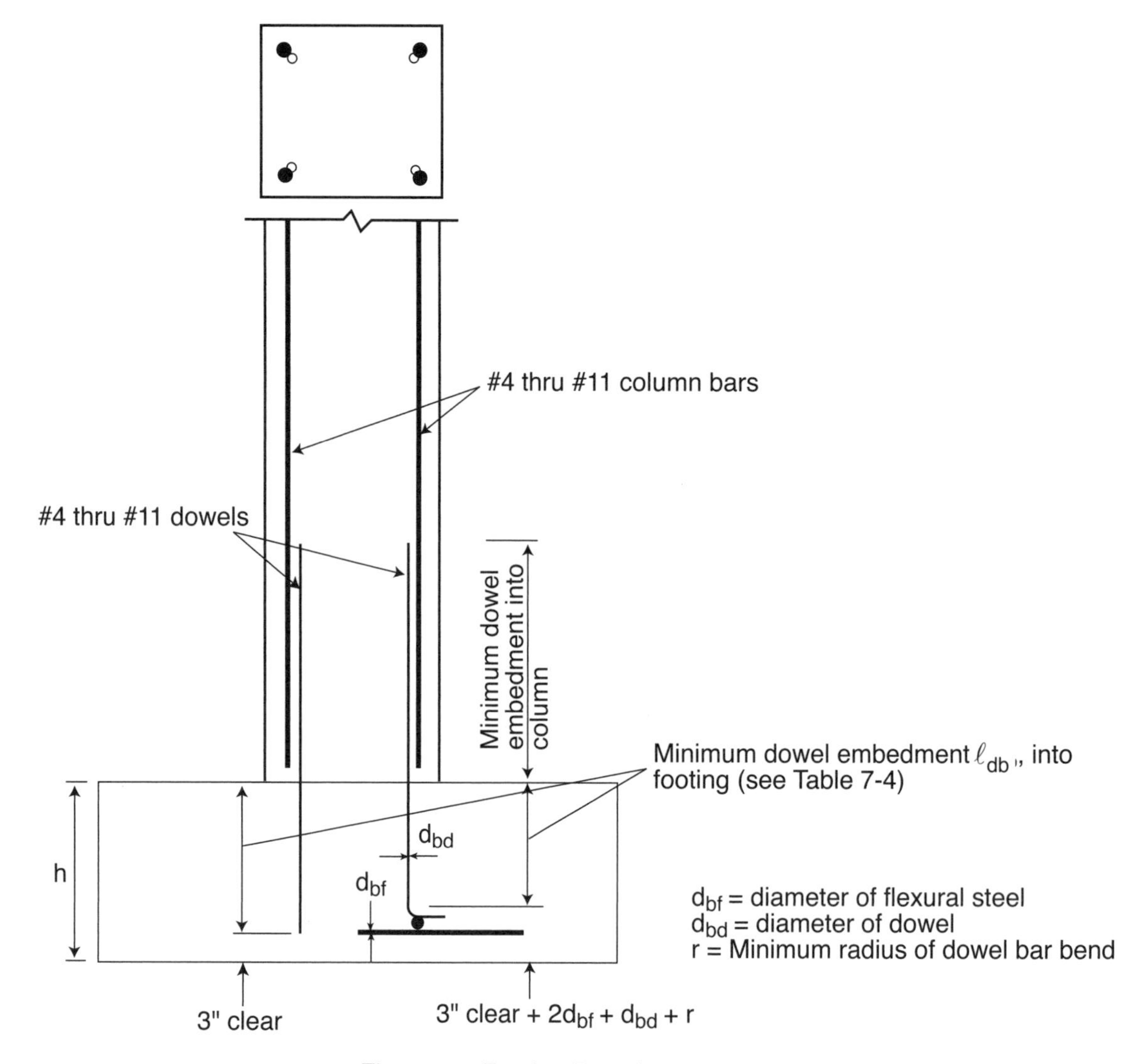

Figure 7-7 Footing Dowels

For the usual case of dowel bars which are smaller in diameter than the column bars, the minimum dowel embedment length into the column must be the larger of the compression development length of the column bar (Table 7-4) or the compression lap splice length of the dowel bar (ACI 12.16.2). The splice length is 0.0005 f_y d_b = 30 d_b for Grade 60 reinforcement, where d_b is the diameter of the dowel bar (ACI 12.16.1). Table 7-5 gives the required splice length for the bar sizes listed. Note that the embedment length into the column is 30 d_b when the dowels are the same size as the column bars.

Table 7-5 Minimum Compression Lap Splice Length for Grade 60 Bars*

Bar Size	Lap Splice Length (in.)
# 4	15
# 5	19
# 6	23
# 7	26
# 8	30
# 9	34
#10	38
#11	42

* $f'_c \geq$ 3000 psi

7.6.2 Horizontal Force Transfer at Base of Column

ACI 11.7.7 permits permanent compression to function like shear friction reinforcement. For all practical cases in ordinary buildings, the column dead load will be more than enough to resist any column shear at the top of the footing.

The horizontal force V_u to be transferred cannot exceed $\phi(0.2f'_cA_c)$ in pounds where A_c is gross area of column(ACI 11.7.5). For a column concrete strength f'_c = 4000 psi, this maximum force is equal to $\phi(800A_c)$. Dowels required to transfer horizontal force must have full tensile anchorage into the footing and into the column (ACI 11.7.8). The values given in Table 8-1 and 8-5 can be modified by the appropriate factors given in ACI 12.2.

7.7 EXAMPLE: REINFORCED FOOTING DESIGN

Design footings for the interior columns of Building No. 2 (5-story flat plate). Assume base of footings located 5 ft below ground level floor slab (see Fig. 7-8). Permissible soil pressure q_a = 4.5 ksf.

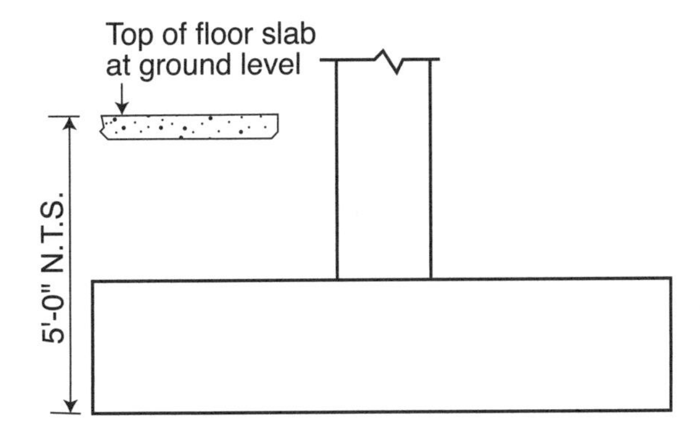

Figure 7-8 Interior Footing for Building No. 2

(1) Design Data:

Service surcharge = 50 psf
Assume weight of soil and concrete above footing base = 130 pcf
Interior columns: 16 in. × 16 in. (see Examples 5.7.1 and 5.7.2)
4-No.8 bars (non-sway frame)
8-No.10 bars (sway frame)

f'_c = 4000 psi (column)
f'_c = 3000 psi (footing)

(2) Load combinations

(a) gravity loads: P_u = 501 kips (Alternate (2))
M_u = 10.6 ft-kips

(b)gravity
+ wind loads: P_u = 423 kips (Alternate (1))
M_u = 65 ft-kips

or P_u = 436 kips
M_u = 113 ft-kips

(3) Base area of footing

Determine footing base area for gravity loads only, then check footing size for gravity plus winds loads.
Total weight of surcharge = (0.130 × 5) + 0.05 = 0.70 ksf
Net permissible soil pressure = 4.5 – 0.70 = 3.8 ksf

$$A_f = \frac{P_u}{1.5q_a} = \frac{501}{1.5(3.8)} = 87.9 \text{ sq ft}^*$$

Try 9 ft-6 in. × 9 ft-6 in. square footing (A_f = 90.25 sq ft)
Check gravity plus wind loading for 9 ft-6 in. × 9 ft-6 in. footing

A_f = 90.25 sq ft
$S_f = bh^2/6 = (9.5)^3/6 = 142.9 \text{ ft}^3$

$$q_u = \frac{P_u}{A_f} + \frac{M_u}{S_f} = \frac{423}{90.25} + \frac{65}{142.9} = 5.1 < 1.5\ (3.8) \quad \text{O.K.}$$

or

$$q_u = \frac{436}{90.25} + \frac{113}{142.9} = 5.6 \text{ ksf} < 1.5\ (3.8) \quad \text{O.K.}$$

* *Neglect small moment due to gravity loads.*

(4) Footing thickness

Footing projection

$$c = [(9.5 - 16/12)]/2 = 4.08 \text{ ft}$$

$$h = 2.2c\sqrt{\frac{P_u}{A_f}} + 4 \text{ in.} = 2.2(4.08)\sqrt{\frac{501}{90.25}} + 4 = 25.2 \text{ in.} > 10 \text{ in.} \quad \text{O.K.}$$

Try h = 27 in. (2 ft-3 in.)
Check if the footing thickness is adequate for shear:

$$d \cong 27 - 4 = 23 \text{ in.}$$

For wide-beam shear, use Fig. 7-3. With $q_u = \dfrac{501}{90.25} = 5.6$ ksf, read d/c ≅ 0.33. Therefore, the minimum d is

$$d = 0.33 \times 4.08 = 1.35 \text{ ft} = 16.2 \text{ in.} < 23 \text{ in.} \quad \text{O.K.}$$

Use Fig. 7-4 for two-way shear:

$$\frac{A_f}{A_c} = \frac{90.25}{(16^2/144)} = 50.8$$

Interpolating between A_f/A_c = 45 and 60, read d/c_1 1.34 for q_u = 5.6 ksf. The minimum d for two-way shear is:

$$d = 1.34 \times 16 = 21.4 \text{ in.} < 23 \text{ in.} \quad \text{O.K.}$$

Therefore, the 27 in. footing depth (d = 23 in.) is adequate for flexure and shear.

(5) Footing reinforcement

$$A_s = 0.022\,h = 0.022(27) = 0.59 \text{ in.}^2\text{/ft}$$

Try No.7 @ 12 in. (A_s = 0.60 in.^{2}/ft; see Table 3-7)
Determine the development length of the No.7 bars (see Fig. 7-6):

cover = 3 in. > $2d_b$ = 2 × 0.875 = 1.8 in.
side cover = 3 in. > 2.5 × 0.875 = 2.2 in.
clear spacing = 12 – 0.875 = 11.1 in. > 5 x 0.875 = 4.4 in.4.4 in.

Since all of the cover and spacing criteria given in Fig. 7-6 are satisfied, Table 7-2 can be used to determine the minimum development length.
For f'_c = 3000 psi: f'_c = 48 in.
Check available development length:

L = 9.5 × 12 = 114 in. > (2 × 48) + 16 + 6 = 118 in. O.K.
Use 9 ft-10 in. × 9 ft-10 in. square footing (L=118 in.)

Total bars required:

$$\frac{118 - 6}{12} = 9.33 \text{ spaces}$$

Use 10-No.7, 9 ft-4 in. long (each way)*

(6) Footing dowels

Footing dowel requirements are different for sway and non-sway frames. For the sway frame, with wind moment transferred to the base of the column, all of the tensile forces produced by the moment must be transferred to the footing by dowels. The number and size of dowel bars will depend on the tension development length of the hooked end of the dowel and the thickness of the footing. The dowel bars must also be fully developed for tension in the column.

For the non-sway frame, subjected to gravity loads only, dowel requirements are determined as follows:**

(a) For 16 × 16 in. column (Table 7-3):

$$\phi P_{nb} = 566 \text{ kips}$$

Minimum dowel area = 1.28 in.^{2}
Since $\phi P_{nb} > P_u = 501$ kips, bearing on concrete alone is adequate for transfer of compressive force.
Use 4-No.6 dowels ($A_s = 1.76$ in.^{2})

(b) Embedment into footing (Table 7-4):

For straight dowel bars,

$h \geq \ell_{db} + 3$ in.
$\ell_{db} = 17$ in. for No.6 dowels with $f'_c = 3000$ psi
$h = 27$ in. $> 17 + 3 = 20$ in. O.K.

For hooked dowel bars,

$h \geq \ell_{db} + r + d_{bd} + 2d_{bf} + 3$ in.
$r = 3d_{bd} = 3 \times 0.75 = 2.25$ in. (ACI Table 7.2)
$h = 27$ in. $> 17 + 2.25 + 0.75 + (2 \times 0.875) + 3 = 24.75$ in. O.K.

(c) Embedment into column:

The minimum dowel embedment length into the column must be the larger of the following:

- compression development length of No.8 column bars ($f'_c = 4000$ psi) = 19 in. (Table 7-4)
- compression lap splice length of No.6 dowel bars = 23 in. (Table 7-5) (governs)

---

* *13-No.6 or 8-No.8 would also be adequate.*

** *The horizontal forces produced by the gravity loads in the first story columns are negligible; thus, dowels required for vertical load transfer will be adequate for horizontal load transfer as well.*

For No.6 hooked dowels, the total length of the dowels is

$$23 + [27 - 3 - (2 \times 0.875)] = 45.25 \text{ in.}$$

Use 4-No.6 dowels × 3 ft-10 in.

Figure 7-9 shows the reinforcement details for the footing in the non-sway frame

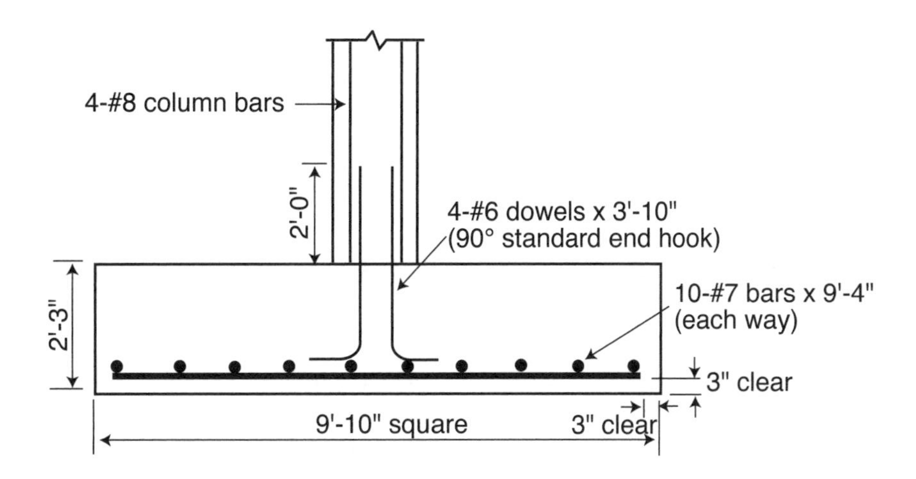

Figure 7-9 Reinforcement Details for Interior Column in Building No.2 (Non-sway Frame)

7.8 ONE-STEP THICKNESS DESIGN FOR PLAIN FOOTINGS

Depending on the magnitude of the loads and the soil conditions, plain concrete footings may be an economical alternative to reinforced concrete footings. Structural plain concrete members are designed according to ACI 318-02, Chapter 22.[7.2]. For plain concrete, the maximum moment design strength is $5\phi\sqrt{f_c'}S$(ACI Eq. 22-2). With $\phi = 0.55$ (ACI 318.1, Section 9.3.5)

A simplified one-step thickness design equation can be derived as follows (see Fig. 7-10):

$$M_u \leq 5\phi\sqrt{f_c'}S$$

For a one-foot design strip:

$$q_u\left(\frac{c^2}{2}\right) \leq 5\phi\sqrt{f_c'}\left(\frac{h^2}{6}\right)$$

$$h_{reqd}^2 = 3q_u\frac{c^2}{5\phi\sqrt{f_c'}} = \frac{0.6P_u c^2}{A_f\phi\sqrt{f_c'}}$$

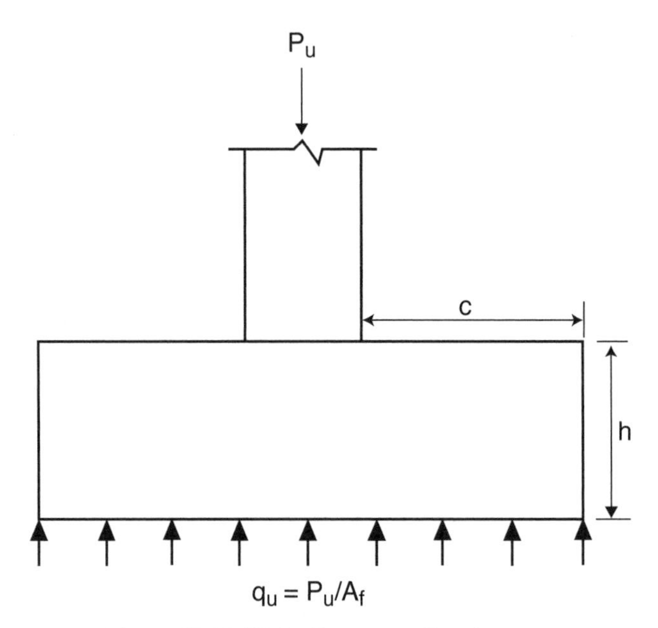

Figure 7-10 Plain Concrete Footing

To allow for unevenness of excavation and for some contamination of the concrete adjacent to the soil, an additional 2 in. in overall thickness is required for plain concrete footings (ACI 318.1, Section 6.3.5); thus,

$$\text{For } f'_c = 3000 \text{ psi:} \quad h = 4.5c\sqrt{\frac{P_u}{A_f}} + 2 \text{ in.}$$

$$\text{For } f'_c = 4000 \text{ psi} \quad h = 4.15c\sqrt{\frac{P_u}{A_f}} + 2 \text{ in.}$$

The above footing thickness equations are in mixed units:

P_u = factored column load, kips
A_f = base area of footing, sq ft
c = greatest distance from face of column to edge of footing, ft (ACI 318.1, Section 7.2.5)
h = overall thickness of footing, in. > 8 in. (ACI 318.1, Section 7.2.4)

Thickness of plain concrete footings will be controlled by flexural strength rather than shear strength for the usual proportions of plain concrete footings. Shear rarely will control. For those cases where shear may cause concern, the nominal shear strength is given in ACI 22.5.4.

7.8.1 Example: Plain Concrete Footing Design

For the interior columns of Building No.2, (Braced Frame), design a plain concrete footing.

From Example 7.7:

A_f = 9 ft-10 in. × 9 ft-10 in. = 96.7 sq ft
P_u = 501 kips
c = footing projection = 4.25 ft

For $f'_c = 3000$ psi:

$$h = 4.5c\sqrt{\frac{P_u}{A_f}} + 2 \text{ in.} = 4.5(4.25)\sqrt{\frac{501}{96.7}} + 2 = 44 + 2 = 45.5 \text{ in.}$$

Bearing on column:
Allowable bearing load $= 0.85\phi f'_c A_g$, $\phi = 0.55$ ACI 318.1, Section 6.2.1
$= 0.85 \times 0.55 \times 4 \times 16^2 = 479$ kips $< P_u$ 501 kips

The excess compression must be transferred to the footing by reinforcement consisting of extended column bars or dowels (calculations not show).

Figure 7-11 illustrates the footing for this case.

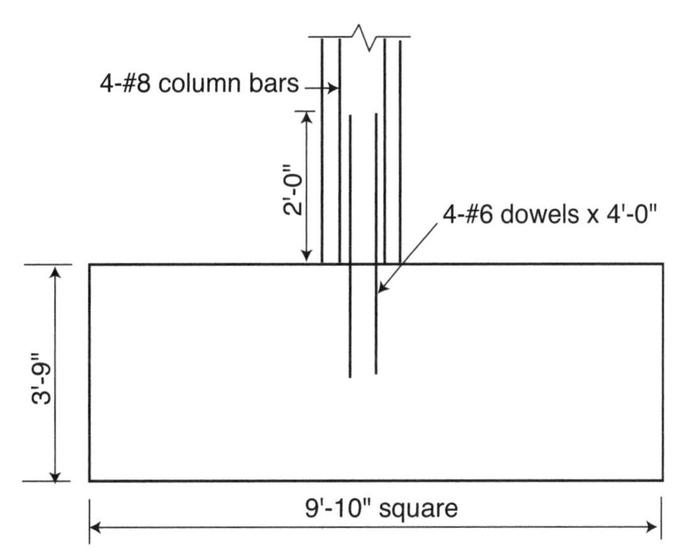

Figure 7-11 Plain Concrete Footing for Interior Column of Building No. 2 (Non-sway Frame)

References

7.1 *Notes on ACI-02*, Chapter 7, Design for Flexure, 8th Edition, EB702, Portland Cement Association, Skokie, Illinois, 2002.

7.2 *Building Code Requirements for Structural Concrete ACI 318-02 and Commentary—ACI 318R-02*, American Concrete Institute, Detroit, Michigan, 2002.

Chapter 8

Structural Detailing of Reinforcement for Economy

8.1 INTRODUCTION

Structurally sound details and proper bar arrangements are vital to the satisfactory performance of reinforced concrete structures. The details and bar arrangements should be practical, buildable, and cost-effective.

Ideally, the economics of reinforced concrete should be viewed in the broad perspective, considering all facets in the execution of a project. While it may be important to strive for savings in materials, many engineers often tend to focus too much on material savings rather than on designing for construction efficiencies. No doubt, savings in material quantities should result from a highly refined "custom design" for each structural member in a building. However, such a savings in materials might be false economy if significantly higher construction costs are incurred in building the custom-designed members.

Trade-offs should be considered in order to minimize the total cost of construction, including the total in-place cost of reinforcement. Savings in reinforcement weight can be traded-off for savings in fabrication, placing, and inspection for overall economy.

8.2 DESIGN CONSIDERATIONS FOR REINFORCEMENT ECONOMY

The following notes on reinforcement selection and placement will usually provide for overall economy and may minimize costly project delays and job stoppages:

(1) First and foremost, show clear and complete reinforcement details and bar arrangements in the Contract Documents. This issue is addressed in Section 1.1 of *ACI Detailing Manual*[8.1]: "...the responsibility of the Engineer is to furnish a clear statement of design requirements; the responsibility of the [Reinforcing Steel] Detailer is to carry out these requirements." ACI 318 further emphasizes that the designer is responsible for the size and location of all reinforcement and the types, locations, and lengths of splices of reinforcement (ACI 1.2.1 and 12.14.1).

(2) Use Grade 60 reinforcing bars. Grade 60 bars are the most widely used and are readily available in all sizes up to and including No.11; No.14 and No.18 bars are not generally inventoried in regular stock. Also, bar sizes smaller than No.6 generally cost more per pound and require more placing labor per pound of reinforcement.

(3) Use straight bars only in flexural members. Straight bars are regarded as standard in the industry. Truss (bent) bars are undesirable from a fabrication and placing standpoint and structurally unsound where stress reversals occur.

(4) In beams, specify bars in single layers only. Use one bar size for reinforcement on one face at a given span location. In slabs, space reinforcement in whole inches, but not at less than 6-in. spacing.

(5) Use largest bar sizes possible for the longitudinal reinforcement in columns. Use of larger bars sizes and fewer bars in other structural members will be restricted by code requirements for development of reinforcement, limits on maximum spacing, and distribution of flexural reinforcement.

(6) Use or specify fewest possible bar sizes for a project.

(7) Stirrups are typically the smaller bar sizes, which usually result in the highest total in-place cost of reinforcement per ton. For overall economy and to minimize congestion of reinforcement, specify the largest stirrup bar size (fewest number of stirrups) and the fewest variations in spacing. Stirrups spaced at the maximum allowable spacing, are usually the most economical.

(8) When closed stirrups are required for structural integrity, and torsion does not govern the design, specify two-piece closed types (conforming to ACI 12.13.5) to facilitate placing unless closed stirrups are required for torsion.

(9) Fit and clearance of reinforcing bars warrant special attention by the Engineer. At beam-column joints, arrangement of column bars must provide enough space or spaces to permit passage of beam bars. Bar details should be properly prepared and reconciled before the bars are fabricated and delivered to the job site. Member joints are far too important to require indiscriminate adjustments in the field to facilitate bar placing.

(10) Use or specify standard reinforcing bar details and practices:

- Standard end hooks (ACI 7.1). Note that the tension development length provisions in ACI 12.5 are only applicable for standard hooks conforming to ACI 7.1.

- Typical bar bends (see ACI 7.2 and Fig. 6 in Ref. 8.1).

- Standard fabricating tolerances (Fig. 4 in Ref. 8.1). More restrictive tolerances must be indicated by the Engineer in the Contract Documents.

- Tolerances for placing reinforcing bars (ACI 7.5). More restrictive tolerances must be indicated by the Engineer in the Contract Documents.

Care must be exercised in specifying more restrictive tolerances for fabricating and placing reinforcing bars. More restrictive fabricating tolerances are limited by the capabilities of shop fabrication equipment. Fabricating and placing tolerances must be coordinated. Tolerances for the formwork must also be considered and coordinated.

(11) Never permit field welding of crossing reinforcing bars for assembly of reinforcement ("tack" welding, "spot" welding, etc.). Tie wire will do the job without harm to the bars.

(12) Avoid manual arc-welded splices of reinforcing bars in the field wherever possible, particularly for smaller projects.

(13) A frequently occurring construction problem is having to make field corrections to reinforcing bars partially embedded in hardened concrete. Such "job stoppers" usually result from errors in placing or fabrication, accidental bending caused by construction equipment, or a design change. Field bending of bars partially embedded in concrete is not permitted except if such bending is shown on the design drawings or authorized by the Engineer (ACI 7.3.2). ACI R7.3 offers guidance on this subject. Further guidance on bending and straightening of reinforcing bars is given in Reference 8.2.

8.3 REINFORCING BARS

Billet-steel reinforcing bars conforming to ASTM A 615, Grade 60, are the most widely used type and grade in the United States. Combining the Strength Design Method with Grade 60 bars results in maximum overall economy. This design practice has made Grade 60 reinforcing bars the standard grade. The current edition of ASTM A 615 reflects this practice, as only bar sizes No.3 through No.6 in Grade 40 are included in the specification. Also listed are Grade 75 bars in sizes No.6 through No.18 only. The larger bar sizes in Grade 75 (No.11, No.14, and No.18) are usually used in columns made of high strength concrete in high-rise buildings. The combination of high strength concrete and Grade 75 bars may result in smaller column sizes, and, thus, more rentable space, especially in the lower levels of a building. It is important to note that Grade 75 bars may not be readily available in all areas of the country; also, as mentioned above, No.14 and No.18 bars are not commonly available in distributors' stock. ACI 3.5.3.2 permits the use of Grade 75 bars provided that they meet all the requirements listed in that section (also see ACI 9.4).

When important or extensive welding is required, or when more bendability and controlled ductility are required (as in seismic construction*), use of low-alloy reinforcing bars conforming to ASTM A 706 should be considered. Note that the specification covers only Grade 60 bars. Local availability should be investigated before specifying A 706 bars.

8.3.1 Coated Reinforcing Bars

Zinc-coated (galvanized) and epoxy-coated reinforcing bars are used increasingly for corrosion-protection in reinforced concrete structures. An example of a structure that might use coated bars is a parking garage where vehicles track in deicing salts.

Zinc-coated (galvanized) reinforcing bars must conform to ASTM A 767; also, the reinforcement to be coated must conform to one of the specifications listed in ACI 3.5.3.1. Bars are usually fabricated before galvanizing. In these cases, the minimum finished bend diameters given in Table 2 of ASTM A 767 must be specified. ASTM A 767 has two classes of coating weights of surface. Class I (3.0 oz./Sq ft for No.3bars and 3.5 oz./sq ft for No.4 and larger bars) is normally specified for general construction. ASTM A 767, requires that sheared ends to be coated with a zinc-rich formulation. Also when bars are fabricated after galvanizing, ASTM A 767

** ACI 21.2.5 specifically requires reinforcing bars complying with ASTM A 706 to be used in frame members and in wall boundary elements subjected to seismic forces. Note that ASTM A 615 Grade 40 and Grade 60 bars are also allowed if they meet all of the requirements in the section.*

requires that the damaged coating be repaired with a zinc-rich formulation. If ASTM A 615 billet-steel bars are being supplied, ASTM A 767 requires that a silicon analysis of each heat of steel be provided. It is recommended that the above pertaining requirements be specified when fabrication after galvanization includes cutting and bending.

Uncoated reinforcing steel (or any other embedded metal dissimilar to zinc) should not be permitted in the same concrete element with galvanized bars, nor in close proximity to galvanized bars, except as part of a cathodic-protection system. Galvanized bars should not be coupled to uncoated bars.

Epoxy-coated reinforcing bars must conform to ASTM A 775, and the reinforcement to be coated must conform to one of the specifications listed in ACI 3.5.3.1. The film thickness of the coating after curing shall be 7 to 12 mils (0.18 to 0.30 mm). Also, there shall not be more than an average of one holidays (pinholes not discernible to the unaided eye) per linear foot of the coated bar.

Proper use of ASTM A 767 and A 775 requires the inclusion of provisions in the project specifications for the following items:

- Compatible tie wire, bar supports, support bars, and spreader bars in walls.
- Repair of damaged coating after completion of welding (splices) or installation of mechanical connections.
- Repair of damaged coating after completion of field corrections, when field bending of coated bars partially embedded in concrete is permitted.
- Minimizing damage to coated bars during handling, shipment, and placing operations; also, limits on permissible coating damage and, when required, repair of damaged coating.

Reference 8.3 contains suggested provisions for preceding items for epoxy-coated reinforcing bars.

8.4 DEVELOPMENT OF REINFORCING BARS

8.4.1 Introduction

The fundamental requirement for development (or anchorage) of reinforcing bars is that a reinforcing bar must be embedded in concrete a sufficient distance on each side of a critical section to develop the peak tension or compression stress in the bar at the section. The development length concept in ACI 318 is based on the attainable average bond stress over the length of embedment of the reinforcement. Standard end hooks or mechanical devices may also be used for anchorage of reinforcing bars, except that hooks are effective for developing bars in tension only (ACI 12.1.1).

8.4.2 Development of Deformed Bars in Tension

The ACI provides two equations for development length calculations:

$$\ell_d = \left(\frac{f_y \alpha \beta \lambda}{25\sqrt{f'_c}} \right) d_b \qquad \text{For No. 6 and smaller bars}$$

$$\ell_d = \left(\frac{f_y \alpha \beta \lambda}{20\sqrt{f'_c}} \right) d_b \qquad \text{For No. 7 and larger bars}$$

where

ℓ_d = development length, in.

d_b = nominal diameter of bar, in.

f_y = specified yield strength for bar, psi

f'_c = specified compressive strength of concrete, psi

α = reinforcement location factor

= 1.3 for horizontal reinforcement so placed that more than 12 in. of fresh concrete is cast below the bar being developed or spliced

= 1.0 for other reinforcement

β = coating factor

= 1.5 for epoxy-coated bars with cover less than $3d_b$ or clear spacing less than $6d_b$

= 1.2 for all other epoxy-coated bars

= 1.0 for uncoated reinforcement

The produce of α and β need not be taken greater than 1.7.

γ = reinforcement size factor

= 0.8 for No. 6 and smaller bars

= 1.0 for No. 7 and larger bars

λ = lightweight aggregate concrete factor

= 1.3 when lightweight aggregate concrete is used, or

= 1.0 for normal weight concrete

The above equations are valid for clear spacing of bars being developed or spliced not less than d_b, clear cover not less than d_b and stirrups or ties not less than the code minimum, or clear spacing not less than $2d_b$ and clear cover not less than d_b (ACI 12.2.2). For cases where the reinforced bars are closely spaced, or the provided cover is less than d_b, the development length must be increased ACI R12.2.

For 4000 psi normal weight concrete and uncoated reinforcing bottom bars:

$\ell_d = 38\ d_b$ for No. 6 and smaller bars

$\ell_d = 48\ d_b$ for No. 7 and larger bars

The development length may be reduced when the provision for excess reinforcement given in ACI 12.2.5 are satisfied. Also ℓ_d must never be taken less than 12 in. (ACI 12.2.1).

Values for tension development length ℓ_d are given in Table 8-1 for grade 60 reinforcing bars. The values in the table are based on bars that are not epoxy-coated and on normal weight concrete. To obtain ℓ_d for top bars (horizontal bars with more than 12 in. of concrete cast below the bars) the tabulated values must be multiplied by 1.3 (ACI 1212.2.4). The cover and clear spacing referred to in the table are depicted in Fig. 8-1.

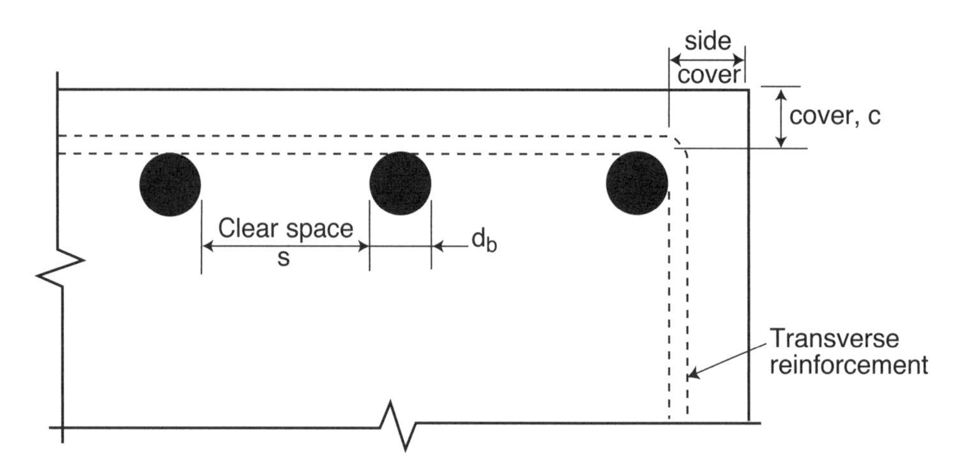

Figure 8-1 Cover and Clear Spacing of the Reinforcement

Table 8-1 Tension Development Length ℓ_d for Grade 60 Bars

	Clear Spacing not less than d_b Clear cover not less than d_b Stirrups throughout A_d not less than minimum or clear spacing not less than $2d_b$ and clear cover not less than d_b		Other spacing and cover conditions	
	f'_c psi		f'_c psi	
Bar size	3000	4000	3000	4000
# 3	16.4	14.2	24.6	21.3
# 4	21.9	19.0	32.9	28.5
# 5	27.4	23.7	41.1	35.6
# 6	32.9	28.5	49.3	42.7
# 7	47.9	41.5	71.9	62.3
# 8	54.8	47.4	82.2	71.2
# 9	61.8	53.5	92.7	80.3
#10	69.6	60.2	104.3	90.4
#11	77.2	66.9	115.8	100.3

Values are based on bars which are not epoxy-coated and on normal weight concrete. For top Bars, multiply tabulated values by 1.3

As can be seen from the table, very long development lengths are required for the larger bar sizes, especially when the cover is less than d_b.

8.4.3 Development of Hooked Bars in Tension

The ACI provides the following equation for development length in tension for bars ending with standard hook:

$$\ell_{dh} = \left(\frac{0.02\beta\lambda f_y}{\sqrt{f_c'}} \right) d_b$$

The values for β and λ are as defined before.

For normal weight concrete with $f_c' = 4000$ psi and uncoated reinforcing bars with $f_y = 60000$ psi the development length $\ell_{dh} = 19\ d_b$. Table 8-2 lists the development length ℓ_{dh} for different bar sizes.

Table 8-2 Minimum Development Lengths ℓ_{dh} for Grade 60 Bars with Standard End Hooks (in.)*

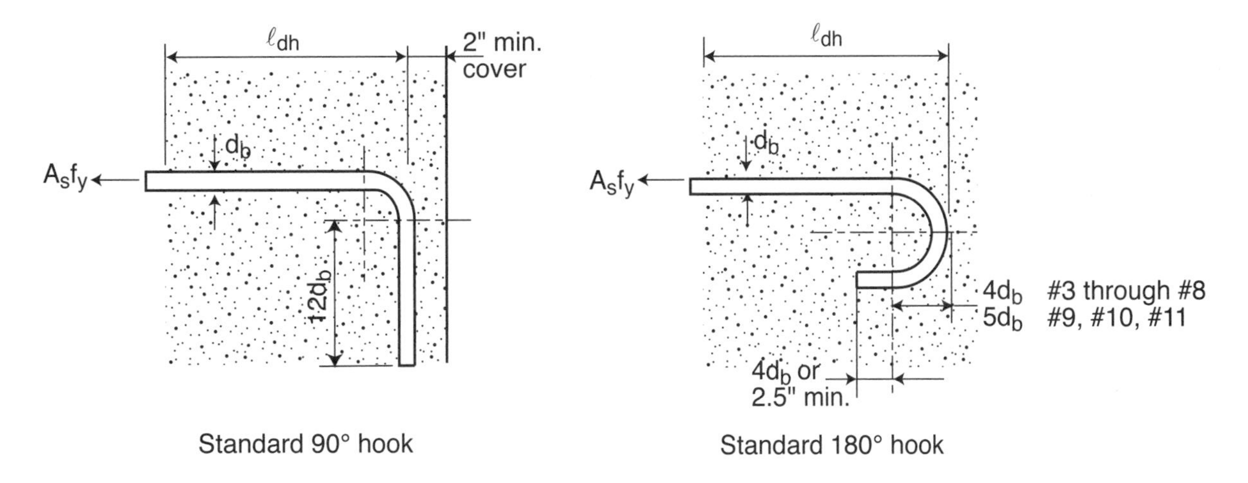

Standard 90° hook

Standard 180° hook

Table 8-2 Minimum Development Lengths ℓ_{dh} for Grade 60 Bars with Standard End Hooks (in.)*

	General use: • Side cover ≥ 2 in. • End cover (90° hooks) ≥ 2 in.		Special confinement: • Side cover ≥ 2 in. • End cover (90° hooks) ≥ 2 in. • Ties or stirrups spaced ≤ 3 d_b	
Bar size	$f_c' = 3000$ psi	$f_c' = 4000$ psi	$f_c' = 3000$ psi	$f_c' = 4000$ psi
# 3	6	6	6	6
# 4	8	7	7	6
# 5	10	9	8	7
# 6	12	10	10	8
# 7	14	12	11	10
# 8	16	14	13	11
# 9	18	15	14	12
#10	20	17	16	14
#11	22	19	18	15

*Values based on normal weight concrete.

The general use development lengths give in Table 8-2 are applicable for end hooks with side cover normal to plane of hook of not less than 2½ in. and end cover (90° hooks only) of not less than 2 in. For these cases, $\ell_{dh} = 0.7\ell_{hb}$, but not less than $8d_b$ or 6 in. (ACI 12.5.1). For hooked bar anchorage in beam-column joints, the hooked beam bars are usually placed inside the vertical column bars, with side cover greater than the 2½-in. minimum required for application of the 0.7 reduction factor. Also, for 90° end hooks with hook extension located inside the column ties, the 2-in. minimum end cover will usually be satisfied to permit the 0.7 reduction factor.

The special confinement condition given in Table 8-2 includes the additional 0.8 reduction factor for confining ties or stirrups (ACI 12.5.3.b). In this case, $\ell_{dh} = (0.7 \times 0.8)\ell_{hb}$, but not less than $8d_b$ or 6 in.

Where development for full f_y is not specifically required, the tabulated values of ℓ_{dh} in Table 8-2 may be further reduced for excess reinforcement (ACI 12.5.3.d). As noted above, ℓ_{dh} must not be less than $8d_b$ or 6 in.

ACI 12.5.4 provides additional requirements for hooked bars terminating at the discontinuous end of members (ends of simply supported beams, free end of cantilevers, and ends of members framing into a joint where the member does not extend beyond the joint). If the full strength of the hooked bar must be developed, and if both the side cover and the top (or bottom) cover over the hook is less than 2½ in., closed ties ore stirrups spaced at $3d_b$ maximum are required along the full development length ℓ_{dh}. The reduction factor in ACI 12.5.3.b must not be used in this case. At discontinuous ends of slabs with confinement provided by the slab continuous on both sides normal to the plane of the hook, the requirements in ACI 12.5.4 for confining ties or stirrups do not apply.

8.4.4 Development of Bars in Compression

Shorter development lengths are required for bars in compression than in tension since the weakening effect of flexural tension cracks in the concrete is not present. The development length for deformed bars in compression is $\ell_{dc} = 0.02 d_b f_y / \sqrt{f'_c}$, but not less than $0.0003 d_b f_y$ or 8 in. (ACI 12.3). For concrete with $f'_c = 4000$ psi and grade 60 reinforcement bars $\ell_{dc} = 19d_b$. The minimum development for bars in compression is 8 in. Table 8-3 lists the development length in compression for grade 60 bars. The values may be reduced by the applicable factors in ACI 12.3.3.

Table 8-3 Minimum Compression Development Lengths ℓ_d for Grade 60 Bars (in.)

Bar size	f'_c = 3000 psi	f'_c = 4000 psi
# 3	9	8
# 4	11	10
# 5	14	12
# 6	17	15
# 7	20	17
# 8	22	19
# 9	25	22
#10	28	25
#11	31	27

8.5 SPLICES OF REINFORCING BARS

Three methods are used for splicing reinforcing bars: 1) lap splices, 2) welded splices, and 3) mechanical connections. The lap splice is usually the most economical splice. When lap splices cause congestion or field placing problems, mechanical connections or welded splices should be considered. The location of construction joints, provision for future construction, and the particular method of construction may also make lap splices impractical. In columns, lapped offset bars may need to be located inside the column above to reduce reinforcement congestion; this can reduce the moment capacity of the column section at the lapped splice location because of the reduction in the effective depth. When the amount of vertical reinforcement is greater than 4%, and/or when large factored moments are present, use of butt splices—either mechanical connections or welded splices—should be considered in order to reduce congestion and to provide for greater nominal moment strength of the column section at the splice location.

Bars in flexural members may be spliced by non-contact lap splices (ACI 12.14.2.3); however, contact lap splices are preferred since the bars are tied and are less likely to displace when the concrete is placed.

Welded splices generally require the most expensive field labor. For projects of all sizes, manual arc-welded splices will usually be the most costly method of splicing due to the costs of inspection.

Mechanical connections are made with proprietary splice devices. Performance information and test data should be obtained directly from the manufacturers. Basic information about mechanical connections and the types of proprietary splice devices currently is available from Reference 8.4. Practical information on splicing and recommendations for the design and detailing of splices are given in Reference 8.5.

8.5.1 Tension Lap Splices

Tension lap splices are classified as Class A or Class B (ACI 12.15.1). The minimum lap length for a Class A splice is $1.0\ell_d$, and for a Class B splice it is $1.3\ell_d$, where ℓ_d is the tension development length of the bars. When calculating the development length ℓ_d the factor in ACI 12.2.5 for excess reinforcement must not be used, since the splice classifications already reflect any excess reinforcement at the splice location.

The minimum lap lengths for Class A splices can be obtained from Table 8-1. For Class B splices, the minimum lap lengths are determined by multiplying the values from Table 8-1 by 1.3. The effective clear spacing between splices bars is illustrated in Fig. 8-2. For staggered splices in slabs or walls, the effective clear spacing is the distance between adjacent spliced bars less the diameters of any intermediate unspliced bars (Fig. 8-2a), The clear spacing to be used for splices in columns with offset bars and for beam bar splices are shown in Figs. 8-2b and 8-2c, respectively.

In general, tension lap splices must be Class B except that Class A splices are allowed when both of the following conditions are met: 1) the area of reinforcement provided is at least twice that required by analysis over the entire length of the splice and 2) one-half or less of the total reinforcement is spliced within the required lap length (ACI 12.15.2). Essentially, Class A splices may be used at locations where the tensile stress is small. It is very important to specify which class of tension splice is to be used, and to show clear and complete details of the splice in the Contract Documents.

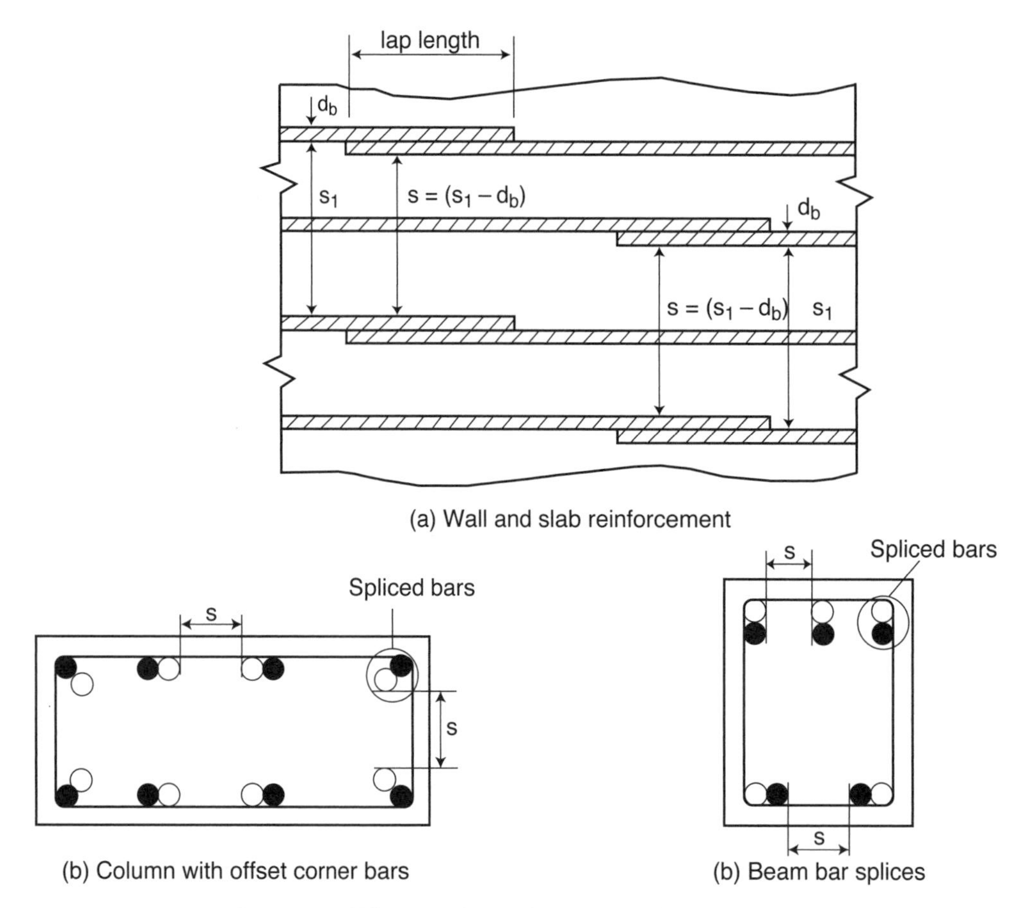

(a) Wall and slab reinforcement

(b) Column with offset corner bars

(b) Beam bar splices

Figure 8-2 Effective Clear Spacing of Spliced Bars

8.5.2 Compression Lap Splices

Minimum lengths for compression lap splices (ACI 12.16.1) for Grade 60 bars in normal weight concrete are given in Table 8-4. The values apply for all concrete strengths greater than or equal to 3000 psi. For Grade 60 bars, the minimum lap splice length is 30 d_b but not less than 12 in. When bars of different size are lap spliced, the splice length shall be the larger of 1) development length of larger bar, or 2) splice length of smaller bar (ACI 12.16.2). For columns, the lap splice lengths may be reduced by a factor of 0.83 when the splice is enclosed throughout its length by ties specified in ACI 12.17.2.4. The 12 in. minimum lap length also applies.

Table 8-4 Minimum Compression Lap Splice Lengths for Grade 60 Bars*

Bar size	Minimum lap length (in.)
# 3	12
# 4	15
# 5	19
# 6	23
# 7	26
# 8	30
# 9	34
#10	38
#11	42

* $f'_c \geq 3000$ psi

8.6 DEVELOPMENT OF FLEXURAL REINFORCEMENT

8.6.1 Introduction

The requirements for development of flexural reinforcement are given in ACI 12.10, 12.11, and 12.12. These sections include provisions for:

- Bar extensions beyond points where reinforcement is no longer required to resist flexure.
- Termination of flexural reinforcement in tension zones.
- Minimum amount and length of embedment of positive moment reinforcement into supports.
- Limits on bar sizes for positive moment reinforcement at simple supports and at points of inflection.
- Amount and length of embedment of negative moment reinforcement beyond points of inflection.

Many of the specific requirements are interdependent, resulting in increased design time when the provisions are considered separately. To save design time and costs, recommended bar details should be used. As was discussed earlier in this chapter, there is potential overall savings in fabrication, placing, and inspection costs when recommended bar details are used.

8.6.2 Recommended Bar Details

Recommended bar details for continuous beams, one-way slabs, one-way joist construction, and two-way slabs (without beams) are given in Figs. 8-3 through 8-6. Similar details can be found in References 8.1 and 8.6. The figures may be used to obtain bar lengths for members subjected to uniformly distributed gravity loads only; adequate bar lengths must be determined by analysis for members subjected to lateral loads. Additionally, Figs. 8-3 through 8-5 are valid for beams, one-way slabs, and one-way joists that may be designed by the approximate method given in ACI 8.3.3.* Fig. 8-6 can be used to determine the bar lengths for two-way slabs without beams.**

8.7 SPECIAL BAR DETAILS AT SLAB-TO-COLUMN CONNECTIONS

When two-way slabs are supported directly by columns (as in flat plates and flat slabs), transfer of moment between slab and column takes place by a combination of flexure and eccentricity of shear (see Chapter 4, Section 4.4.1). The portion of the unbalanced moment transferred by flexure is assumed to be transferred over a width of slab equal to the column width c plus 1.5 times the slab thickness h on either side of the column. For edge and interior columns, the effective slab width is (c+3h), and for corner columns it is (c+1.5h). An adequate amount of negative slab reinforcement is required in this effective slab width to resist the portion of the unbalanced moment transferred by flexure (ACI 13.5.4). In some cases, additional reinforcement must be concentrated over the column to increase the nominal moment resistance of the section. Note that minimum bar spacing requirements must be satisfied at all locations in the slab (ACI 13.3.2). Based on recommendations in Reference 8.7, examples of typical details at edge and corner columns are shown in Figs. 8-7 and 8-8.

---

* *Under normal conditions, the bar lengths give in Figs. 8-3 through 8-5 will be satisfactory. However, for special conditions, a more detailed analysis will be required. In any situation, it is the responsibility of the engineer to ensure that adequate bar lengths are provided.*

** *To reduce placing and inspection time, all of the top bars in the column strip of a two-way slab system can have the same length at a particular location (either 0.30 ℓ_n for flat plates or 0.33 ℓ_n for flat slabs), instead of the two different lengths shown in Figs. 8-6(a) and 8-6(c).*

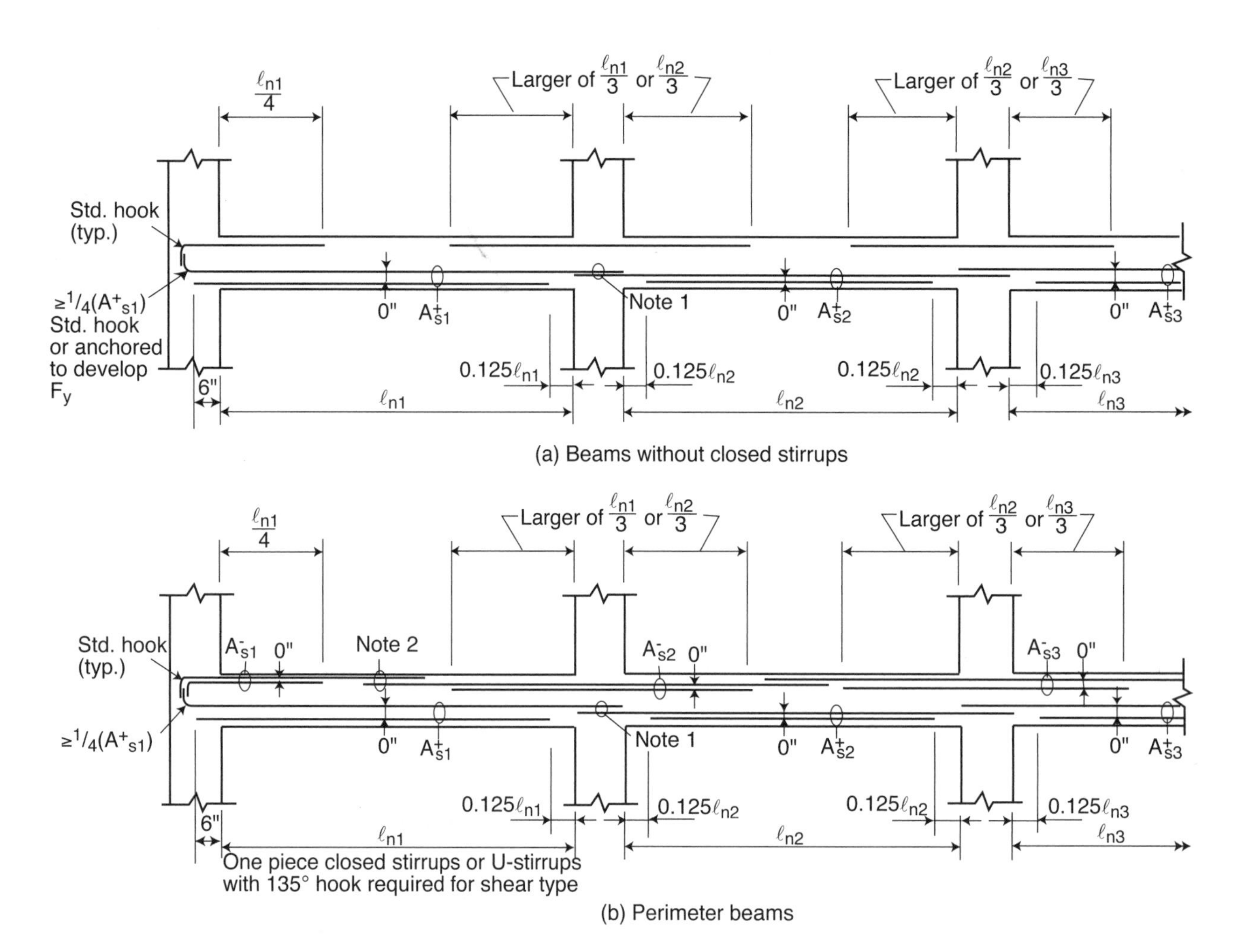

(a) Beams without closed stirrups

(b) Perimeter beams

Notes: (1) Larger of $^1/_4(A^+_{s1})$ or $^1/_4(A^+_{s2})$ but not less than two bars continuous or spliced with Class A splices or mechanical or welded splice (ACI 7.13.2.2 and 7.13.2.3)
(2) Larger of $^1/_6(A^-_{s1})$ or $^1/_6(A^-_{s2})$ but not less than two bars continuous or spliced with Class A splices or mechanical or welded splice (ACI 7.13.2.2)

Figure 8-3 Recommended Bar Details for Beams

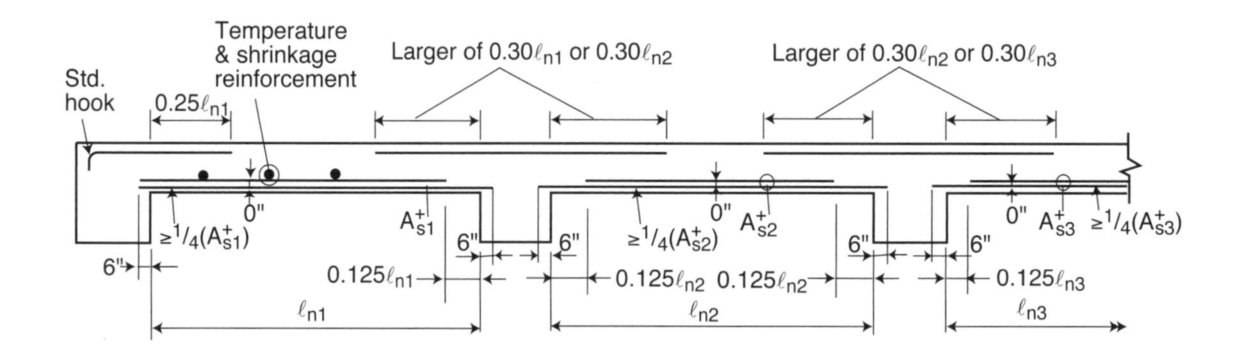

Figure 8-4 Recommended Bar Details for One-Way Slabs

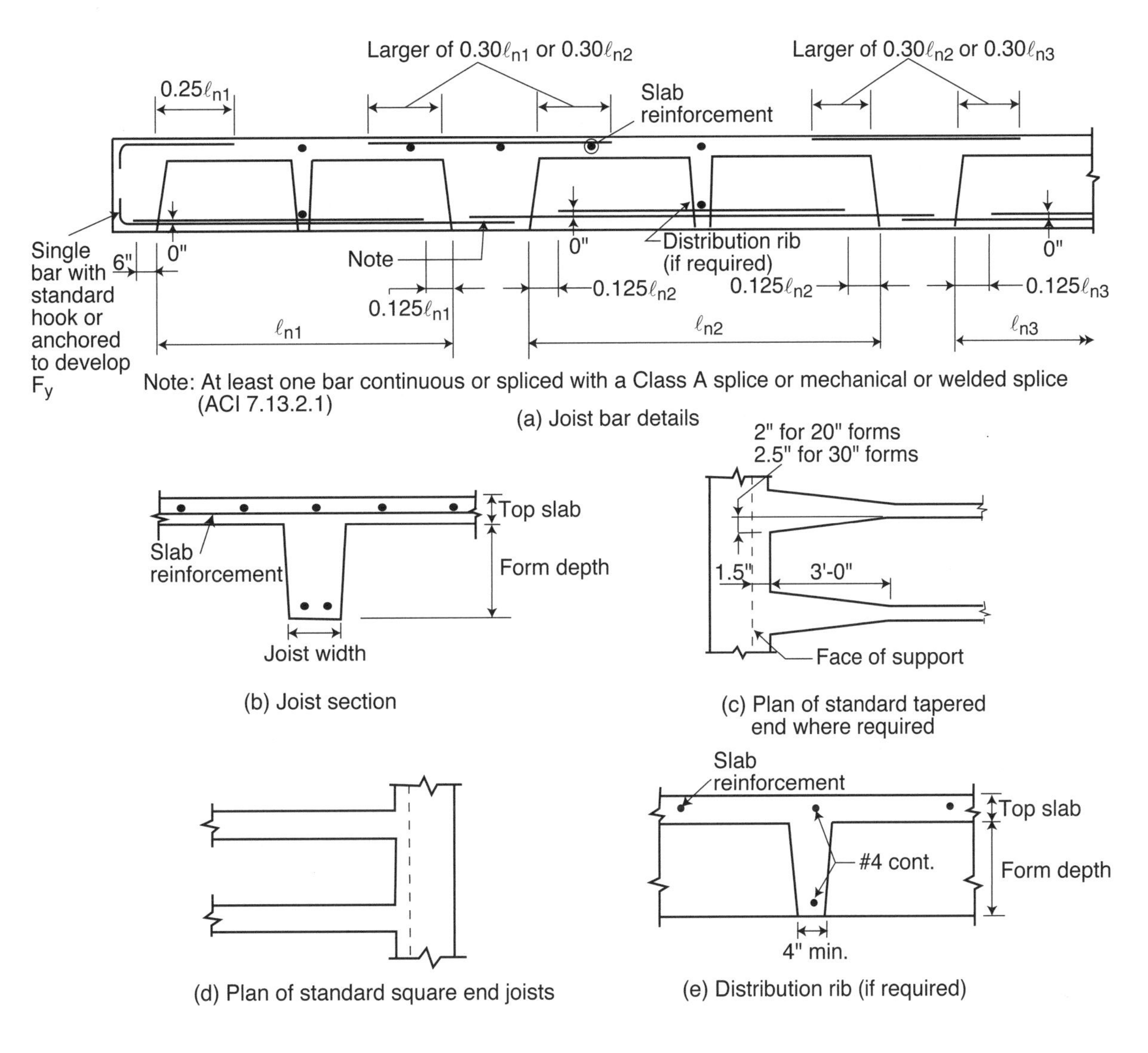

Figure 8-5 Recommended Bar Details for One-Way Joist Construction

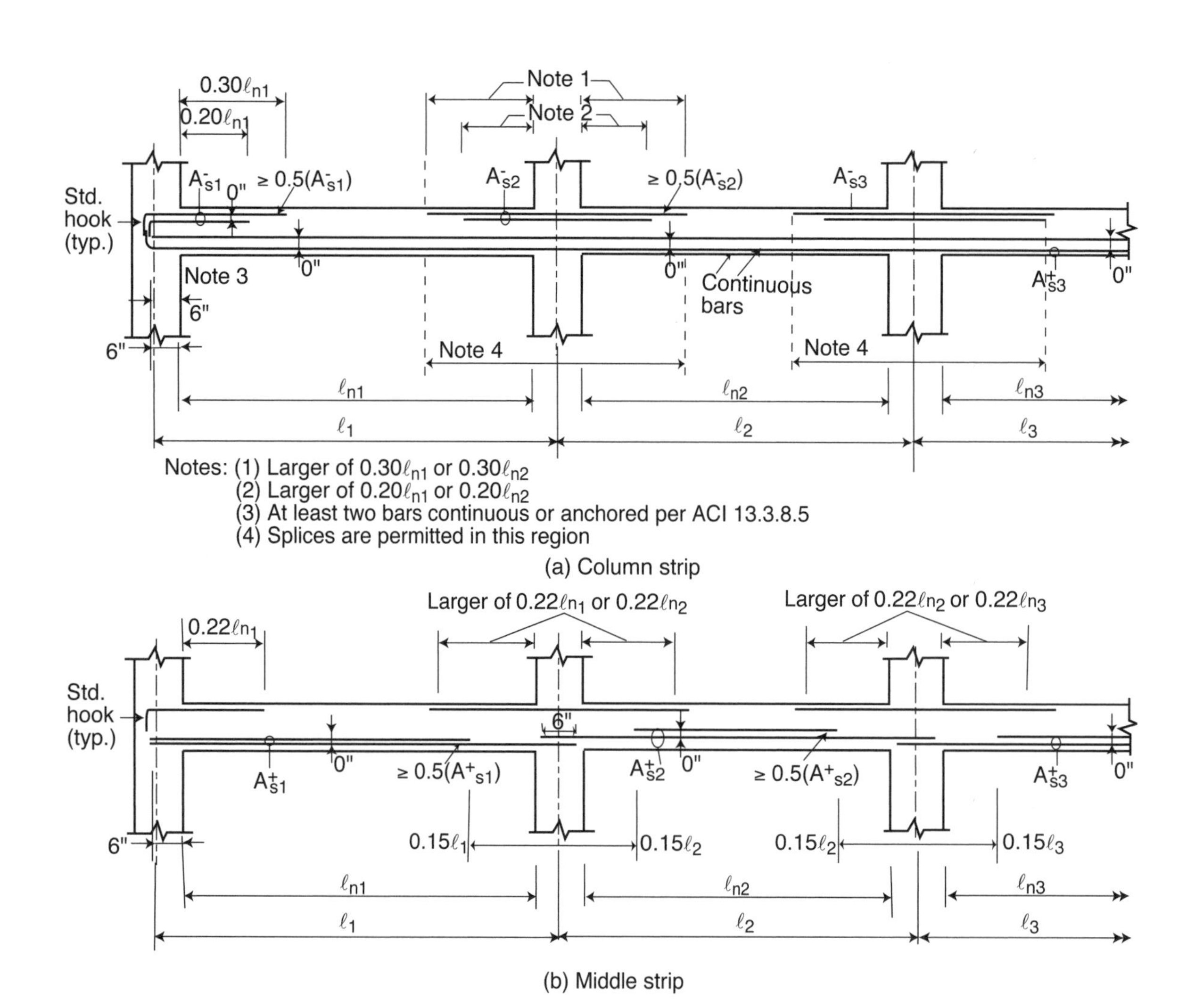

Figure 8-6 Recommended Bar Details for Two-Way Slabs (Without Beams)

Start 1st bar @ column centerline (if uniformly spaced bar is on centerline, start additional bars @ 3" on each side). Provide 3" min. spacing from uniformly spaced bars (if possible).

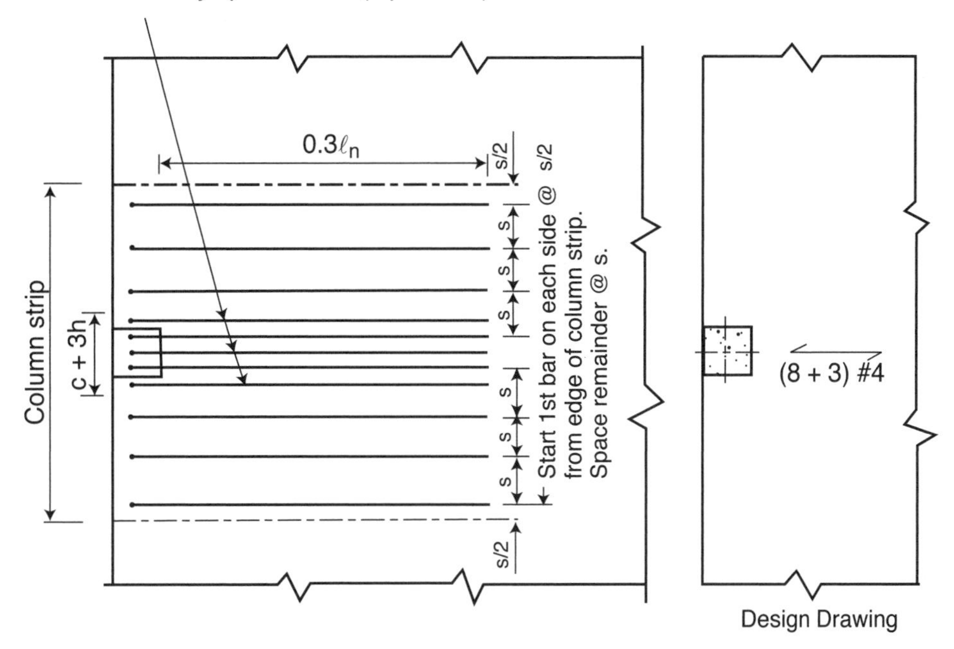

Notes: (1) Maximum spacings s = 2 x slab thickness ≤ 18 in. (ACI 13.4.2)

(2) Where additional top bars are required, show the total number of bars on the design drawing as (8 + 3) #4 where 8 indicates the number of uniformly spaced bars and 3 indicates the number of additional bars.

Figure 8-7 Example of a Typical Detail for Top Bars at Edge Columns (Flat Plate)

Start 1st additional bar @ 6" from edge. Space remainder @ 3" (if possible). Provide 3" min. spacing from uniformly spaced bars (if possible).

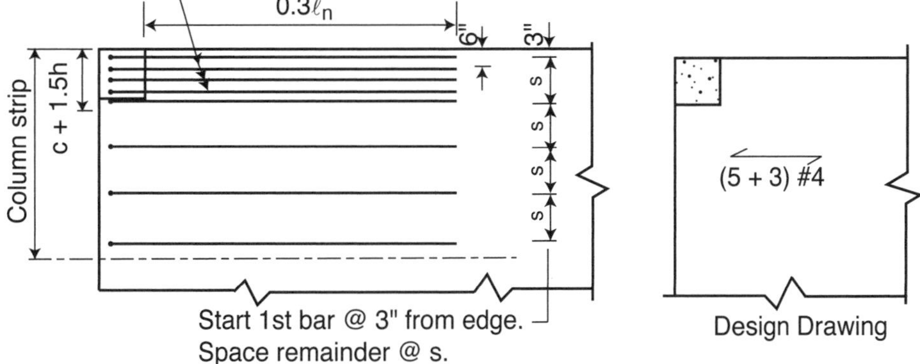

Figure 8-8 Example of a Typical Detail for Top Bars at Corner Columns (Flat Plate)

8.8 SPECIAL SPLICE REQUIREMENTS FOR COLUMNS

8.8.1 Construction and Placing Considerations

For columns in multistory buildings, one-story high preassembled reinforcement cages are usually used. It is common practice to locate the splices for the vertical column bars just above the floor level. In certain situations, it may be advantageous to use two-story high cages since this will reduce the number of splices and, for lap splices, will reduce the amount of reinforcing steel. However, it is important to note that two-story high cages are difficult to brace; the required guy wires or projecting bars may interfere with other construction operations such as the movement of cranes for transporting equipment and material. Also, it is more difficult and time-consuming to place the beam or girder bars at the intermediate floor level since they have to be threaded through the column steel. These two reasons alone are usually more than sufficient to offset any expected savings in steel that can be obtained by using two-story high cages. Thus, one-story high cages are usually preferred.

8.8.2 Design Considerations

Special provisions for column splices are given in ACI 12.17. In general, column splices must satisfy requirements for all load combinations for the column. For example, column design will frequently be governed by the gravity load combination (all bars in compression). However, the load combination, which includes wind loads, may produce tensile stresses in some of the bars. In this situation, a tension splice is required even though the load combination governing the column design did not produce any tensile stresses.

When the bar stress due to factored loads is compressive, lap splices, butt-welded splices, mechanical connections, and end-bearing splices are permitted. Table 8-4 may be used to determine the minimum compression lap splice lengths for Grade 60 bars. Note that these lap splice lengths may be multiplied by 0.83 for columns with the minimum effective area of ties (throughout the splice length) given in ACI 12.17.2.4. In no case shall the lap splice length be less than 12 in. Welded splices and mechanical connectors must meet the requirements of ACI 12.14.3. A full welded splice, which is designed to develop in tension, at least 1.25 A_bf_y (A_b = area of bar) will be adequate for compression as well. A full mechanical connection must develop in compression (or tension) at least 1.25 A_bf_y. End-bearing splices transfer the compressive stresses by bearing of square cut ends of the bars held in concentric contact by a suitable device (ACI 12.16.4). These types of splices may be used provided the splices are staggered or additional bars are provided at splice locations (see ACI 12.17.4 and the following discussion).

A minimum tensile strength is required for all compression splices. A compression lap splice with a length greater than or equal to the minimum value given in ACI 12.16.1 has a tensile strength of at least 0.25 A_bf_y. As noted above, full welded splices and full mechanical connectors develop at least 1.25 A_bf_y in tension. For end-bearing splices, the continuing bars on each face of the column must have a tensile strength of 0.25 A_sf_y where A_s is the total area of steel on the face of the column. This implies that not more than three-quarters of the bars can be spliced on each face of the column at any one location. Consequently, to ensure minimum tensile strength, end-bearing splices must be staggered or additional bars must be added if more than three-quarters of the bars are to be spliced at any one location.

Lap splices, welded splices, and mechanical connections are permitted when the bar stress is tensile; end-bearing splices must not be used (ACI 12.16.4.1). According to ACI 12.14.3, full welded splices and full mechanical connections must develop in tension at least 1.25 A_bf_y. When the bar stress on the tension face of the column is less than or equal to 0.5 f_y, lap splices must be Class B if more than one-half of the bars are spliced at any section, or Class A if half or fewer of the bars are spliced and alternate splices are staggered by the tension development length ℓ_d (ACI 12.17.2). Class B splices must be used when the bar stress is greater than 0.5 f_y.

Lap splice requirements for columns are illustrated in Fig. 8-9. For factored load combinations in Zone 1, all column bars are in compression. In Zone 2, the bar stress f_s on the tension face of the column varies from zero to 0.5 f_y in tension. For load combinations in Zone 3, f_s is greater that 0.5 f_y. The load combination that produces the greatest tensile stress in the bars will determine which type of lap splice is to be used. Load-moment design charts (such as the one in Figs. 5-16 through 5-23 in Chapter 5) can greatly facilitate the design of lap splices for columns.

Typical lap splice details for tied columns are shown in Fig. 8-10. Also given in the figure are the tie spacing requirements of ACI 7.8 and 7.10.5 (see Chapter 5). When a column face is offset 3 in. or more, offset bent longitudinal bars are not permitted (ACI 7.8.1.5). Instead, separate dowels, lap spliced with the longitudinal bars adjacent to the offset column faces must be provided. Typical splice details for footing dowels are given in Chapter 7, Fig. 7-7.

8.8.3 Example: Lap Splice Length for an Interior Column of Building #2, Alternate (2) Slab and Column Framing with Structural Walls (Non-sway Frame)

In this example, the required lap splice length will be determined for an interior column in the 2nd story; the splice will be located just above the 9 in. floor slab at the 1st level.

(1) Column Size and Reinforcement

In Example 5.7.2, a 16 × 16 in. column size was established for the entire column stack. It was determined that 4-No.8 bars were required in both the 1st and 2nd floor columns.

(2) Lap Splice Length

Since the columns carry only gravity loads, all of the column bars will be in compression (Zone 1 in Fig. 8-9). Therefore, a compression lap splice is sufficient.

From Table 8-4, the minimum compression lap splice length required for the No.8 bars is 30 in. In this situation, No.3 ties are required @ 16 in. spacing.

According to ACI 12.17.2.4, the lap splice length may be multiplied by 0.83 if ties are provided with an effective area of 0.0015hs throughout the lap splice length.

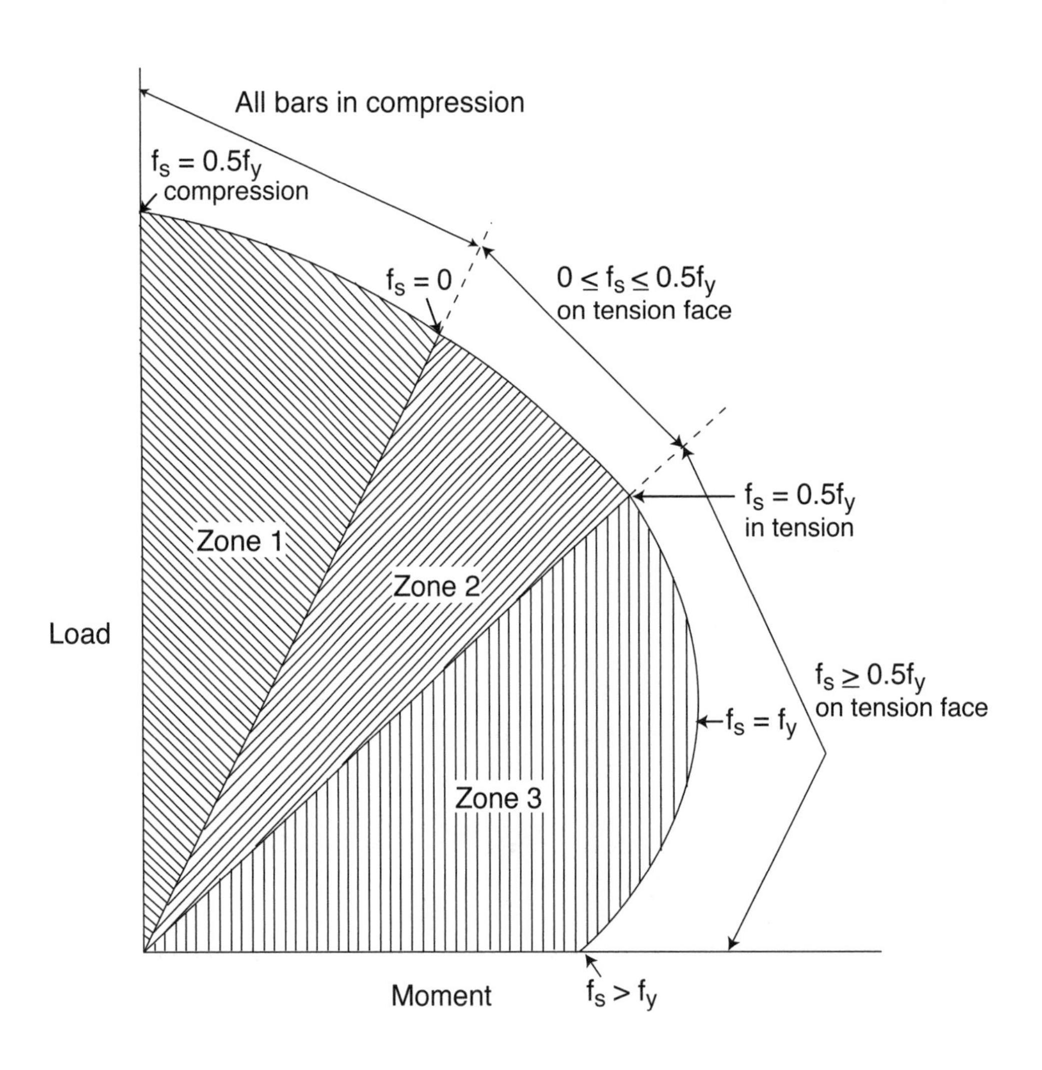

Figure 8-9 Special Splice Requirements for Columns

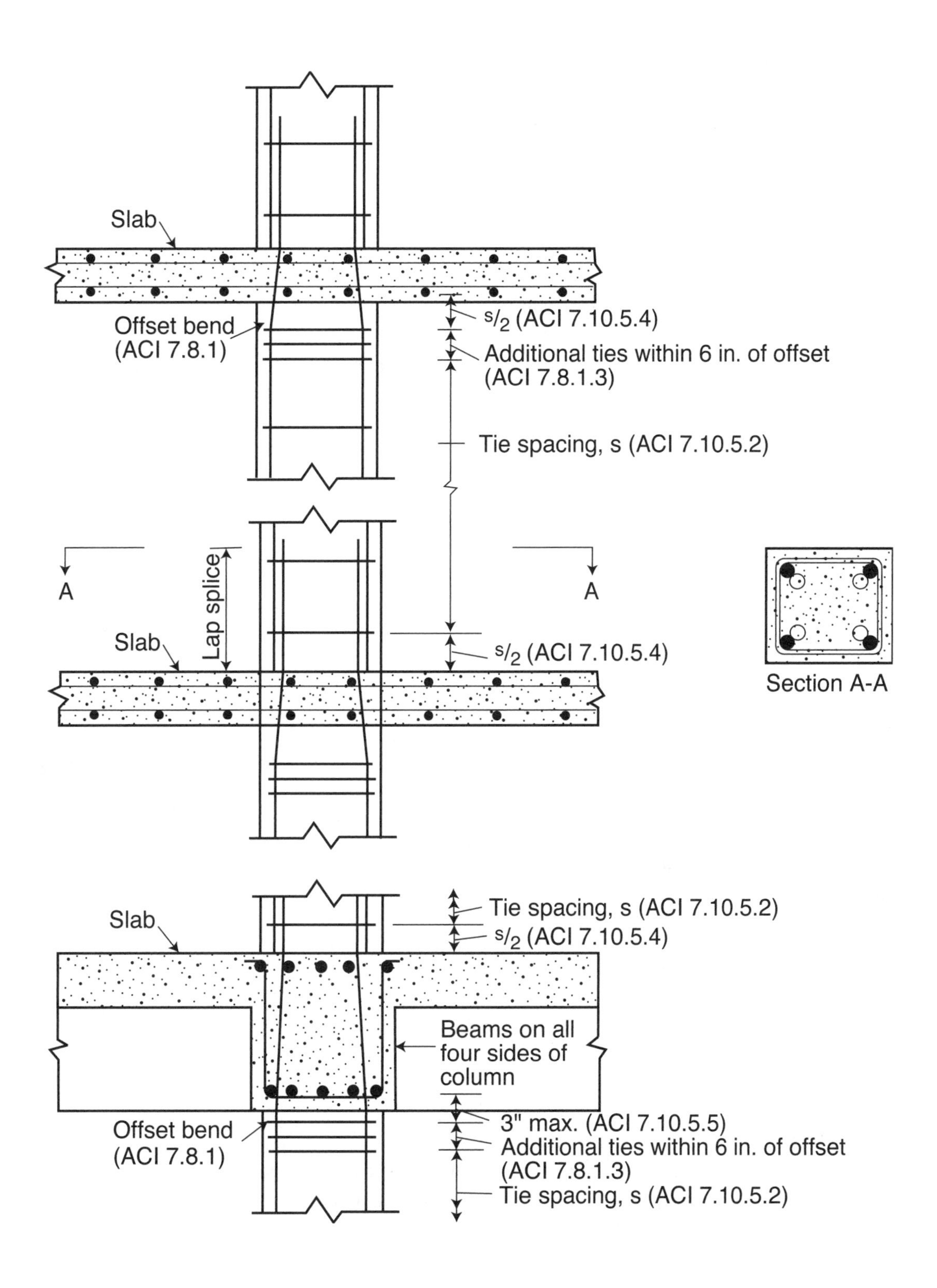

Figure 8-10 Column Splice and Tie Details

Two legs of the No.3 ties are effective in each direction; thus, the required spacing s can be determined as follows:

$2 \times 0.11 = 0.22 \text{ in.2} = 0.0015 \times 16 \times s$

or, $s = 9.2$ in.

Splice length $= 0.83 \times 30 = 24.9$

Use a 2 ft-1 in. splice length with No.3 ties @ 9 in. throughout the splice length.

8.8.4 Example: Lap Splice Length for an Interior Column of Building #.2, Alternate (1) Slab and Column Framing Without Structural Walls (Sway Frame)

As in Example 8.8.3, the required lap splice length will be determined for an interior column in the 2nd story, located just above the slab at the 1st level.

(1) Load Data

In Example 5.7.1, the following load combinations were obtained for the 2nd story interior columns:

Gravity loads:	$P_u = 399$ kips	ACI Eq. (9-2)
	$M_u = 13$ ft-kips	
Gravity + Wind loads:	$P_u = 399$ kips	ACI Eq. (9-3)
	$M_u = 47$ ft-kips	
	$P_u = 371$ kips	ACI Eq. (9-4)
	$M_u = 77$ ft-kips	
	$P_u = 243$ kips	ACI Eq. (9-6)
	$M_u = 68$ ft-kips	

(2) Column Size and Reinforcement

A 16 × 16 in. column size was established in Example 5.7.1 for the 1st story columns. This size is used for the entire column stack; the amount of reinforcement can be decreased in the upper levels. It was also shown that the 1st story columns were slender, and the 8-No.10 bars were required for the factored axial loads and magnified moments at this level.

The reinforcement for the 2nd story columns can be determined using the procedure outlined in Chapter 5. As was the case for the 1st story columns, the 2nd story columns are slender (use k = 1.2; see Chapter 5):

$$\frac{k\ell_u}{r} = \frac{1.2\,[(12 \times 12) - 8.5]}{0.3 \times 16} = 34 > 22$$

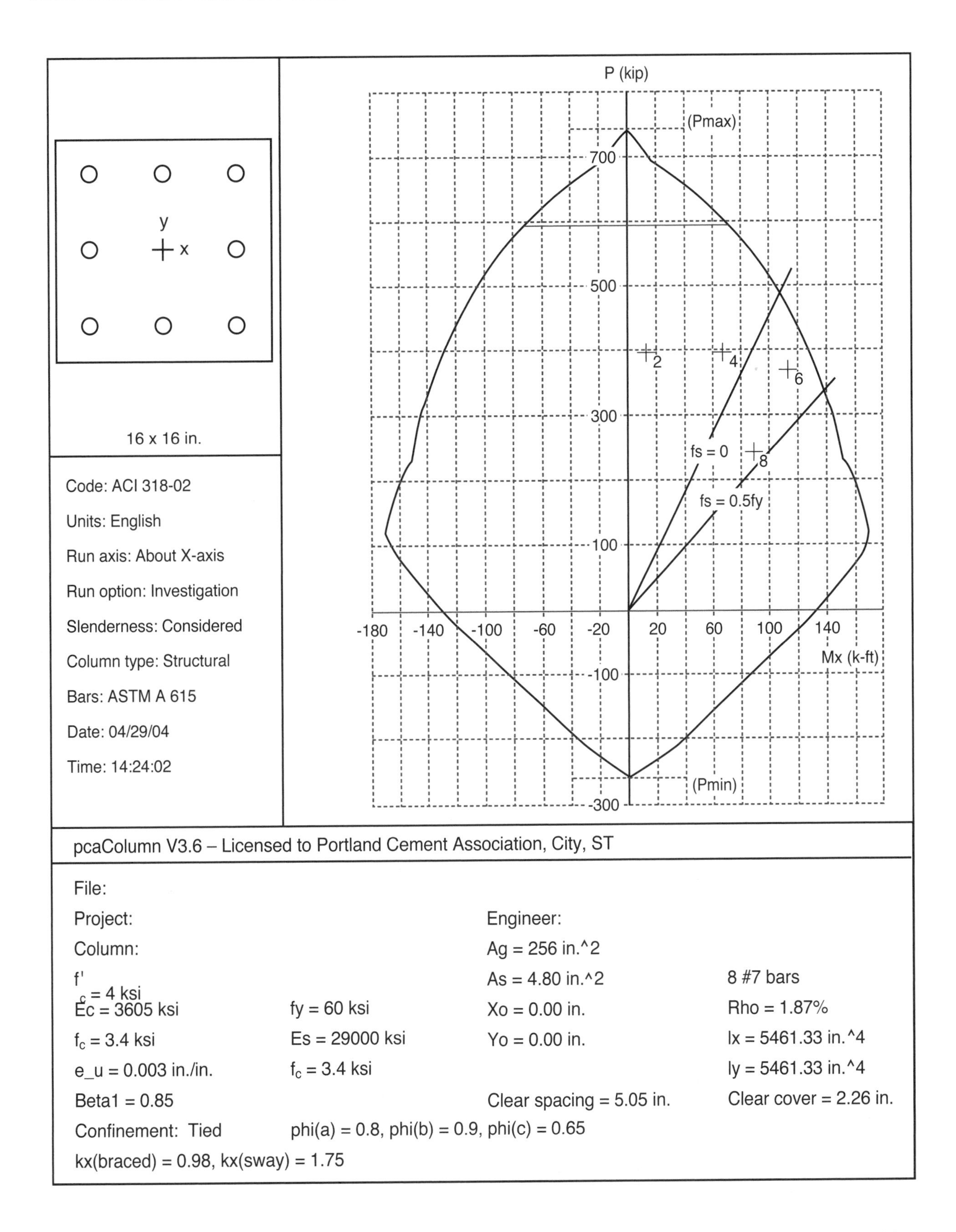

Figure 8-11 Interaction Diagram for an Interior column in the 2nd Story of building No. 2, Alternate (1)

Figure 8-11 is the output from PCACOLUMN for the section reinforced with 8-No.7 bars.*

(3) Lap Splice Length

The load combination represented by point 8 (ACI Eq. (9-6)) in Fig. 8-11 governs the type of lap splice to be used, since it is the combination that produces the greatest tensile stress f_s in the bars. Note that the load combination represented by point 6 (ACI Eq. (9-4)) which governed the design of the column does not govern the design of the splice. Since $0 < f_s < 0.5\ f_y$ at point 8, a Class B splice must be used (all the bars spliced ACI 12.17.2.2).

Required splice length = 1.3 ℓ_d where ℓ_d is the tension development length of the No.10 bars (of the lower column).

Clear bar spacing ≅ 3.5 in. = 2.8 d_b (see Fig. 8-2)

Cover > d_b = 1.27 in.

From Table 8-1, ℓ_d = 60.2 in.

$1.3\ell_d = 1.3 \times 60.2 = 78.3$ in.

Thus, a 6 ft-6 in. splice length would be required which is more than one-half of the clear story height.

Decreasing the bar size in the 1st story columns would result in slightly smaller splice lengths; however, the reinforcement ratio would increase from 4% (8-No.10) to 4.7% (12-No.9). Also, labor costs would increase since more bars would have to be placed and spliced.

One possible alternative would be to increase the column size. For example, a 18 × 18 in. column would require about 8-No.8 bars in the 1st story. It is important to note that changing the dimensions of the columns would change the results from the lateral load analysis, affecting all subsequent calculations; a small change, however, should not significantly alter the results.

References

8.1 *ACI Detailing Manual – 1994,* SP-66(94), American Concrete Institute, Detroit, Michigan, 1994, (PCA LT185).

8.2 Stecich, J.P., Hanson, J.M., and Rice, P.F., "Bending and Straightening of Grade 60 Reinforcing Bars," *Concrete International: Design & Construction,* VOl. 6, No. 8, August 1984, pp. 14-23.

8.3 "Suggested Project Specifications Provisions for Epoxy-Coated Reinforcing Bars," *Engineering Data Report No. 19,* Concrete Reinforcing Steel Institute, Schaumburg, Illinois, 1984.

8.4 *Mechanical Connections of Reinforcing Bars,* ACI Committee 439, American Concrete Institute, Detroit, Michigan, 1991(Reapproved 1999), 16 pp.

* *25% of the gross moment of inertia of the column strip slab was used to account for cracking.*

8.5 *Reinforcement: Anchorages and Splices,* 4th Edition, Concrete Reinforcing Steel Institute, Schaumburg, Illinois, 1997, 100 pp.

8.6 *CRSI Handbook,* Concrete Reinforcing Steel Institute, Schaumburg, Illinois, 9th Edition, 2002, (PCA LT185).

8.7 "Design of Reinforcement for Two-Way Slab-to-Column Frames Laterally Braced or Not Braced," *Structural Bulletin No. 9,* Concrete Reinforcing Steel Institute, Schaumburg, Illinois, June 1993.

Chapter 9

Design Considerations for Economical Formwork

9.1 INTRODUCTION

Depending on a number of factors, the cost of formwork can be as high as 60% of the total cost of a cast-in-place concrete structure. For this reason, it is extremely important to devise a structural system that will minimize the cost of formwork. Basic guidelines for achieving economical formwork are given in Reference 9.1, and are summarized in this chapter.

Formwork economy should initially be considered at the conceptual stage or the preliminary design phase of a construction project. This is the time when architectural, structural, mechanical, and electrical systems are conceived. The architect and the engineer can help reduce the cost of formwork by following certain basic principles while laying out the plans and selecting the structural framing for the building. Additional savings can usually be achieved by consulting a contractor during the initial design phases of a project.

Design professionals, after having considered several alternative structural framing systems and having determined those systems that best satisfy loading requirements as well as other design criteria, often make their final selections on the concrete framing system that would have the least amount of concrete and possibly the least amount of reinforcing steel. This approach can sometimes result in a costly design. Complex structural frames and nonstandard member cross sections can complicate construction to the extent that any cost savings to be realized from the economical use of in-place (permanent) materials can be significantly offset by the higher costs of formwork. Consequently, when conducting cost evaluations of concrete structural frames, it is essential that the costs of formwork be included.

9.2 BASIC PRINCIPLES TO ACHIEVE ECONOMICAL FORMWORK

There is always the opportunity to cut costs in any structural system. The high cost of formwork relative to the costs of the other components makes it an obvious target for close examination. Three basic design principles that govern formwork economy for all site-cast concrete structures are given below.

9.2.1 Standard Forms

Since most projects do not have the budget to accommodate custom forms, basing the design on readily available standard form sizes is essential to achieve economical formwork. Also, designing for actual dimensions of standard nominal lumber will significantly cut costs. A simplified approach to formwork carpentry means less sawing, less piecing together, less waste, and less time; this results in reduced labor and material costs and fewer opportunities for error by construction workers.

9.2.2 Repetition

Whenever possible, the sizes and shapes of the concrete members should be repeated in the structure. By doing this, the forms can be reused from bay to bay and from floor to floor, resulting in maximum overall savings. The relationship between cost and changes in depth of horizontal construction is a major design consideration. By standardizing the size or, if that is not possible, by varying the width and not the depth of beams, most requirements can be met at a lowered cost, since the forms can be reused for all floors. To accommodate load and span variations, only the amount of reinforcement needs to be adjusted. Also, experience has shown that changing the depth of the concrete joist system from floor to floor because of differences in superimposed loads actually results in higher costs. Selecting different joist depths and beam sizes for each floor may result in minor savings in materials, but specifying the same depth for all floors will achieve major savings in forming costs.

9.2.3 Simplicity

In general, there are countless variables that must be evaluated and then integrated into the design of a building. Traditionally, economy has meant a time-consuming search for ways to cut back on quantity of materials. As noted previously, this approach often creates additional costs—quite the opposite effect of that intended.

An important principle in formwork design is simplicity. In light of this principle, the following questions should be considered in the preliminary design stage of any project:

(1) Will custom forms be cost-effective? Usually, when standard forms are used, both labor and materials costs decrease. However, custom forms can be as cost-effective as standard forms if they are required in a quantity that allows mass production.

(2) Are deep beams cost-effective? As a rule, changing the beam depth to accommodate a difference in load will result in materials savings, but can add considerably to forming costs due to field crew disruptions and increased potential for field error. Wide, flat beams are more cost-effective than deep narrow beams.

(3) Should beam and joist spacing be uniform or vary with load? Once again, a large number of different spacings (closer together for heavy loads, farther apart for light) can result in material savings. However, the disruption in work and the added labor costs required to form the variations may far exceed savings in materials.

(4) Should column size vary with height and loading? Consistency in column size usually results in reduced labor costs, particularly in buildings of moderate height. Under some rare conditions, however, changing the column size will yield savings in materials that justify the increased labor costs required for forming.

(5) Are formed surface tolerances reasonable? Section 3.4 of ACI Standard 347[9.2] provides a way of quantitatively indicating tolerances for surface variations due to forming quality. The suggested tolerances for formed cast-in-place surfaces are shown in Table 9-1 (Table 3.1 of ACI 347). The following simplified guidelines for specifying the class of formed surface will usually minimize costs: a) Class A finish should be specified for surfaces prominently exposed to public view, b) Class B finish should be specified for surfaces less prominently exposed to public view, c) Class C finish should be specified for all noncritical or unexposed surfaces, and d) Class D finish should be specified for concealed surfaces where roughness is not objectionable. If a more stringent class of surface is specified than is necessary for a particular formed surface, the increase in cost may become disproportionate to the increase in quality; this is illustrated in Fig. 9-1.

Table 9-1 Permitted Irregularities in Formed Surfaces Checked with a 5-ft Template[9.1]

Class of surface			
A	B	C	D
1/8 in.	1/4 in.	1/2 in.	1 in.

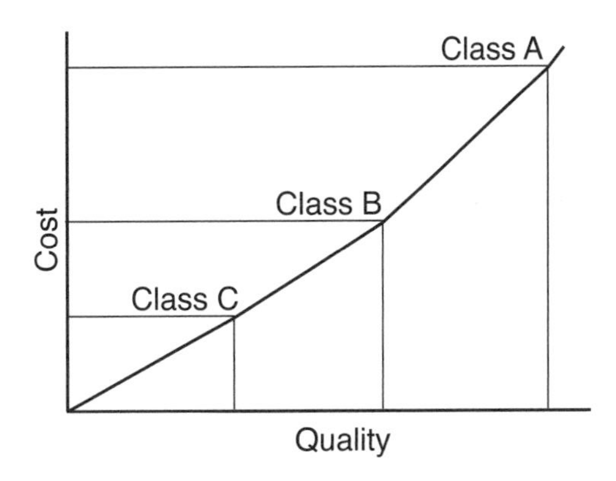

Figure 9-1 Class of Surface Versus Cost

9.3 ECONOMICAL ASPECTS OF HORIZONTAL FRAMING

Floors and the required forming are usually the largest cost component of a concrete building structure. The first step towards achieving maximum economy is selecting the most economical floor system for a given plan layout and a given set of loads. This will be discussed in more detail below. The second step is to define a regular, orderly progression of systematic shoring and reshoring. Timing the removal of the forms and requiring a minimum amount of reshoring are two factors that must be seriously considered since they can have a significant impact on the final cost.

Figures 1-5 and 1-6 show the relative costs of various floor systems as a function of bay size and superimposed load. Both figures are based on a concrete strength f'_c = 4000 psi. For a given set of loads, the slab system that is optimal for short spans is not necessarily optimal for longer spans. Also, for a given span, the slab system that is optimal for lighter superimposed loads is not necessarily optimal for heavier loads. Reference 9.3 provides material and cost estimating data for various floor systems. It is also very important to consider the fire resistance of the floor system in the preliminary design stage (see Chapter 10). Required fire resistance ratings can dictate the type of floor system to specify in a particular situation.

The relationship between span length, floor system, and cost may indicate one or more systems to be economical for a given project. If the system choices are equally cost-effective, then other considerations (architectural, aesthetic, etc.) may become the determining factor.

Beyond selection of the most economical system for load and span conditions, there are general techniques that facilitate the most economical use of the chosen system.

9.3.1 Slab Systems

Whenever possible, avoid offsets and irregularities that cause a "stop and start" disruption of labor and require additional cutting (and waste) of materials (for example, breaks in soffit elevation). Depressions for terrazzo, tile, etc. should be accomplished by adding concrete to the top surface of the slab rather than maintaining a constant slab thickness and forming offsets in the bottom of the slab. Cross section (a) in Fig. 9-2 is less costly to form than cross section (b).

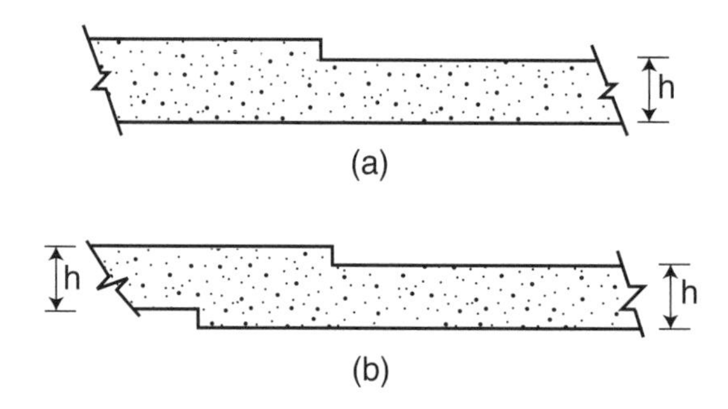

Figure 9-2 Depressions in Slabs

When drop panels are used in two-way systems, the total depth of the drop h_1 should be set equal to the actual nominal lumber dimension plus 3/4-in. for plyform (see Fig. 9-3). Table 9-2 lists values for the depth h_1 based on common nominal lumber sizes. As noted above, designs which depart from standard lumber dimensions are expensive.

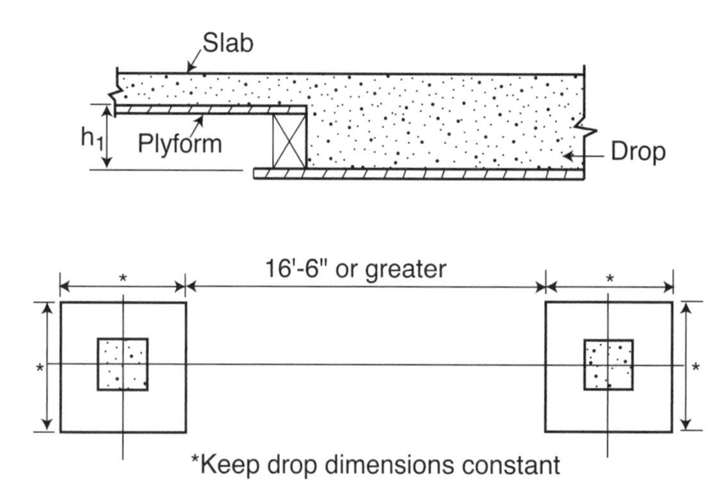

Figure 9-3 Formwork for Drop Panels

Table 9-2 Drop Panel Depth, h_1

Nominal lumber size	Actual lumber size (in.)	Plyform thickness (in.)	h_1 (in.)
2X	1½	¾	2¼
4X	3½	¾	4¼
6X	5½	¾	6¼
8X	7¼	¾	8

Whenever possible, a minimum 16 ft (plus 6 in. minimum clearance) spacing between drop panel edges should be used (see Fig. 9-3). Again, this permits the use of 16 ft long standard lumber without costly cutting of material. For maximum economy, the plan dimensions of the drop panel should remain constant throughout the entire project.

9.3.2 Joist Systems

Whenever possible, the joist depth and the spacing between joists should be based on standard form dimensions (see Table 9-3).

The joist width should conform to the values given in Table 9-3 also. Variations in width mean more time for interrupted labor, more time for accurate measurement between ribs, and more opportunities for jobsite error; all of these add to the overall cost.

Table 9-3 Standard Form Dimensions for One-Way Joist Construction (in.)

Width	Depth	Flange width	Width of joist
20	8, 10, 12	7/8, 2 ½	5, 6
30	8, 10, 12, 14, 16, 20	7/8, 3	5, 6, 7
53	16, 20	3½	7, 8, 9, 10
66	14, 16, 20	3	6, 7, 8, 9, 10

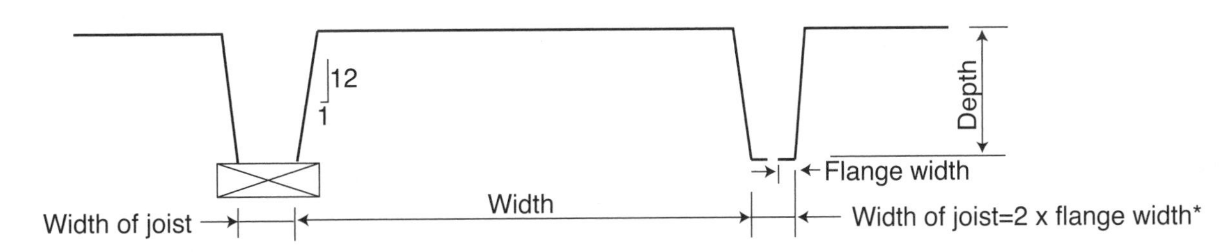

*Applies to flange widths > 7/8 in.

It is extremely cost-effective to specify a supporting beam with a depth equal to the depth of the joist. By doing this, the bottom of the entire floor system can be formed in one horizontal plane. Additionally, installation costs for utilities, partitions, and ceilings can all be reduced.

9.3.3 Beam-Supported Slab Systems

The most economical use of this relatively expensive system relies upon the principles of standardization and repetition. Of primary importance is consistency in depth and of secondary importance is consistency in width. These two concepts will mean a simplified design; less time spent interpreting plans and more time for field crews to produce.

9.4 ECONOMICAL ASPECTS OF VERTICAL FRAMING

9.4.1 Walls

Walls provide an excellent opportunity to combine multiple functions in a single element; by doing this, a more economical design is achieved. With creative layout and design, the same wall can be a fire enclosure for stair or elevator shafts, a member for vertical support, and bracing for lateral loads. Walls with rectangular cross-sections are less costly then nonrectangular walls.

9.4.2 Core Areas

Core areas for elevators, stairs, and utility shafts are required in many projects. In extreme cases, the core may require more labor than the rest of the floor. Standardizing the size and location of floor openings within the core will reduce costs. Repeating the core framing pattern on as many floors as possible will also help to minimize the overall costs.

9.4.3 Columns

Although the greatest costs in the structural frame are in the floor system, the cost of column formwork should not be overlooked. Whenever possible, use the same column dimensions for the entire height of the building. Also, use a uniform symmetrical column pattern with all of the columns having the same orientation. Planning along these general lines can yield maximum column economy as well as greater floor framing economy because of the resulting uniformity in bay sizes.

9.5 GUIDELINES FOR MEMBER SIZING

9.5.1 Beams

- For a line of continuous beams, keep the beam size constant and vary the reinforcement from span to span.

- Wide flat beams (same depth as joists) are easier to form than beams projecting below the bottom of the joints (see Fig, 9-4).

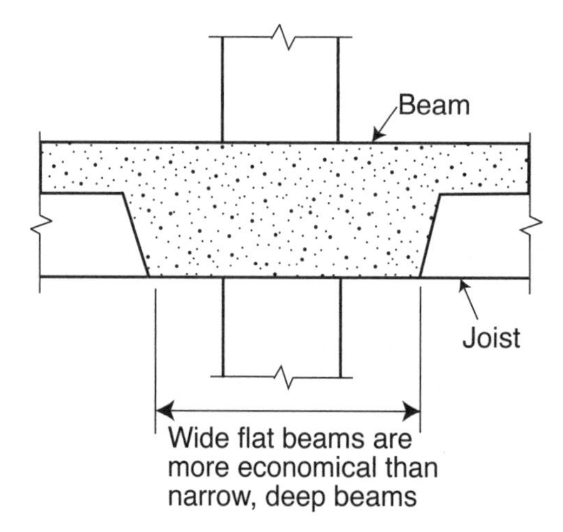

Figure 9-4 One-Way Joist Floor System

- Spandrel beams are more cost intensive than interior beams due to their location at the edge of a floor slab or at a slab opening. Fig. 9-5 lists some of the various aspects to consider when designing these members.

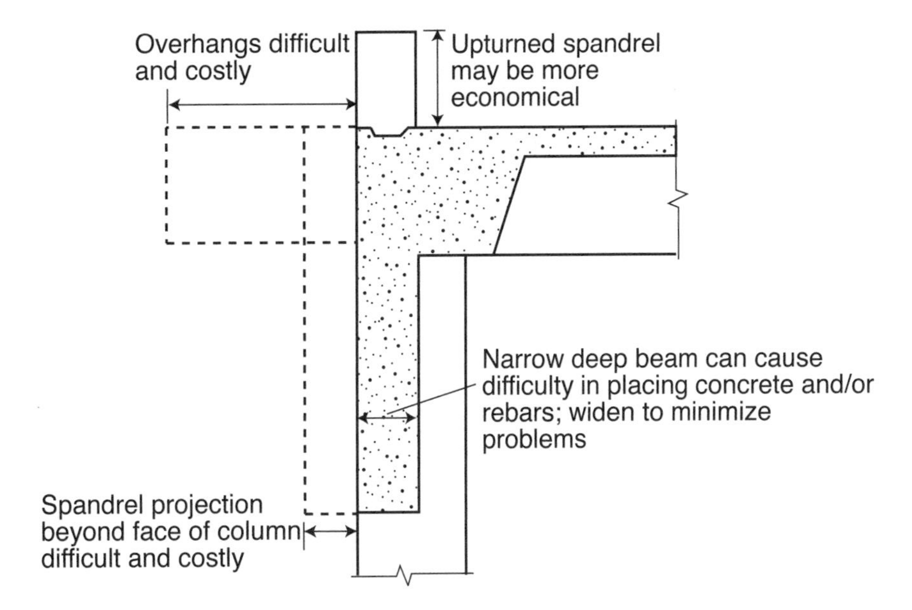

Figure 9-5 Spandrel Beams

- Beams should be as wide as, or wider than, the columns into which they frame (see Fig. 9-6). In addition for formwork economy, this also alleviates some of the reinforcement congestion at the intersection.

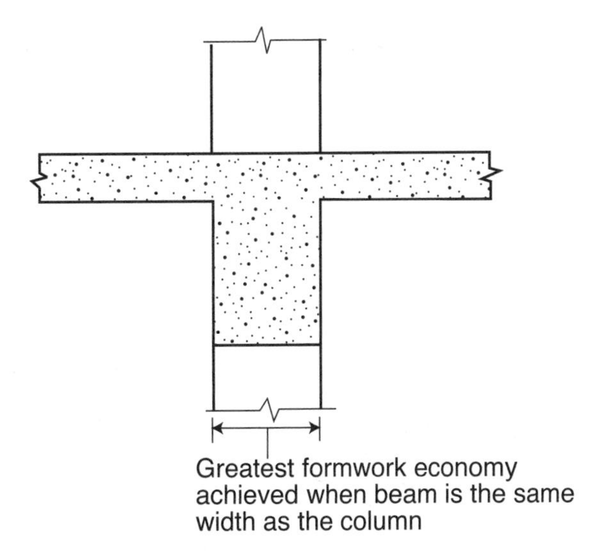

Figure 9-6 Beam-Column Intersections

- For heavy loading or long spans, a beam deeper than the joists may be required. In these situations, allow for minimum tee and lugs at sides of beams as shown in Fig. 9-7. Try to keep difference in elevation between bottom of beam and bottom of floor system in modular lumber dimensions.

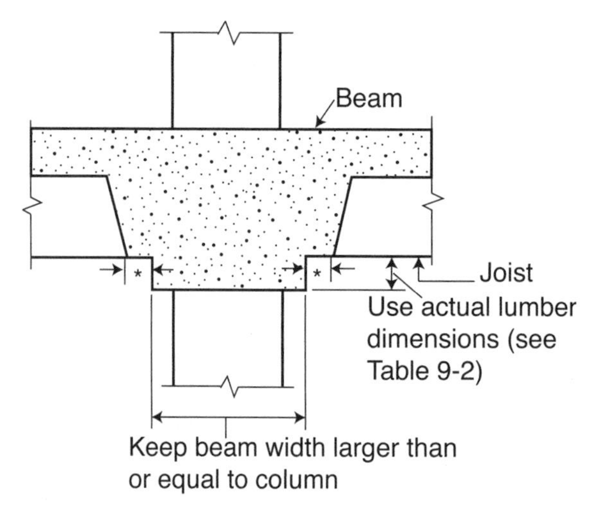

Figure 9-7 One-Way Joist Floor System with Deep Beams

9.5.2 Columns

- For maximum economy, standardize column location and orientation in a uniform pattern in both directions (see Fig, 9-8).

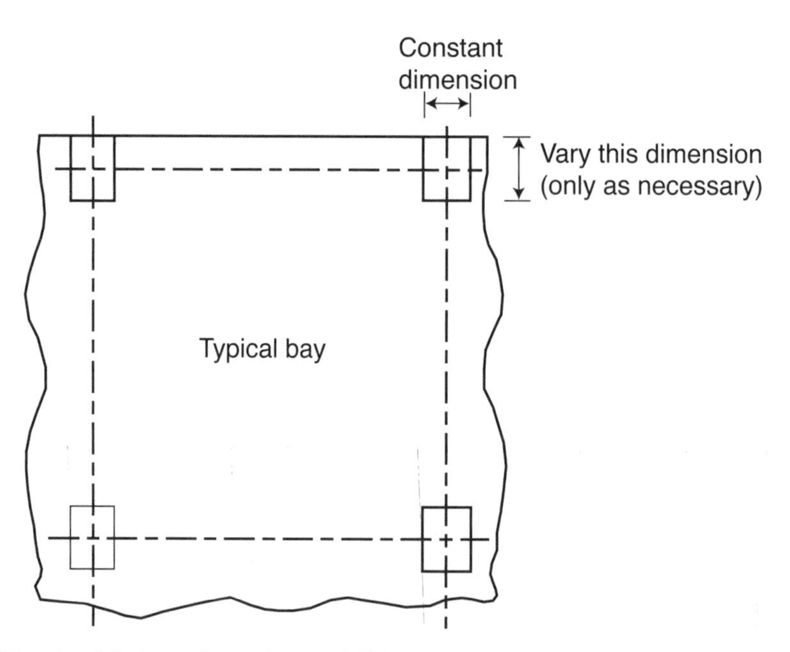

Figure 9-8 Standard Column Location and Orientation for a Typical Bay

- Columns should be kept the same size throughout the building. If size changes are necessary, they should occur in 2 in. increments, one side at a time (for example, a 22 × 22 in. column should go to a 24 x 22 in., then to a 24 × 24 in., etc.) Gang forming can possibly be used when this approach to changing columns sizes is utilized. When a flying form system is used, the distance between column faces and the flying form must be held constant. Column size changes must be made parallel to the flying form.

- Use the same shape as often as possible throughout the entire building. Square, rectangular, or round columns are the most economical; use other shapes only when architectural requirements so dictate.

9.5.3 Walls

- Use the same wall thickness throughout a project if possible; this facilitates the reuse of equipment, ties, and hardware. In addition, this minimizes the possibilities of error in the field. In all cases, maintain sufficient wall thickness to permit proper placing and vibrating of concrete.

- Wall openings should be kept to a minimum number since they can be costly and time-consuming. A few larger openings are more cost-effective than many smaller openings. Size and location should be constant for maximum reuse of formwork.

- Brick ledges should be kept at a constant height with a minimum number of steps. Thickness as well as height should be in dimensional units of lumber, approximately as closely as possible those of the masonry to be placed. Brick ledge locations and dimensions should be detailed on the structural drawings.

- Footing elevations should be kept constant along any given wall if possible. This facilitates the use of wall gang forms from footing to footing. If footing steps are required, use the minimum number possible.

- For buildings of moderate height, pilasters can be used to transfer column loads into the foundation walls. Gang forms can be used more easily if the pilaster sides are splayed as shown in Fig. 9-9.

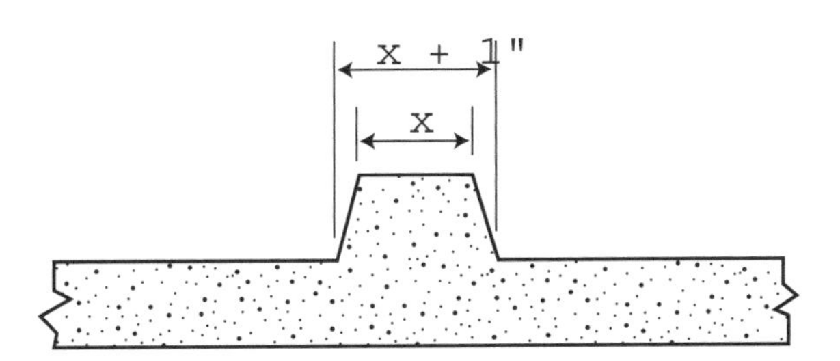

Figure 9-9 Pilasters

9.6 OVERALL STRUCTURAL ECONOMY

While it has been the primary purpose of this chapter to focus on those considerations that will significantly impact the costs of the structural system relative to formwork requirements, the 10-step process below should be followed during the preliminary and final design phases of the construction project as this will lead to overall structural economy:

(1) Study the structure as a whole.

(2) Prepare freehand alternative sketches comparing all likely structural framing systems.

(3) Establish column locations as uniformly as possible, keeping orientation and size constant wherever possible.

(4) Determine preliminary member sizes from available design aids (see Section 1.8).

(5) Make cost comparisons based on sketches from Step 2 quickly, roughly, but with an adequate degree of accuracy.

(6) Select the best balance between cost of structure and architectural/mechanical design considerations.

(7) Distribute prints of selected framing scheme to all design and building team members to reduce unnecessary future changes.

(8) Plan your building. Visualize how forms would be constructed. Where possible, keep beams and columns simple without haunches, brackets, widened ends or offsets. Standardize concrete sizes for maximum reuse of forms.

(9) During final design, place most emphasis on those items having greatest financial impact on total structural frame cost.

(10) Plan your specifications to minimize construction costs and time by including items such as early stripping time for formwork and acceptable tolerances for finish.

Reference 9.4 should be consulted for additional information concerning formwork.

References

9.1 *Concrete Buildings, New Formwork Perspectives*, Ceco Industries, Inc., 1985.

9.2 *Guide to Formwork for Concrete*, ACI 347-01, American Concrete Institute, Detroit, Michigan, 2002, 33 pp.

9.3 *Concrete Floor Systems—Guide to Estimating and Economizing*, SP041, Portland Cement Association, Skokie, Illinois, 2000, 41 pp.

9.4 Hurd, M.K., *Formwork for Concrete*, (prepared under direction of ACI Committee 347, Formwork for Concrete), SP-4, 6th Ed., American Concrete Institute, Detroit, Michigan, 1995.

Chapter 10

Design Considerations for Fire Resistance

10.1 INTRODUCTION

State and municipal building codes throughout the country regulate the fire resistance of the various elements and assemblies comprising a building structure. Structural frames (columns and beams), floor and roof systems, and load bearing walls must be able to withstand the stresses and strains imposed by fully developed fires and carry their own dead loads and superimposed loads without collapse.

Fire resistance ratings required of the various elements of construction by building codes are a measure of the endurance needed to safeguard the structural stability of a building during the course of a fire and to prevent the spread of fire to other parts of the building. The determination of fire rating requirements in building codes is based on the expected fire severity (fuel loading) associated with the type of occupancy and the building height and area.

In the design of structures, building code provisions for fire resistance are sometimes overlooked and this may lead to costly mistakes. It is not uncommon, for instance, to find that a concrete slab in a waffle slab floor system may only require a 3 to 4-1/2 in. thickness to satisfy ACI 318 strength requirements. However, if the building code specifies a 2-hour fire resistance rating for that particular floor system, the slab thickness may need to be increased to 3-1/2 to 5 in., depending on type of aggregate used in the concrete. Indeed, under such circumstances and from the standpoint of economics, the fire-resistive requirements may indicate another system of construction to be more appropriate, say, a pan-joist or flat slab/plate floor system. Simply stated, structural members possessing the fire resistance prescribed in building codes may differ significantly in their dimensional requirements from those predicated only on ACI 318 strength criteria. Building officials are required to enforce the stricter provisions.

The purpose of this chapter is to make the reader aware of the importance of determining the fire resistance requirements of the governing building code before proceeding with the structural design.

The field of fire technology is highly involved and complex and it is not the intent here to deal with the chemical or physical characteristics of fire, nor with the behavior of structures in real fire situations. Rather, the goal is to present some basic information as an aid to designers in establishing those fire protection features of construction that may impact their structural design work.

The information given in this chapter is fundamental. Modern day designs, however, must deal with many combinations of materials and it is not possible here to address all the intricacies of construction. Rational methods of design for dealing with more involved fire resistance problems are available. For more comprehensive discussions on the subject of the fire resistive qualities of concrete and for calculation methods used in solving design problems related to fire integrity, the reader may consult Reference 10.1.

10.2 DEFINITIONS

Structural Concrete:

- Siliceous aggregate concrete: concrete made with normal weight aggregates consisting mainly of silica or compounds other than calcium or magnesium carbonate.

- Carbonate aggregate concrete: concrete made with aggregates consisting mainly of calcium or magnesium carbonate, e.g., limestone or dolomite.

- Sand-lightweight concrete: concrete made with a combination of expanded clay, shale, slag, or slate or sintered fly ash and natural sand. Its unit weight is generally between 105 and 120 pcf.

- Lightweight aggregate concrete: concrete made with aggregates of expanded clay, shale, slag, or slate or sintered fly ash, and weighing 85 to 115 pcf.

Insulating Concrete:

- Cellular concrete: a lightweight insulating concrete made by mixing a preformed foam with Portland cement slurry and having a dry unit weight of approximately 30 pcf.

- Perlite concrete: a lightweight insulating concrete having a dry unit weight of approximately 30 pcf made with perlite concrete aggregate produced from volcanic rock that, when heated, expands to form a glass-like material or cellular structure.

- Vermiculite concrete: a lightweight insulating concrete made with vermiculite concrete aggregate, a laminated micaceous material produced by expanding the ore at high temperatures. When added to Portland cement slurry the resulting concrete has a dry unit weight of approximately 30 pcf.

Miscellaneous Insulating Materials:

- Glass fiber board: fibrous glass roof insulation consisting of inorganic glass fibers formed into rigid boards using a binder. The board has a top surface faced with asphalt and kraft reinforced with glass fibers.

- Mineral board: a rigid felted thermal insulation board consisting of either felted mineral fiber or cellular beads of expanded aggregate formed into flat rectangular units.

10.3 FIRE RESISTANCE RATINGS

10.3.1 Fire Test Standards

The fire-resistive properties of building components and structural assemblies are determined by standard fire test methods. The most widely used and nationally accepted test procedure is that developed by the American Society of Testing and Materials (ASTM). It is designated as ASTM E 119, Standard Methods of Fire Tests of Building Construction and Materials. Other accepted standards, essentially alike, include the National Fire Protection Association Standard Method No. 251; Underwriters Laboratories; U.L. 263; American National

Standards Institute's ANSI A2-1; ULC-S101 from the Underwriters Laboratories of Canada; and Uniform Building Code Standard No. 43-1.

10.3.2 ASTM E 119 Test Procedure

A standard fire test is conducted by placing an assembly in a test furnace. Floor and roof specimens are exposed to controlled fire from beneath, beams from the bottom and sides, walls from one side, and columns from all sides. The temperature is raised in the furnace over a given period of time in accordance with ASTM E 119 standard time-temperature curve shown in Fig. 10-1.

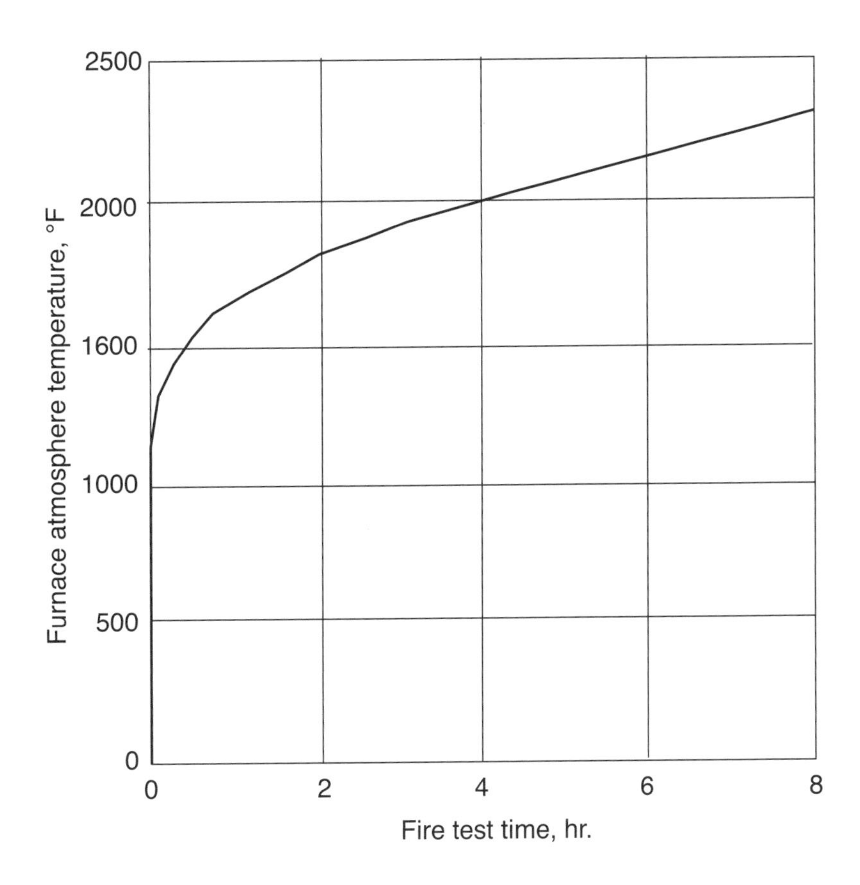

Figure 10-1 Standard Time-Temperature Relationship of Furnace Atmosphere (ASTM E 119)

This specified time-temperature relationship provides for a furnace temperature of 1000°F at five minutes from the beginning of the test, 1300°F at 10 minutes, 1700°F at one hour, 1850°F at two hours, and 2000°F at four hours. The end of the test is reached and the fire endurance of the specimen is established when any one of the following conditions first occur:

(1) For walls, floors, and roof assemblies the temperature of the unexposed surface rises an average of 150°F above its initial temperature of 325°F at any location. In addition, walls achieving a rating classification of one hour or greater must withstand the impact, erosion and cooling effects of a hose steam test.

(2) Cotton waste placed on the unexposed side of a wall, floor, or roof system is ignited through cracks or fissures developed in the specimen.

(3) The test assembly fails to sustain the applied load.

(4) For certain restrained and all unrestrained floors, roofs and beams, the reinforcing steel temperature rises to 1100°F.

Though the complete requirements of ASTM E 119 and the conditions of acceptance are much too detailed for inclusion in this chapter, experience shows that concrete floor/roof assemblies and walls usually fail by heat transmission (item 1); and columns and beams by failure to sustain the applied loads (item 3), or by beam reinforcement failing to meet the temperature criterion (item 4).

Fire rating requirements for structural assemblies may differ from code to code; therefore, it is advisable that the designer take into account the building regulations having jurisdiction over the construction rather than relying on general perceptions of accepted practice.

10.4 DESIGN CONSIDERATIONS FOR FIRE RESISTANCE

10.4.1 Properties of Concrete

Concrete is the most highly fire-resistive structural material used in construction. Nonetheless, the properties of concrete and reinforcing steel change significantly at high temperatures. Strength and the modulus of elasticity are reduced, the coefficient of expansion increases, and creep and stress relaxations are considerably higher.

Concrete strength, the main concern in uncontrolled fires, remains comparatively stable at temperatures ranging up to 900°F for some concretes and 1200°F for others. Siliceous aggregate concrete, for instance, will generally maintain its original compressive strength at temperatures up to 900°F, but can lose nearly 50% of its original strength when the concrete reaches a temperature of about 1200°F. On the other hand, carbonate aggregate and sand-lightweight concretes behave more favorably in fire, their compressive strengths remaining relatively high at temperatures up to 1400°F, and diminishing rapidly thereafter. These data reflect fire test results of specimens loaded in compression to 40% of their original compressive strength.

The temperatures stated above are the internal temperatures of the concrete and are not to be confused with the heat intensity of the exposing fire. As an example, in testing a solid carbonate aggregate slab, the ASTM standard fire exposure after 1 hour will be 1700°F, while the temperatures within the test specimen will vary throughout the section: about 1225°F at 1/4 in. from the exposed surface, 950°F at 3/4 in., 800°F at 1 in., and 600°F at 1-1/2 in.; all within the limit of strength stability.

It is to be realized that the strength loss in concrete subjected to intense fire is not uniform throughout the structural member because of the time lag required for heat penetration and the resulting temperature gradients occurring across the concrete section. The total residual strength in the member will usually provide an acceptable margin of safety.

This characteristic is even more evident in massive concrete building components such as columns and girders. Beams of normal weight concrete exposed to an ASTM E 119 fire test will, at two hours when the exposing fire is at 1850°F, have internal temperatures of about 1200°F at 1 in. inside the beam faces and less than 1000°F at 2 in. Obviously, the dimensionally larger concrete sections found in main framing systems will suffer far less net loss of strength (measured as a percentage of total cross-sectional area) than will lighter assemblies.

Because of the variable complexities and the unknowns of dealing with the structural behavior of buildings under fire as total multidimensional systems, building codes continue to specify minimum acceptable levels of fire endurance on a component by component basis—roof/floor assemblies, walls, columns, etc. It is known, for instance, that in a multi-bay building, an interior bay of a cast-in-place concrete floor system subjected to fire will be restrained in its thermal expansion by the unheated surrounding construction. Such restraint increases the structural fire endurance of the exposed assembly by placing the heated concrete in compression. The restraining forces developed are large and, under elastic behavior, would cause the concrete to exceed its original compressive strength were it not for stress relaxations that occur at high temperatures. According to information provided in Appendix X3 of ASTM E 119, cast-in-place beams and slab systems are generally considered restrained (see Table 10-5 in Section 10.4.3).

In addition to the minimum acceptable limits given in the building codes, the use of calculation methods for determining fire endurance are also accepted, depending on the local code adoptions (see Reference 10.1).

10.4.2 Thickness Requirements

Test findings show that fire resistance in concrete structures will vary in relation to the type of aggregate used. The differences are shown in Table 10-1 and 10-2.

Table 10-1 Minimum Thickness for Floor and Roof Slabs and Cast-in-Place Walls, in. (Load-Bearing and Nonload-Bearing)

	Fire resistance rating				
Concrete type	1 hr.	1½ hr.	2 hr.	3 hr.	4 hr.
Siliceous aggregate	3.5	4.3	5.0	6.2	7.0
Carbonate aggregate	3.2	4.0	4.6	5.7	6.6
Sand-lightweight	2.7	3.3	3.8	4.6	5.4
Lightweight	2.5	3.1	3.6	4.4	5.1

Table 10-2 Minimum Concrete Column Dimensions, in.

	Fire resistance rating				
Concrete type	1 hr.	1½ hr.	2 hr.	3 hr.	4 hr.
Siliceous aggregate	8	8	10	12	14
Carbonate aggregate	8	8	10	12	12
Sand-lightweight	8	8	9	10.5	12

In studying the tables above it is readily apparent that there may be economic benefits to be gained from the selection of the type of concrete to be used in construction. The designer is encouraged to evaluate the alternatives.

10.4.3 Cover Requirements

Another factor to be considered in complying with fire-resistive requirements is the minimum thickness of concrete cover for the reinforcement. The concrete protection specified in ACI 318 for cast-in-place concrete will generally equal or exceed the minimum cover requirements shown in the following tables, but there are a few exceptions at the higher fire ratings and these should be noted.

The minimum thickness of concrete cover to the positive moment reinforcement is given in Table 10-3 for one-way or two-way slabs with flat undersurfaces.

The minimum thickness of concrete cover to the positive moment reinforcement (bottom steel) in reinforced concrete beams is shown in Table 10-4.

Table 10-3 Minimum Cover for Reinforced Concrete Floor or Roof Slabs, in.

	Restrained Slabs*				Unrestrained Slabs*			
	Fire resistance rating				Fire resistance rating			
Concrete type	1 hr.	1 ½ hr.	2 hr.	3 hr.	1 hr.	1 ½ hr.	2 hr.	3 hr.
Siliceous aggregate	¾	¾	¾	¾	¾	¾	1	1¼
Carbonate aggregate	¾	¾	¾	¾	¾	¾	¾	1¼
Sand-lightweight or lightweight	¾	¾	¾	¾	¾	¾	¾	1¼

*See Table 10-5

Table 10-4 Minimum Cover to Main Reinforcing Bars in Reinforced Concrete Beams, in.
(Applicable to All Types of Structural Concrete)

		Fire resistance rating				
Restrained or unrestrained*	Beam width, in.**	1 hr.	1½ hr.	2 hr.	3 hr.	4 hr.
Restrained	5	¾	¾	¾	1	1¼
Restrained	7	¾	¾	¾	¾	¾
Restrained	≥ 10	¾	¾	¾	¾	¾
Unrestrained	5	¾	1	1¼	-	-
Unrestrained	7	¾	¾	¾	1¾	3
Unrestrained	≥ 10	¾	¾	¾	1	1¾

*See Table 10-5
**For beam widths between the tabulated values, the minimum cover can be determined by interpolation.

The minimum cover to main longitudinal reinforcement in columns is shown in Table 10-6.

Table 10-5 Construction Classification, Restrained and Unrestrained
(Table X3.1 from ASTM E 119)

Construction	Classification
I. Wall bearing	
Single span and simply supported end spans of multiple bays:[A]	
(1) Open-web steel joists or steel beams, supporting concrete slab, precast units, or metal decking	unrestrained
(2) Concrete slabs, precast units, or metal decking	unrestrained
Interior spans of multiple bays:	
(1) Open-web steel joists, steel beams or metal decking, supporting continuous concrete slab	restrained
(2) Open-web steel joists or steel beams, supporting precast units or metal decking	unrestrained
(3) Cast-in-place concrete slab systems	restrained
(4) Precast concrete where the potential thermal expansion is resisted by adjacent construction[B]	restrained
II. Steel framing:	
(1) Steel beams welded, riveted, or bolted to the framing members	restrained
(2) All types of cast-in-place floor and roof systems (such as beam-and-slabs, flat slabs, pan joists, and waffle slabs)where the floor or roof system is secured to the framing members	restrained
(3) All types of prefabricated floor or roof systems where the structural members are secured to the framing members and the potential thermal expansion of the floor or roof system is resisted by the framing system or the adjoining floor or roof construction[B]	restrained
III. Concrete framing:	
(1) Beams securely fastened to the framing members	restrained
(2) All types of cast-in-place floor or roof systems (such as beam-and-slabs, pan joists, and waffle slabs) where the floor system is cast with the framing members	restrained
(3) Interior and exterior spans of precast systems with cast-in-place joints resulting in restraint equivalent to that which would exist in condition III (1)	restrained
(4) All types of prefabricated floor or roof systems where the structural members are secured to such systems and the potential thermal expansion of the floor or roof systems is resisted by the framing system or the adjoining floor or roof construction[B]	restrained
IV. Wood construction:	
All Types	unrestrained

[A]Floor and roof systems can be considered restrained when they are tied into walls with or without tie beams, the walls being designed and detailed to resist thermal thrust from the floor or roof system.

[B]For example, resistance to potential thermal expansion is considered to be achieved when:

(1) Continuous structural concrete topping is used,
(2) The space between the ends of precast units or between the ends of units and the vertical face of supports is filled with concrete or mortar, or
(3) The space between the ends of precast units and the vertical faces of supports, or between the ends of solid or hollow core slab units does not exceed 0.25% of the length for normal weight concrete members or 0.1% of the length for structural lightweight concrete members.

Table 10-6 Minimum Cover for Reinforced Concrete Columns, in.

	Fire resistance rating				
Concrete type	1 hr.	1½ hr.	2 hr.	3 hr.	4 hr.
Siliceous aggregate	1½	1½	1½	1½	2
Carbonate aggregate	1½	1½	1½	1½	1½
Sand-lightweight	1½	1½	1½	1½	1½

10.5 MULTICOURSE FLOORS AND ROOFS

Symbols: Carb = carbonate aggregate concrete
Sil = siliceous aggregate concrete
SLW = sand-lightweight concrete

10.5.1 Two-Course Concrete Floors

Figure 10-2 gives information on the fire resistance ratings of floors that consist of a base slab of concrete with a topping (overlay) of a different type of concrete.

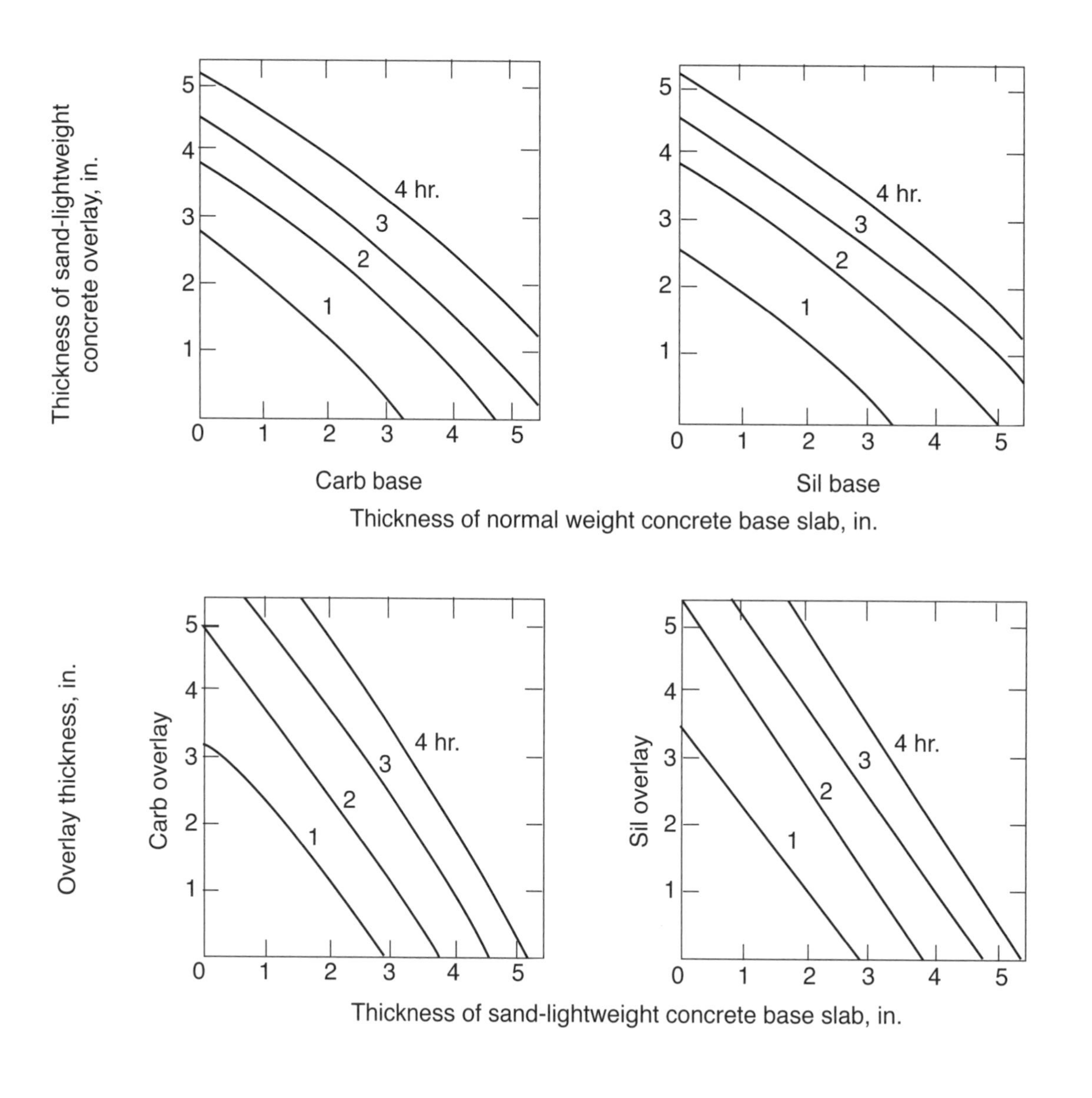

Figure 10-2 Fire Resistance Ratings for Two-Course Floor Slabs

10.5.2 Two-Course Concrete Roofs

Figure 10-3 gives information on the fire resistance ratings of roofs that consist of a base slab of concrete with a topping (overlay) of an insulating concrete; the topping does not include built-up roofing. For the transfer of heat, three-ply built-up roofing contributes 10 minutes to the fire resistance ratings; thus, 10 minutes may be added to the values shown in the figure.

10.5.3 Concrete Roofs with Other Insulating Materials

Figure 10-4 gives information on the fire resistance ratings of roofs that consist of a base slab of concrete with an insulating board overlay; the overlay includes standard 3-ply built-up roofing.

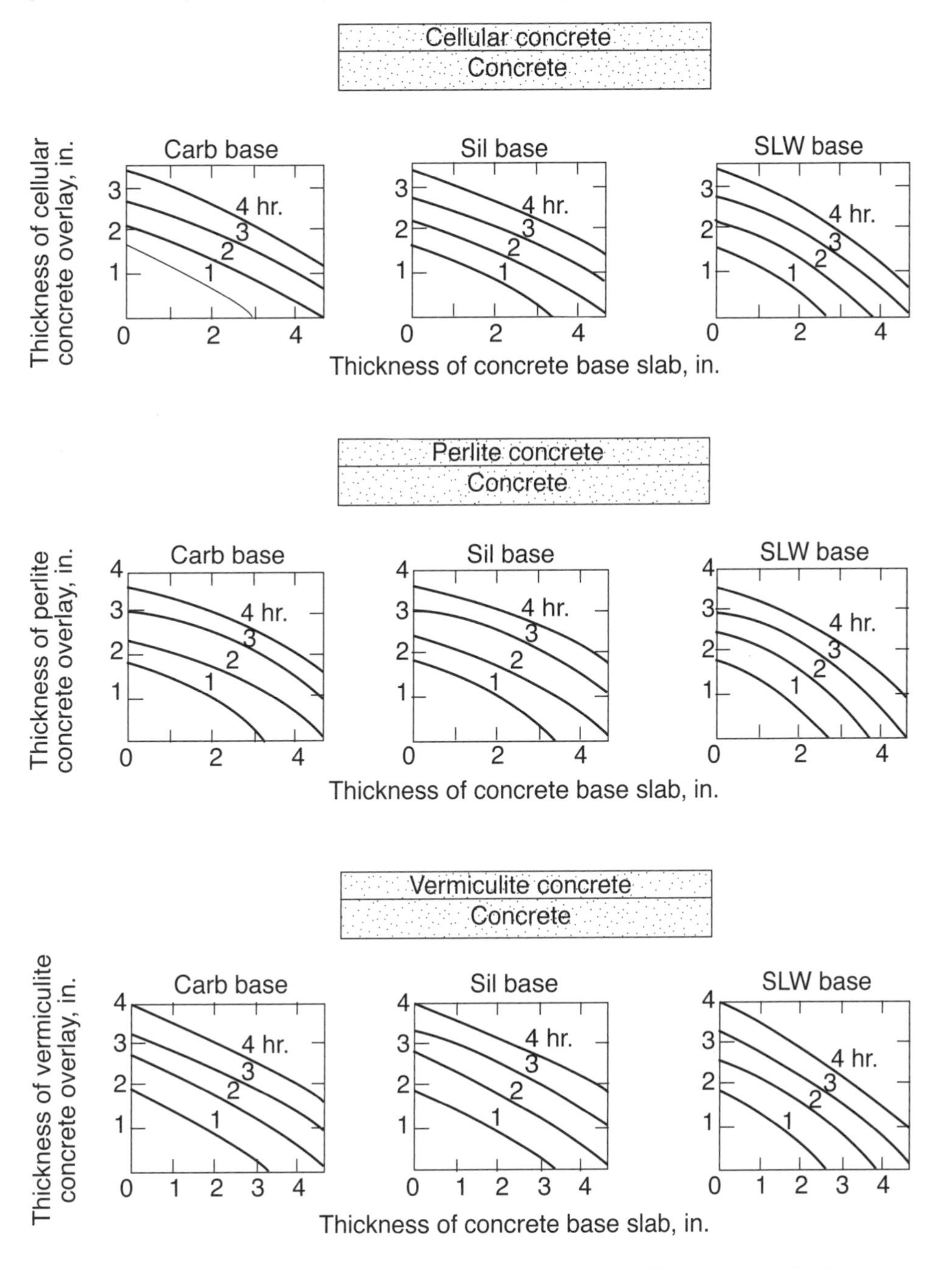

Figure 10-3 Fire Resistance Ratings for Two-Course Roof Slabs

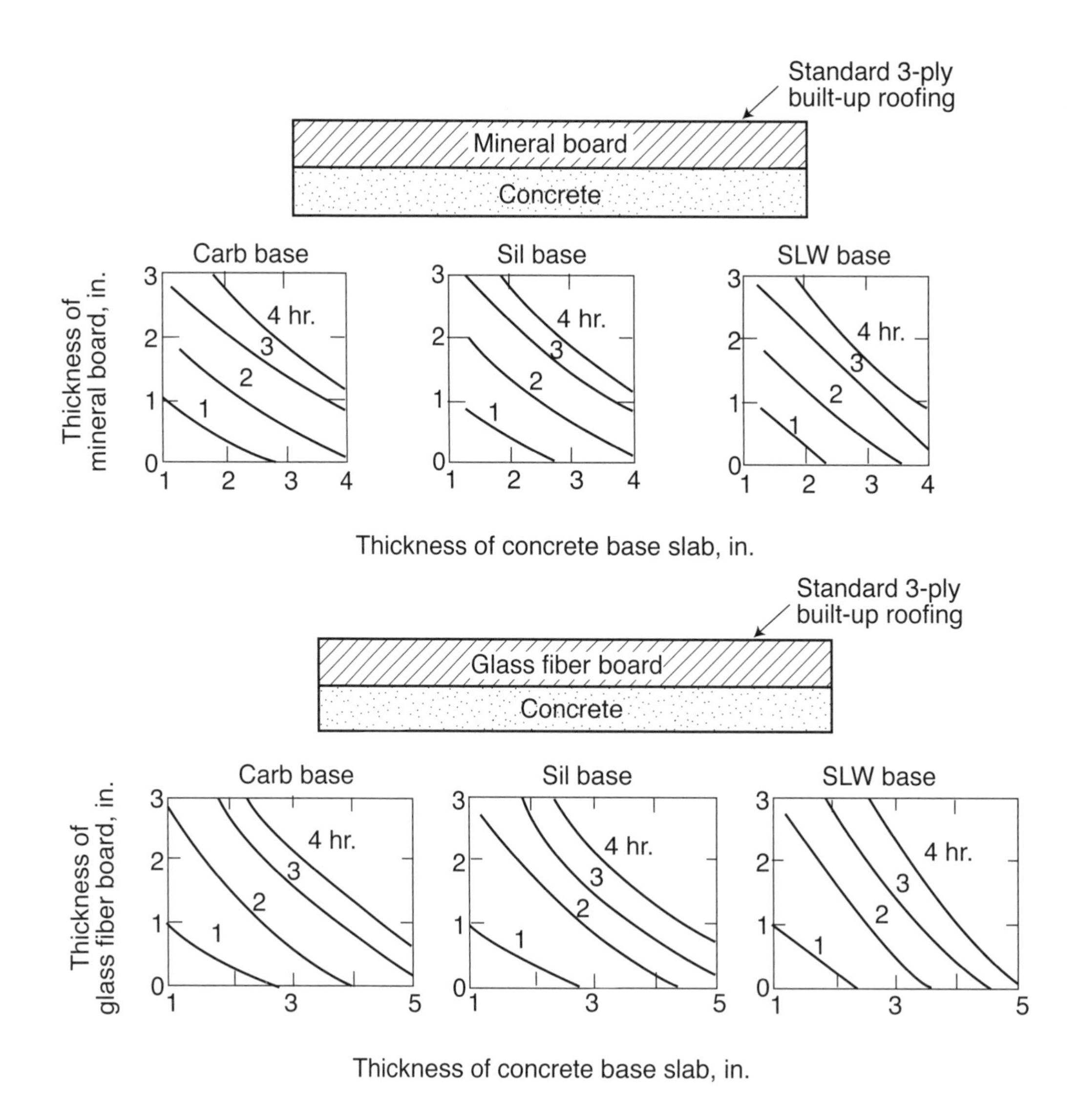

Figure 10-4 Fire Resistance Ratings for Roof Slabs with Insulating Overlays and Standard 3-Ply Built-Up Roofing

Reference

10.1 *Reinforced Concrete Fire Resistance*, Concrete Reinforcing Steel Institute, Schaumburg, Illinois, 256 pp.

Chapter 11

Design Considerations for Earthquake Forces

11.1 INTRODUCTION

The objective of this chapter is to introduce the basic seismic design provisions of the International Building Code (2003 IBC) that apply to the structures intended to be within the scope of this publication where seismic forces are resisted entirely by moment frame or shearwalls. The material in this chapter does not cover structures with plan or vertical irregularities (1616.5) and assumes rigid diaphragm typical for cast-in-place concrete floor systems. Examples of structural systems with plan or vertical irregularities are shown in Figures 11-1 and 11-2 respectively. For comprehensive background on seismic design and detailing requirements for all cases, refer to References 11.1, 11.2 and 11.3.

The 2003 IBC refers or adopts with modification provisions from ASCE 7-02[2]. Reference to both documents will be made throughout this chapter as applicable. The 2003 IBC contains contour maps for the maximum considered earthquake (MCE) spectral response accelerations (5 percent of critical damping) at periods of 0.2 second (S_s) and 1.0 second (S_1). The mapped values of S_s and S_1 are based on Site Class B (see below for site class definition). The earthquake effects that buildings and structures are proportioned to resist are based on what is called Design Basis Earthquake (DBE). The design seismic forces prescribed in the 2003 IBC are generally less than the elastic inertia forces induced by the DBE (Reference 11.3). Structures subjected to seismic forces must resist collapse when subjected to several cycles of loading in the inelastic range. Therefore, critical regions of certain members must be designed and detailed to safely undergo sufficient inelastic deformability. The building code contains structural detailing requirements to enable the structure and members to dissipate seismic energy by inelastic deformation in order to prevent collapse.

11.2 SEISMIC DESIGN CATEGORY (SDC)

The 2003 IBC requires that a Seismic Design Category (SDC) (A, B, C, D, E, or F) be assigned to each structure (1616.3). A SDC is used to determine:

(1) Permissible structural systems

(2) Level of detailing

(3) Limitations on height and irregularity

(4) The components of the structure that must be designed and detailed for seismic resistance

(5) Structures exempt from seismic design requirements

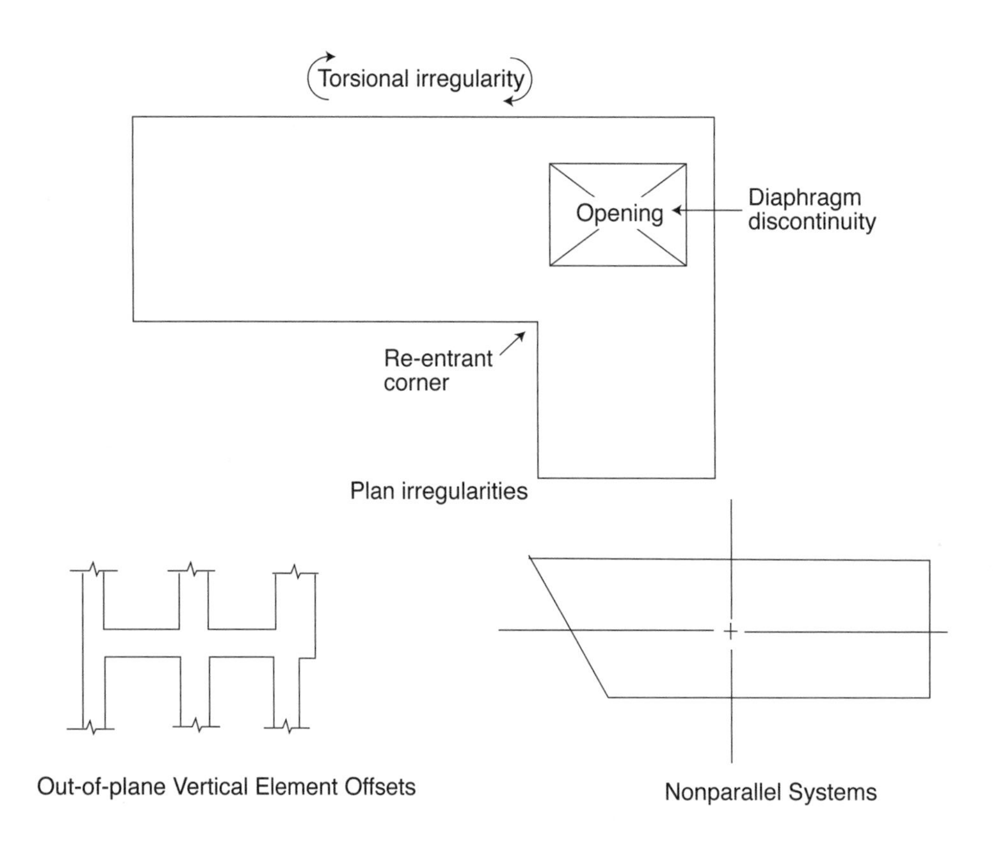

Figure 11-1 Structures with Plan Irregularities

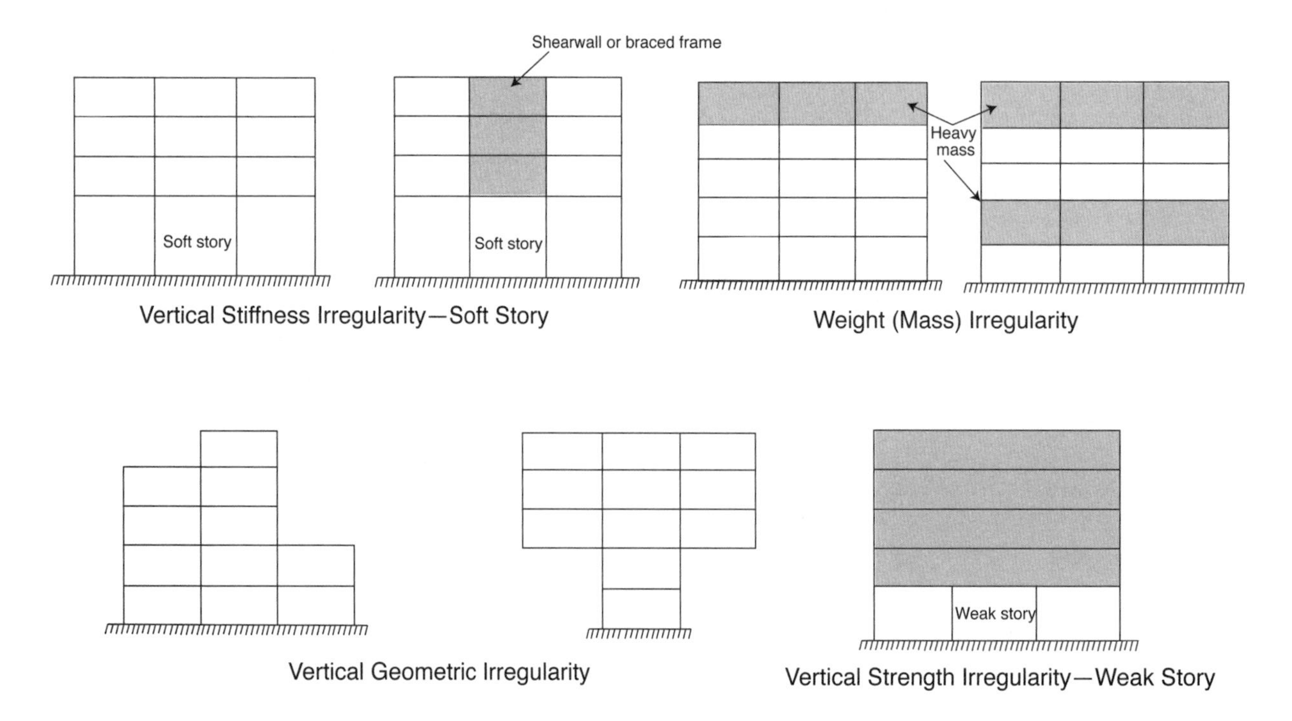

Figure 11-2 Structures with Vertical Irregularities

(6) The types of lateral force analysis that must be performed

The SDC is a function of two parameters 1) the Seismic Use Group I, II, or III (1616.2), which depends on the occupancy and use of the building, and 2) the design spectral response acceleration. The following procedure summarizes the determination of the SDC:

(1) From 2003 IBC Figures 1615(1) through 1615(10), determine the maximum considered earthquake spectral response acceleration at short period (0.2 second) S_s and at one second and S_1. These values can alternatively be obtained from the software prepared by the United States Geological Survey (USGS) by inputting the zip code of the building location. This software is distributed with the 2003 IBC.

(2) Depending on the soil properties at the site, determine the site class definition (A, B, C, D, E, or F), from 1615.1.1. In the absence of sufficient details on site soil properties it is allowed to assume site class D unless the building official specifies higher site class (E or F).

(3) Based on the magnitude of the maximum considered earthquake spectral response accelerations (S_s and S_1) from step 1 and the site class from step 2 determine the site coefficients F_a and F_v from Tables 1615.1.2(1) and 1615.1.2(2).

(4) Calculate the adjusted (for the site class effects) maximum considered earthquake spectral response acceleration S_{MS} and S_{M1} as follows:

$$S_{MS} = F_a S_s$$
$$S_{M1} = F_v S_1$$

(5) Calculate the five percent damped design spectral response acceleration at short period S_{DS}, and at one second period S_{D1} (1615.1.3):

$$S_{DS} = 2/3\ S_{MS}$$
$$S_{D1} = 2/3\ S_{M1}$$

(6) SDC (A, B, C, D, E, or F) is determined based on the Seismic Use Group (I, II, or III) as defined in 1616.2 and the short period design spectral response accelerations S_{DS}. Another seismic design category is determined based on the seismic use group and S_{D1} (2003 IBC Table 1616.3(1) and Table 1616.3(2)). The more critical of the two categories shall be used.

It is important to make a distinction between the Seismic Use Group (I, II, and III) and the classification of the buildings based on occupancy categories (I, II, III, and IV).

The 2003 IBC allows the SDC to be determined based on S_{DS} alone (from table 1616.3(1)) provided that all the following conditions apply (1616.3):

a) The approximate fundamental period of the structure, T (see Section 11.6.1 for T calculations), in each of the two orthogonal directions is less than 0.8 T_s where $T_s = S_{D1}/S_{DS}$.

b) The seismic response coefficient, C_s (see Section 11.6.1) is determined based on S_{DS}.

c) The diaphragms are rigid as defined in 2003 IBC Section 1602.

Applicable to buildings of moderate height, this exception can make a substantial difference in the SDC of the building which impacts the type of detailing required and ultimately plays a key role in the building economy. The utilization of this exception is illustrated in the examples given in this chapter.

11.3 REINFORCED CONCRETE EARTHQUAKE-RESISTING STRUCTURAL SYSTEMS

The basic reinforced concrete seismic force resisting systems are shown in Figure 11-3. The permitted structural system, height limitations, and reinforcement detailing depend on the determined SDC. A brief description of each system follows:

(1) **Bearing Wall System:** Load bearing walls provide support for most or all of the gravity loads. Resistance to lateral forces is provided by the same walls acting as shearwalls.

(2) **Building Frame System:** A structural system with essentially complete space frame provides support for the gravity loads. Resistance to lateral forces is provided by shearwalls.

(3) **Moment-Resisting Frame System:** An essentially complete space frame provides support for the gravity loads and resistance to lateral loads at the same time (simultaneously).

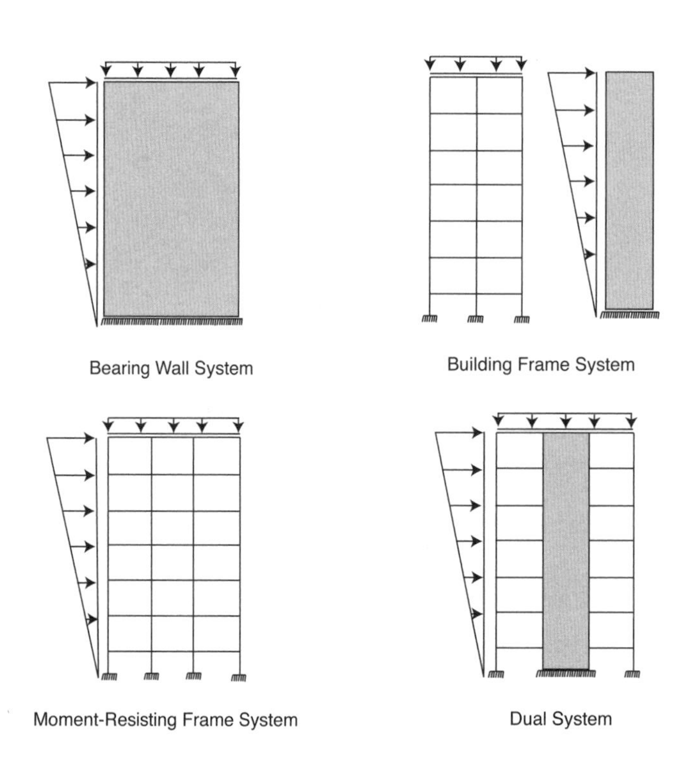

Figure 11-3 Earthquake-Resisting Structural Systems of Reinforced Concrete

(4) **Dual System:** A structural system with space frames and walls to provide support for the gravity loads. Resistance to earthquake loads is provided by shearwalls and moment-resisting frames. The shearwalls and moment-resisting frames are designed to resist the design base shear in proportion to their relative rigidities. The moment resisting frame must be capable of resisting at least 25 percent of the design base shear (1617.6.1).

Based on 2003 IBC 1617.6 Table 11-1 shows the permitted system and height limitation for each SDC. The values of R and C_d are taken from ASCE Table 9.5.2.2. The table could be used to select the required seismic-force-resisting systems for a specific SDC. The building frame system and the moment-resisting frame system are commonly used and are suitable for buildings of regular shape and moderate height. Bearing wall and dual system buildings are, therefore, not discussed in this publication.

11.4 STRUCTURES EXEMPT FROM SEISMIC DESIGN REQUIREMENTS

The 2003 IBC allows certain structures to be exempt from the seismic design requirements (1614.1). These structures include: detached one- and two-family dwellings with SDC A, B, or C or located where the mapped short-period spectral response acceleration S_s is less than 0.4g. Agricultural storage structures intended only for incidental human occupation are also included in the exemption.

Also structures located at the following locations are assigned SDC A:

a) Where $S_S \leq 0.15g$ and $S_1 \leq 0.04g$

b) Where $S_{DS} \leq 0.167g$ and $S_{D1} \leq 0.067g$

11.5 EARTHQUAKE FORCES

The IBC code requires a structural analysis procedure to determine the magnitude and distribution of earthquake forces (1616.6). The permitted analysis procedure for certain structures depends on the seismic design category and the structure characteristics (seismic use group, structural irregularity, height, and location). The 2003 IBC refers to ASCE 7-02[11.2] (Minimum Design Loads for Buildings and other Structures) for the permitted analysis procedures for different structures (ASCE Table 9.5.2.5.1). Therefore, the provisions of ASCE 7-02 will be used from here on. Six analytical procedures are introduced in ASCE 7-02 (index force, simplified analysis, equivalent lateral force, modal response spectrum, linear response history, and nonlinear response history). The index force procedure is applied for structures with SDC A; the lateral force considered at each floor level for this procedure is equal to one percent (1%) of the gravity (W) load at that floor level. The simplified analysis method is used for light-framed construction not exceeding three stories or building with any construction with flexible diaphragm not exceeding two stories, this procedure is not covered in this publication. For buildings within the scope of this publication, the equivalent lateral forces procedure (ASCE 9.5.5) provides the most suitable approach. The following section presents the equivalent lateral force method. For other analytical procedures References 11.1 and 11.2 should be consulted.

Table 11-1 Permitted Building Systems for Different SDC (ASCE Table 9.5.2.2)

Basic Seismic Force Resisting System			SDC						R	C_d
			A	B	C	D	E	F		
Bearing Wall	Special Reinforced Concrete Shearwall					160 ft	160 ft	100 ft	5	5
	Ordinary Reinforced Concrete Shearwall								4	4
Building Frame	Special Reinforced Concrete Shearwall					160 ft	160 ft	100 ft	6	5
	Ordinary Reinforced Concrete Shearwall								5	4.5
Moment Resisting Frame	Special Frame								8	5.5
	Intermediate Frame								5	4.5
	Ordinary Frame								3	2.5
Dual System	With Special Moment Frame	Special Shearwall							8	6.5
		Ordinary Shearwall							7	6
	With Intermediate Moment Frame	Special Shearwall				160 ft	100 ft	100 ft	6	5
		Ordinary Shearwall							5.5	4.5

Permitted with the indicated height limit (if any)

Not permitted

11.6 EQUIVALENT LATERAL FORCE PROCEDURE

For reinforced concrete structures with or without irregularities and assigned to SDC A, B or C, the equivalent lateral force procedure can be used. Also the equivalent lateral force method can be used for regular structures assigned to SDC D, E, or F provided that the fundamental period of the structure $T < 3.5\ T_s$ where $T_s = S_{D1}/S_{DS}$ (see below for calculation of T). Limitations on the applicability of this method to irregular structures assigned to SDC D, E, or F are given in ASCE (9.5.2.5.1).

11.6.1 Design Base Shear

The seismic base shear, V, in a given direction is a fraction of the dead weight of the structure. V is calculated from:

$$V = C_S W$$

where:

C_S = seismic response coefficient
W = the effective seismic weight of the structure which includes the total dead load and the loads listed below (ASCE 9.5.3):

(1) In areas used for storage, a minimum of 25% of the floor live load (floor live load in public garages and open parking structures need not be included)

(2) Where an allowance for partition load is included in the floor load design, the actual partition weight or a minimum weight of 10 psf of floor area, whichever is greater

(3) Total operating weight of permanent equipment

(4) 20% of flat roof snow load where flat roof snow load exceeds 30 psf

The seismic response coefficient is calculated as follows:

$$C_s = \frac{S_{D1}}{\left(\frac{R}{I_E}\right)T}$$

The value of C_S need not exceed:

$$C_s = \frac{S_{DS}}{\left(\frac{R}{I_E}\right)}$$

Also the value of C_S should not be taken less than:

$$C_s = 0.004 S_{DS} I_E$$

For structures assigned to SDC E or F, C_S shall not be taken less than:

$$C_s = \frac{0.5 S_1}{R / I_E}$$

where:

R = response modification factor depending on the basic seismic-force-resisting system, from ASCE Table 9.5.2.2. Table 11-1 lists the values for R for different reinforced concrete seismic force resisting systems.

I_E = occupancy importance factor depending on the nature of occupancy, from ASCE 9.1.4
T = the fundamental period of the structure in seconds. T can be calculated from the following equation:

$$T = C_t h_n^x$$

where:

C_t = building period coefficient
C_t = 0.016 for concrete moment resisting frames
C_t = 0.02 for other concrete systems
h_n = building height in feet
x = 0.9 for concrete moment resisting frames
x = 0.75 for other concrete systems

Figure 11-4 shows the fundamental period for concrete building systems for different heights.

For concrete moment resisting frame buildings with less than 12 floors and story height of 10 feet minimum, T calculations can be further simplified to T = 0.1 N where N is the number of stories

For concrete shearwall structures the fundamental period can be approximated using ASCE Eq. 9.5.5.3 2-2

11.6.2 Vertical Distribution of Seismic Forces

In the equivalent lateral force method the design base shear V is distributed at different floor levels as follows (ASCE 9.5.5.4):

$$F_x = C_{vx} V$$

$$C_{vx} = \frac{w_x h_x^k}{\sum_{i=1}^{n} w_i h_i^k}$$

where:

F_x = lateral force at floor level x
C_{vx}= vertical distribution factor for level x
k = a distribution exponent related to the building period T
= 1 for T ≤ 0.5 second
= 2 for T ≥ 2.5 second
= 2 for 0.5 second < T < 2.5 second, or linear interpolation between 1 and 2

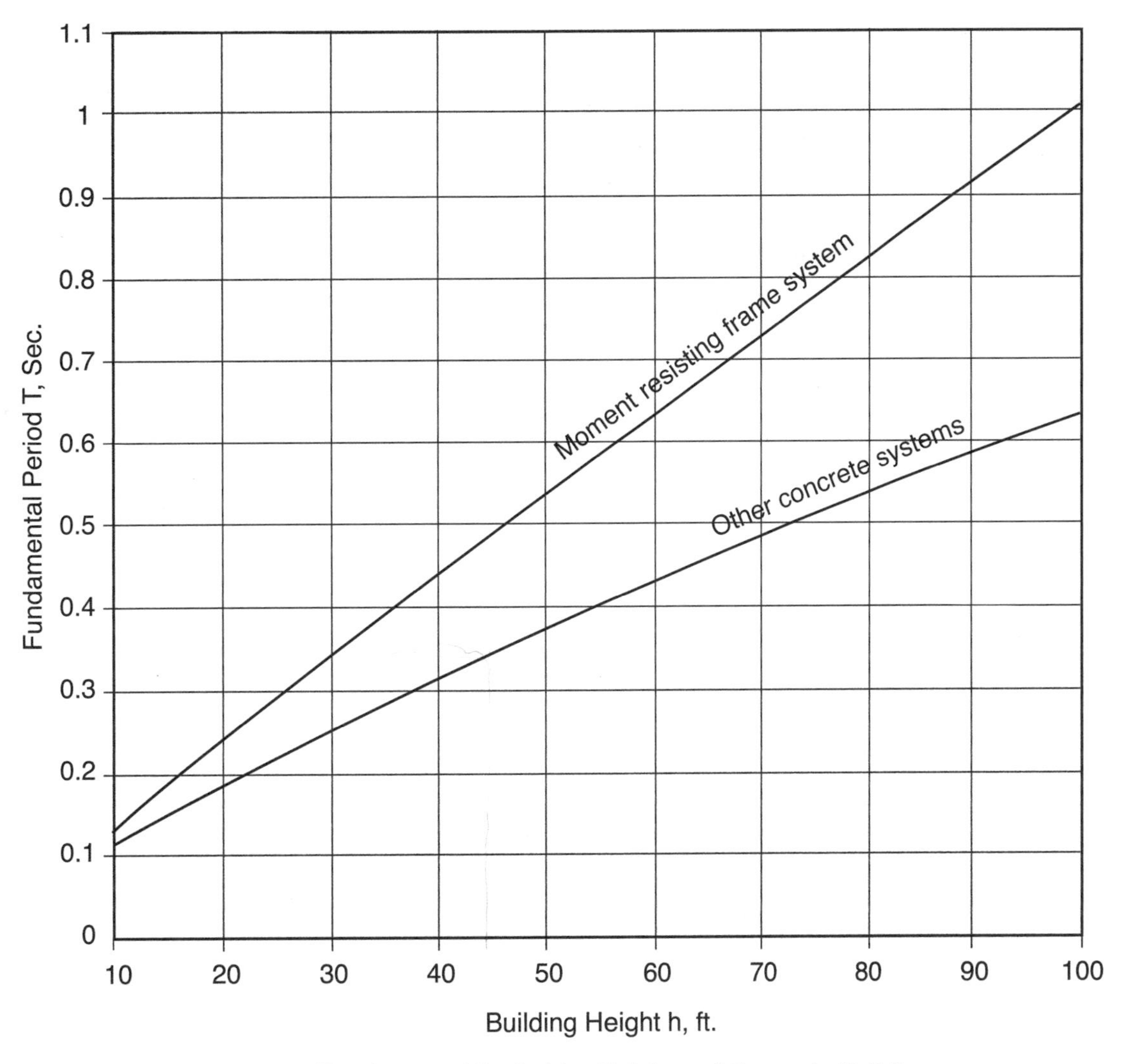

Figure 11-4 Fundamental Period for Reinforced Concrete Buildings

h_i and h_x = the height in feet from the base level to i or x
w_i and w_x = the portion of W assigned to level i or x

For most of the structures covered in this publication, the fundamental period T is less than 0.5 second. For this case the above equation simplifies to:

$$C_{vx} = \frac{w_x h_x}{\sum_{i=1}^{n} w_i h_i}$$

11.6.2.1 Distribution of Seismic Forces to Vertical Elements of the Lateral Force Resisting System

The seismic design story shear V_x in any story x is the sum of the lateral forces acting at that story in addition to the lateral forces acting on all the floor levels above (ASCE Eq. 9.5.5.5):

$$V_x = \sum_{i=x}^{n} F_i$$

Figure 11-5 shows the vertical distribution of the seismic force F_x and the story shear V_x in buildings with $T \leq 0.5$.

The lateral shear force V_x is typically transferred to the lateral force resisting elements (shearwalls or frames) by the roof and floors acting as diaphragms. At each level the floor diaphragm distributes the lateral forces from above to the shearwalls and frames below. The distribution of the lateral force to the lateral force resisting elements (shearwalls or frames) depends on the relative rigidity of the diaphragm and the lateral force resisting elements. For analysis purposes the diaphragms are typically classified as rigid, semi-rigid, and flexible. Cast-in-place concrete floor systems are considered and modeled as rigid diaphragms. In rigid diaphragms, the lateral force V_x is distributed to the shearwalls and frames in proportion to their relative stiffnesses.

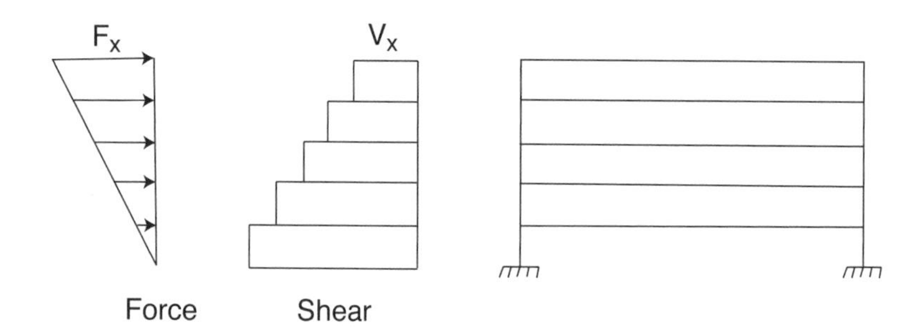

Figure 11-5 Vertical Distribution of Seismic Base Shear in Low-rise Buildings ($T \leq 0.5$ sec)

For building frame system (consisting of shearwalls and frames) the shearwalls are designed to resist the entire story shear V_x. For SDC D, E, and F the frames must be designed to resist the effects caused by the lateral deflections, since they are connected to the walls through the floor slab (1617.6.2.4).

The seismic design story shear V_x is considered to act at the center of mass of the story. The center of mass is the location where the mass of an entire story may be assumed to be concentrated. The location of the center of mass can be determined by taking the moment of the components weights about two orthogonal axes x and y. The distribution of V_x to the walls and frames depends on the relative location of the center of mass with respect to the location of the center of rigidity of the floor. The center of rigidity is the point where the equivalent lateral story stiffness (for frames or walls) may be considered to be located.

When the centers of mass and rigidity coincide, the lateral load resisting elements in the story displace an equal distance horizontally (translate) and the story shear V_x is distributed to the lateral force resisting elements in proportion to their relative stiffnesses. If the center of mass does not coincide with the center of rigidity the lateral load resisting elements in the story displace unequally (translate and rotate) and the story shear V_x is distributed to the lateral force resisting elements depending on their location and their relative stiffnesses. Figure 11-6 illustrates the two cases. ASCE 9.5.5.5.2 requires that the horizontal force acting on each element be increased due to accidental torsional moment. Such moment is the result of an assumed offset between the center of mass and the center of rigidity of the seismic force resisting elements. The assumed offset is 5 percent of the building plan dimension perpendicular to the force direction at each level and is intended to account for inaccuracies in building weight and stiffness estimate.

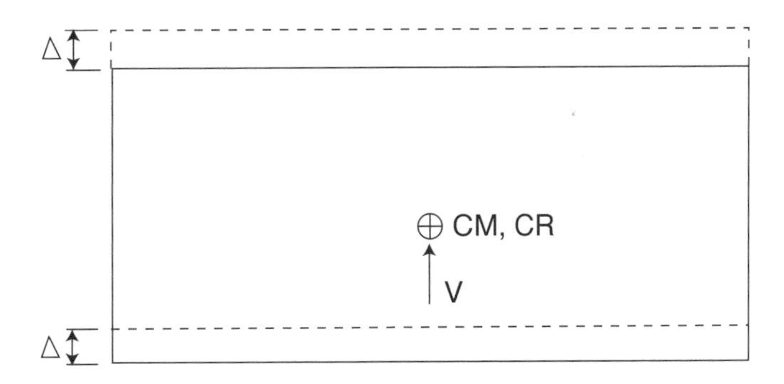

(a) Centers of mass and rigidity at the same point

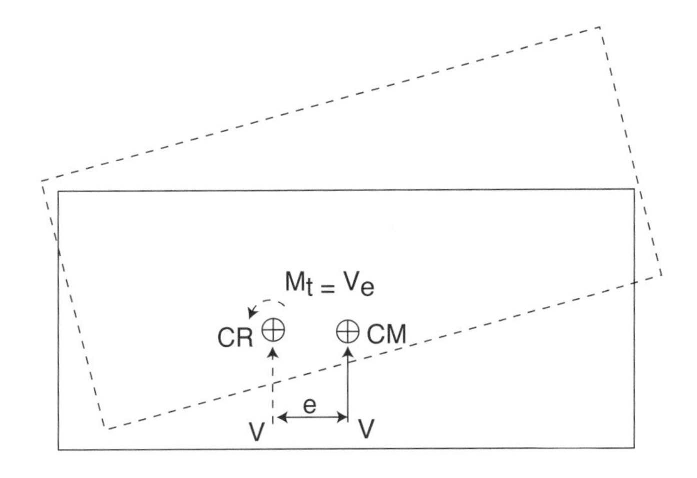

(b) Centers of mass and rigidity at distinct points

Figure 11-6 Floor Displacements

The location of the center of rigidity for a story can be determined by calculating the coordinates $\bar{x}_r$ and $\bar{y}_r$ from arbitrary located origin as follows:

$$\bar{x}_r = \frac{\Sigma (k_i)_y \, x_i}{\Sigma (k_i)_y}$$

$$\bar{y}_r = \frac{\Sigma (k_i)_x \, y_i}{\Sigma (k_i)_x}$$

where:

$(k_i)_y$ = lateral stiffness of lateral load resisting element (wall or frame) in the y-direction
$(k_i)_x$ = lateral stiffness of lateral load resisting element (wall or frame) in the x-direction
x_i and y_i = coordinates measured from arbitrary located origin to the centroid of lateral force resisting element i.

The distribution of the seismic shear force V_x (at floor x) to the different lateral force resisting element (shear-walls or frames) can be calculated from the following equations:

For force seismic force V_x applied in x direction

$$(V_i)_x = \frac{(k_i)_x}{\Sigma(k_i)_x} V_x + \frac{y_i(k_i)_x}{J_r} V_x e_y$$

For force seismic force V_x applied in y direction

$$(V_i)_y = \frac{(k_i)y}{\Sigma(k_i)_y} V_x + \frac{x_i(k_i)_y}{J_r} V_x e_x$$

where:

x_i , y_i = perpendicular distances from the lateral force resisting element i to the center of rigidity parallel to x and y axes respectively

J_r = rotational stiffness for all lateral force resisting elements in the story

$$= \Sigma x_i^2 (k_i)_y + \Sigma y_i^2 (k_i)_x$$

e_x , e_y = perpendicular distance from the center of mass to the center of rigidity or assumed eccentricities parallel to x and y axes respectively

$(k_i)_x$, $(k_i)_y$ = stiffnesses of lateral force resisting element in x and y direction, respectively

11.6.2.2 Direction of Seismic Load

To determine the seismic force effects on different structural members, a structural analysis for the building needs to be preformed. The design seismic forces should be applied in the direction which produces the most critical load effect in each structural component. Provisions on application of loading are given in ASCE 9.5.2.5.2 as a function of the SDC and the irregularity of the structure. Applications of these provisions for irregular structures are beyond the scope of this publication.

11.6.3 Load Combinations for Seismic Design

The seismic forces effect E in the load combinations introduced in Chapter 2 (Section 2.2.3) is the combined effect of horizontal and vertical earthquake induced forces and is calculated as follows:

For load combination ACI Equation 9-5:

$$E = \rho Q_E + 0.2\, S_{DS} D$$

For load combination ACI Equation 9-7:

$$E = \rho Q_E - 0.2\, S_{DS} D$$

where ρ is a redundancy coefficient = 1 for structures assigned SDC A, B, or C and is the largest ρ_x calculated at each story from the following formula for other SDC:

$$\rho_x = 2 - \frac{20}{r_{max_x} \sqrt{A_x}} \geq 1.0$$

where:

A_x = the floor area in square feet of the diaphragm level above the story

r_{max_x} is a factor calculated as follows:

(1) For moment resisting frame, r_{max_x} = maximum of the sum of the shears in any two adjacent columns divided by the story shear

(2) For shear walls in building frame system, r_{max_x} = (maximum wall shear $\times$ $10/\ell_w$)/total story shear

(3) For dual systems (not discussed in this publication) refer to ASCE 9.5.2.4

11.7 OVERTURNING

A building must be designed to resist the overturning effects caused by the seismic forces (ASCE 9.5.5.6). The overturning moment (M_x) at any level x is determined from the following equation:

$$M_x = \sum_{i=x}^{n} F_i(h_i - h_x)$$

where

F_i	= the portion of the base shear V, induced at level i.
h_i and h_x	= the height in feet from the base to level i or x.

11.8 STORY DRIFT

ASCE 7-02 specifies maximum allowable limits for story drift Δ resulting from the design earthquake (ASCE Table 9.5.2.8). Drift control is important to limit damage to partitions, shafts and stair enclosures, glass and other fragile nonstructural elements. The design story drift Δ is the difference of the lateral deflection δ_x (resulting from the design earthquake) at the floor level x at the top and bottom of story under consideration (Figure 11-7). The design story drift Δ_x is calculated as follows:

$$\Delta_x = \delta_x - \delta_{x-1}$$

where δ_x and δ_{x-1} are the magnified lateral displacement at the top and bottom of the story considered (Figure 11-7). The magnified lateral displacement is calculated from the following equation:

$$\delta_x = C_d \delta_{xe} / I_E$$

where

C_d	= deflection amplification factor presented in ASCE Table 9.5.2.2
δ_{xe}	= elastic lateral deflection (in.) due to the code prescribed seismic forces
I_E	= occupancy importance factor defined in 11.6.1

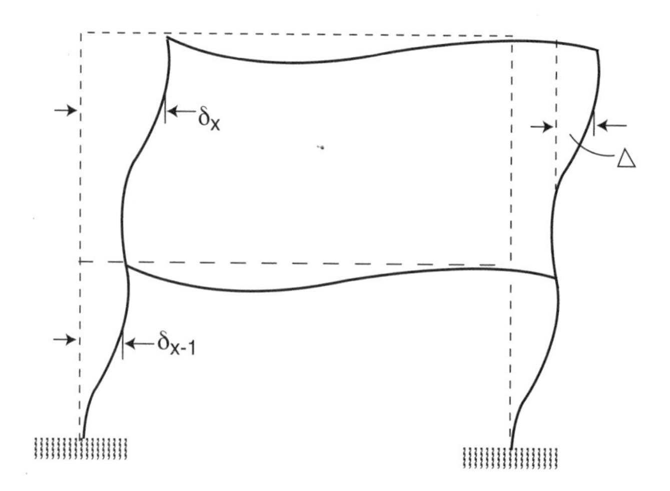

Figure 11-7 Interstory Drift, Δ

11.9 P-Δ EFFECT

Seismic forces cause the structure to deflect laterally. As a result, secondary moments are induced in the structural members due to the displaced gravity load as shown in Figure 11-8. This secondary moment effect is known as the P-Δ effect. P-Δ effects are not required to be considered if the stability index θ is equal to or less than 0.10. The stability index θ is calculated as follows (ASCE Eq. 9.5.5.7.2-1):

$$\theta = \frac{P_x \Delta}{V_x h_{sx} C_d}$$

where:

P_x = total unfactored vertical design load at and above level x (kips)
Δ = design story drift (inches) occurring simultaneously with V_x.
V_x = seismic shear force (kips) acting between level x and x-1
h_{sx} = story height (feet) below level x
C_d = deflection amplification factor defined in 11.8

The stability coefficient, θ, shall not exceed θ_{max} calculated as follows:

$$\theta_{max} = \frac{0.5}{\beta C_d} \leq 0.25$$

where:

β = the ratio of shear demand to shear capacity for the story between level x and x-1 which may be conservatively taken equal to 1.0.

For cases when the stability index θ is greater than 0.10 but less than or equal to θ_{max}, the drift and element forces shall be calculated including P-Δ effects. To obtain the story drift including the P-Δ effect, the design story drift shall be multiplied by $1.0/(1.0-\theta)$. Where θ is greater than θ_{max} the structure is potentially unstable and shall be redesigned to provide the needed stiffness to control drift.

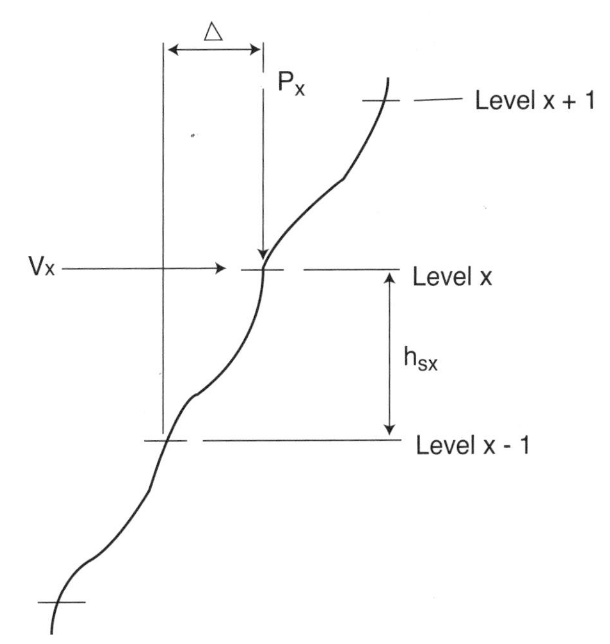

Figure 11-8 P-Δ Effects

11.10 DESIGN AND DETAILING REQUIREMENTS

The magnitudes of the design seismic forces determined by the analysis procedure (Equivalent Lateral Force Procedure) are reduced from the magnitudes of the actual forces that an elastic structure may experience during an earthquake by the response modification factor R. It is uneconomical and unnecessary to design a structure to respond elastically when subjected to the anticipated ground motion resulting from an earthquake. Traditionally, structures and their components are designed to yield under the code prescribed seismic forces. However, the yielding members are expected to undergo substantial additional deformation beyond the yield point while retaining strength capacity. This demonstrates a minimum level of ductility to prevent collapse that is suitable for the seismic design category assigned to the structure. In addition to proportioning the structural members' dimensions and reinforcement for the seismic force effects, structures must be properly detailed so that they are able to dissipate the earthquake energy through inelastic deformation and provide the required ductility. The IBC requires compliance with the requirements of the ACI 318 Code for design of reinforced concrete structures to achieve the required ductility. Design and detailing requirements for ordinary structure members are presented in Chapters 1 through 18 in the ACI 318 Building Code and are as given in this book. No additional requirements are required for ordinary shear walls and ordinary moment resisting frames. Additional requirements for the design and detailing for special reinforced concrete shear walls, intermediate moment frame, and special moment frame are presented in Chapter 21 in the ACI 318 Building Code. Table 11-2 presents a summary of ACI 318-02 sections need to be satisfied for different cast-in-place concrete frames and walls.

Table 11-2 ACI Detailing Requirements for Seismic Design

Structural element	ACI 318-02 requirements
Ordinary moment frame	Chapters 1 through 18
Intermediate moment frame	Chapter 1 through 18 Section 21.2.2.3 Section 21.12
Special moment frame	Chapters 1 through 18 Sections 21.2 through 21.5
Ordinary reinforced concrete wall	Chapters 1 through 18
Special reinforced concrete wall	Chapters 1 through 18 Section 21.2 Section 21.7

11.11 EXAMPLES

Two examples are provided to illustrate the application of the equivalent lateral force procedure. The two examples provide design calculations for shearwalls (Example 1) and typical frame flexural member (Example 2). For comprehensive coverage for other seismic design requirement see Reference 11.3.

11.11.1 Example 1 – Building # 2 Alternate (2) Shearwalls

Seismic design data

Assuming that the building is located in the Midwest with the maximum considered earthquake spectral response accelerations for short period (0.2 second) and one second period determined from IBC maps (2003 IBC Figures 1615(1) through 1615(10)) as follows:

$S_S = 0.26$ g
$S_1 = 0.12$ g

Soil site class definition is D

Seismic use group I

Determination of the Seismic Design Category SDC

(1) Based on the values of S_s and S_1 and site class D, determine the site coefficients F_a and F_v from 2003 IBC Tables 1615.2(1) and 1615.2(2). Notice that linear interpolation is performed.

$F_a = 1.59$
$F_v = 2.32$

(2) Calculate the adjusted maximum considered earthquake spectral response acceleration:

$S_{MS} = F_a\ S_s = 1.59 \times 0.26 = 0.41g$
$S_{M1} = F_v\ S_s = 2.32 \times 0.12 = 0.28g$

(3) Calculate the design earthquake spectral response accelerations:

$S_{DS} = 2/3\ S_{MS} = 2/3(0.41) = 0.28g$
$S_{D1} = 2/3\ S_{M1} = 2/3(0.28) = 0.19g$

(4) For $S_{DS} = 0.28g$ 2003 IBC Table 1616.3(1) shows that the SDC is B

For $S_{D1} = 0.19g$ 2003 IBC Table 1616.3(2) shows that the SDC is C

Use the more severe of the two:

SDC = C

Considering the exception in IBC 1616.3 (see Section 11.2) check the building SDC:

$T_s = S_{D1}/S_{DS} = 0.19/0.28 = 0.68$
Building T (see below) $= 0.45 < 0.8(T_s) = 0.8(0.68) = 0.54$ O.K.

The SDC can be determined based on Table 1616.3(1) and the building assigned SDC B. For illustrative purposes only this building example will continue using SDC C.

Table 11-1 shows that the building frame system with ordinary shear wall can be used for SDC C.

Fundamental Period

The approximate natural period of the structure can be calculated as follows:

$T = C_t h_n^x$

For building frame system $C_t = 0.02$ and $x = 0.75$

$T = 0.02(63)^{0.75} = 0.45$ second (also T can be obtained from Fig. 11-4)

Seismic Response Coefficient:

$$C_s = \frac{S_{D1}}{\left(\frac{R}{I_E}\right)T}$$

For building frame system with ordinary shear walls, the modification factor R = 5 Table 11-1 (ASCE Table 9.5.2.2). The occupancy importance factor $I_E = 1$ (ASCE 9.1.4).

$$C_s = \frac{0.19}{\left(\frac{5}{1}\right)0.45} = 0.083$$

C_s should not be taken less than:

$$C_s = 0.044 S_{DS} I_E = 0.044 \times 0.28 \times 1 = 0.012$$

C_s need not exceed:

$$C_s = \frac{S_{DS}}{\left(\frac{R}{I_E}\right)} = \frac{0.28}{\left(\frac{5}{1}\right)} = 0.055 \quad (5.5\%)$$

Use $C_s = 0.055$

Effective Seismic Weight

The effective seismic weight for this case includes the total dead load and the partition weight. The effective seismic loads for different floors are calculated as follows:

For the first floor:

Slab	= (121 ft)(61 ft)(0.142 ksf)	=1048 kips
Interior columns (8 columns)	= 8 (1.33 ft)(1.33 ft)(13.5 ft)(0.15 kcf)	= 28.7 kips
Exterior columns (12 columns)	= 12(1 ft)(1 ft)(13.5 ft)(0.15 kcf)	= 24.3 kips
Walls	= 2[(20.66 ft)(0.66 ft)+2(7.33 ft)(0.66 ft)](13.5 ft)(0.15 kcf)	= 94.4 kips
Effective seismic weight (1st floor)	= 1048 + 28.7 + 24.3 + 94.4	= 1195 kips

For second to fourth floor:

Slab	= (121 ft)(61 ft)(0.142 ksf)	=1048 kips
Interior columns (8 columns)	= 8 (1.33 ft)(1.33 ft)(12 ft)(0.15 kcf)	= 25.5 kips
Exterior columns (12 columns)	= 12(1 ft)(1 ft)(12 ft)(0.15 kcf)	= 21.6 kips
Walls	= 2[(20.66 ft)(0.66 ft)+2(7.33 ft)(0.66 ft)](12 ft)(0.15 kcf)	= 83.9 kips
Effective seismic weight	= 1048 + 25.5 + 21.6 + 83.9	= 1179 kips

For the roof:

Slab	= (121 ft)(61 ft)(0.122 ksf)	= 900 kips
Interior columns (8 columns)	= 8 (1.33 ft)(1.33 ft)(6 ft)(0.15 kcf)	= 12.8 kips
Exterior columns (12 columns)	= 12(1 ft)(1 ft)(6 ft)(0.15 kcf)	= 10.8 kips
Walls	= 2[(20.66 ft)(0.66 ft) + 2(7.33 ft)(0.66 ft)](6 ft)(0.15 kcf)]	= 42 kips
Effective seismic weight (roof)	= 900 + 12.8 + 10.8 + 42	= 966 kips

The total effective seismic weight:

$$W = 1195 + 3(1179) + 966 = 5698 \text{ kips}$$

Seismic Base Shear

$$V = C_s W$$

$$V = 0.055 \times 5698 = 313.4 \text{ kips}$$

Vertical Distribution of the Base Shear

The vertical distribution of the base shear V can be calculated using ASCE Eqs. 9.5.5.4-1 and 9.5.5.4-2 (Section 11.6.2). For T = 0.45 second the distribution exponent k = 1. The calculations for the lateral forces and story shear are shown in Table 11-3:

Table 11-3 Forces and Story Shear Calculations

Level	Height h_x (ft)	Story Weight w_x (kips)	$w_x h_x^k$	Lateral Force F_x (kips)	Story Shear V_x (kips)
5	63	966	60,858	88	88
4	51	1,179	60,129	87	176
3	39	1,179	45,981	67	242
2	27	1,179	31,833	46	288
1	15	1,195	17,925	26	314
	Σ	5,698	216,726	314	

For building frame system the lateral force are carried by the shear walls. Gravity loads (dead load and live load) are carried by the frames.

Distribution of the Seismic Forces to the Shear Walls in N-S Direction

For north south direction, consider that the lateral forces are carried by the two 248 in. by 8 in. shear walls. For the symmetrical floor layouts the center of rigidity coincides with the center of mass at the geometric center of the 120ft by 60ft floor area.

The shear force transmitted to each of the two shear walls must be increased due to code required displacement of the center of mass (point of application of lateral force) 5 percent of the building plan dimension perpendicular to the force direction. For the N-S direction the displacement considered = 0.05 × 120' = 6'. To calculate the force in the shear wall the equation $(V_i)_y = \frac{(k_i)_y}{\Sigma(k_i)_y} V_x + \frac{x_i(k_i)_y}{J_r} V_x e_x$ simplifies due to symmetry in geometry and element stiffness to $(V_i)_y = \frac{1}{2} V_x + \frac{1}{120} V_x e_x$ for this case.

The shear forces and moments for the shearwall at each floor level are shown in Table 11-4:

Table 11-4 Shear Forces and Moments at each shearwall (N-S)

Level	Height h_x (ft)	Lateral Force F_x (kips)	Story Shear V_x (kips)	$V_x e_x$ (kip-ft)	$(V_i)_y$ (kips)	Moment (kip-ft)
5	63	88	88	529.8	48.6	583
4	51	87	176	1053	96.6	1741
3	39	67	242	1454	133.2	3340
2	27	46	288	1731	158.7	5244
1	15	26	314	1887	173.0	7839

Load Combinations

For lateral force resisting (shear walls in this case) the following load combinations apply (Chapter 2 Table 2-6):

$U = 1.4D$	Eq. (9-1)	
$U = 1.2D + 1.6L + 0.5L_r$	Eq. (9-2)	
$U = 1.2D + 1.6L_r + 0.5L$	Eq. (9-3)	
$U = 1.2D + 1.6L_r \pm 0.8W$		
$U = 1.2D \pm 1.6W + 0.5L + 0.5L_r$	Eq. (9-4)	
$U = 1.2D \pm E + 0.5L$	Eq. (9-5)	(seismic)
$U = 0.9D \pm 1.6W$	Eq. (9-6)	
$U = 0.9D \pm 1.0E$	Eq (9-7)	(seismic)

The seismic load effect E in equations (9-5) and (9-7) includes the effect of the horizontal and vertical earthquake induced forces (see Section 11.6.3). For SDC C the redundancy coefficient $\rho = 1$

$E = Q_E + 0.2S_{DS}D = Q_E + 0.2(0.28)D = Q_E + 0.056D$ For Eq (9-5)

$E = Q_E - 0.2S_{DS}D = Q_E - 0.2(0.28)D = Q_E - 0.056D$ For Eq (9-7)

Calculations of Gravity Loads on the Shearwall

Based on dead loads calculated in Chapter 6 Section 6.5.1

Total wall weight	= 3.53(63)	= 222 kips
Roof dead load	= 0.122(480)	= 59 kips
Four floors dead load	= 0.142(480)(4)	= 273 kips
Total dead load	= 222 + 59 + 273	= 554 kips

Proportion total dead load between wall segments:

Two 8 ft segments: (2 × 96 in.)	= 192/440	= 0.44
One 20 ft-8 in. segment: (248 in.)	= 248/440	= 0.56
For two 8 ft segments: Dead load	= 0.44(554)	= 244 kips
One 20 ft-8 in. segment	= 0.56(554)	= 310 kips
Roof live load	= 20(480)/1000	= 10 kips

Four floors live load = (29.5 + 24.5 + 22.5 + 21) (480)/1000 = 47 kips (see Section 5.7.1)

Proportion of live load between wall segments:

Roof live load

For two 8 ft segments	= 0.44(10)	= 4.4 kips
One 20 ft-8 in. segment	= 0.56(10)	= 5.6 kips

4 floors live load

For 2-8 ft segments:	= 0.44(47)	= 21 kips
1-20 ft-8 in. segment	= 0.56(47)	= 26 kips

Recall wind load analysis (see Chapter 2, Section 2.2.1.1):

For wind load N-S direction

$$M = [(16.2 \times 63) + (31.6 \times 51) + (30.6 \times 39) + (29.2 \times 27) + (30.7 \times 15)]/2 = 2537 \text{ ft-kip}$$

The axial force, bending moment, and shear acting on the base of the 20'-8" long shear wall resisting lateral loads in the N-S direction are summarized in Table 11-5. Table 11-6 shows the factored axial force, bending moment and shear. It is clear from the table that seismic forces will govern the design of the shearwalls in this example.

Story Drift and P-Δ Effect

To check that the maximum allowable limits for story drift are not exceeded (ASCE Table 9.5.2.8) the displacement at each floor level δ_{xe} need to be calculated and amplified (See Section 11.8). Then the stability index θ should be calculated and checked against 0.1 where the P-Δ effects are not required to be considered.

Table 11-5 Forces at the Base of Shear Wall N-S Direction

Load Case	Axial Force (kips)	Bending Moment (ft-kips)	Shear Force (kips)
Dead load D	310	0	
Roof live load L_r	5.6	0	
Live load L	26	0	
Wind W	0	2537	69.2
Earthquake E	0	7839	157

Table 11-6 Factored Axial Forces, Moment and Shear

Load Combinations		P_u (kips)	M_u (ft-kips)	V_u (kips)
Eq. (9-1)	U = 1.4D	434	0	0
Eq. (9-2)	U = 1.2D + 1.6L + 0.5L_r	416.4	0	0
Eq. (9-3)	U = 1.2D + 1.6L_r + 0.5L	393.96	0	0
	U = 1.2D + 1.6L_r + 0.8W	380.96	2029.6	55.36
	U = 1.2D + 1.6L_r - 0.8W	380.96	-2029.6	-55.36
Eq. (9-4)	U = 1.2D + 1.6W + 0.5L+ 0.5L_r	387.8	4059.2	110.72
	U = 1.2D - 1.6W + 0.5L+ 0.5L_r	387.8	-4059.2	-110.72
Eq. (9-5)	U = 1.2D + E + 0.5L	385	7839	173
	U = 1.2D - E + 0.5L	385	-7839	-173
Eq. (9-6)	U = 0.9D + 1.6W	279	4059.2	110.72
	U = 0.9D - 1.6W	279	-4059.2	-110.72
Eq (9-7)	U = 0.9D + 1.0E	279	7839	173
	U = 0.9D - 1.0E	279	-7839	-173

Design for shear:

The maximum factored shear force From Table 11-6:

$V_u = 173$ kips

Determine ϕV_c and maximum allowable ϕV_n

From Table 6-5 for 8 in. wall $\phi V_c = 7.3 \times 20.67 = 150.9$ kips

maximum $\phi V_n = 36.4 \times 20.67 = 752.4$ kips

Wall cross section is adequate ($V_u < \phi V_n$); however, shear reinforcement must be provided ($V_u > \phi V_c$).

Determine the required horizontal shear reinforcement

$\phi V_s = V_u - \phi V_c = 173 - 150.9 = 22.1$ kips

$\phi V_s = 22.1/20.67 = 1.1$ kips/ft length of wall

Select horizontal bars from Table 6-4

For No.3 @ 18 in., $\phi V_s = 2.6$ kips/ft > 1.1 kips/ft

Minimum horizontal reinforcement (of the gross area) for reinforcing bars not greater than No. 5, $\rho = 0.002$ (ACI 14.3.3) $= 0.002(8)(18) = 0.29$ in.^{2} > area of No. 3

Use No. 4 @ 12 in spacing $\rho = 0.2/(8 \times 12) = 0.0021$

$s = 12$ in. $< s_{max} = 18$ in. O.K.

Determine required vertical shear reinforcement (ACI Eq. 11-32):

$$\rho_v = 0.0025 + 0.5(2.5 - h_w/\ell_w)(\rho_h - 0.0025)$$

$$= 0.0025 + 0.5(2.5 - 3.05)(0.0021 - 0.0025)$$

$$= 0.0026$$

where $h_w/\ell_w = 63/20.62 = 3.05$

Required $A_v/s_1 = \rho_v h = 0.0026 \times 8 = 0.021$ in.^{2}/in.

For No. 4 bars: $s_1 = 0.2/0.021 = 9.5$ in. < 18 in. O.K.

Use No. 4 @ 9 in. vertical reinforcement.

Design for flexure:

Check moment strength for required vertical shear reinforcement No. 4 @ 9 in

ρ_v = 0.2/(9 × 8) = 0.0028.

For 1-20 ft-8 in. wall segment at first floor level:

Consider P_u = 279 kips
M_u = 7839 ft-kips
ℓ_w = 248 in.
h = 8 in.

For No.4 @ 9 in.:

A_{st} = 0.0028 × 20.67(12)(8) = 5.5 in.^{2}

$$\omega = \left(\frac{5.5}{248 \times 8}\right)\frac{60}{4} = 0.042$$

$$\alpha = \frac{279}{248 \times 8 \times 4} = 0.035$$

$$\frac{c}{\ell_w} = \frac{0.042 + 0.035}{2(0.042) + 0.72} = 0.095$$

$$M_n = 0.5 \times 5.5 \times 60 \times 248\left(1 + \frac{279}{5.5 \times 60}\right)(1\ \ 0.095)/12 = 5699 \text{ ft-kips}$$

$$\phi M_n = 0.9(5699) = 5129 \text{ ft-kips} < M_u = 7839 \text{ ft-kips.}$$

Use No. 6 @ 6 in spacing r = 0.44/(6X12) = 0.0065

A_{st} = 0.0065 × 20.67(12)(8) = 12.81 in.^{2}

$$\omega = \left(\frac{12.81}{248 \times 8}\right)\frac{60}{4} = 0.097$$

$$\alpha = \frac{279}{248 \times 8 \times 4} = 0.035$$

$$\frac{c}{\ell_w} = \frac{0.035 + 0.097}{2(0.097) + 0.72} = 0.144$$

$$M_n = 0.5 \times 12.81 \times 60 \times 248\left(1 + \frac{279}{12.81 \times 60}\right)(1 - 0.144)/12 = 9267 \text{ ft-kips}$$

$$\phi M_n = 0.9(9267) = 8340 \text{ ft-kips} > M_u = 7839 \text{ ft-kips. O.K.}$$

Distribution of the Seismic Forces to the Shear Walls in E-W Direction

For east west direction, the lateral forces are carried by the four 96" by 8" shear walls. The torsional stiffness of the two N-S direction segments (248 in. each and 20 feet apart) is much larger than that of the four walls. Assuming that the N-S walls resist all the torsion and neglecting the contribution of the E-W walls, the shear force transmitted to each of the four shear walls is $(V_i)_x = \frac{1}{4} V_x$ for this case.

The shear forces and moments for each shearwall at each floor level are shown in Table 11-7:

Table 11-7 Shear Forces and Moments at each Shear wall (E-W)

Level	Height h_x (ft)	Lateral Force F_x (kips)	Story Shear V_x (kips)	$(V_i)_x$ (kips)	Moment (kip-ft)
5	63	88	88	22	264
4	51	87	176	44	792
3	39	67	242	61	1518
2	27	46	288	72	2382
1	15	26	314	79	3560

For each two 8 ft segment:

Dead load	= 244/2	= 122 kips
Roof live load for one 8 ft segments	= 4.4/2	= 2.2 kips
Four floors live load for one 8 ft segments	= 21/2	= 10.5 kips

From wind load analysis (see Chapter 2, Section 2.2.1.1):

For wind load E-W direction

$V = (6.9 + 13.4 + 12.9 + 12.2 + 12.6)$ = 14.5 kips

$M = [(6.9 \times 63) + (13.4 \times 51) + (12.9 \times 39) + (12.2 \times 27) + (12.6 \times 15)]/4$ = 535 ft-kip

The axial force, bending moment, and shear acting on the base of the shear wall resisting lateral loads in the E-W direction are summarized in Table 11-8. Table 11-9 shows the factored axial force, bending moment and shear. It is clear from the table that seismic forces will govern the design of the shearwalls in this example.

Table 11-8 Forces at the Base of Shear Walls E-W Direction

Load Case	Axial Force (kips)	Bending Moment (ft-kips)	Shear Force (kips)
Dead load D	122	0	
Roof live load L_r	2.2	0	
Live load L	10.5	0	
Wind W	0	535	14.5
Earthquake E	0	3560	79

Table 11-9 Factored Axial Forces, Moment and Shear (E-W direction)

Load Combinations		P_u (kips)	M_u (ft-kips)	V_u (kips)
Eq. (9-1)	U = 1.4D	170.8	0	0
Eq. (9-2)	U = 1.2D + 1.6L + 0.5L_r	164.3	0	0
Eq. (9-3)	U = 1.2D + 1.6L_r + 0.5L	155.17	0	0
	U = 1.2D + 1.6L_r + 0.8W	149.92	428	11.6
	U = 1.2D + 1.6L_r - 0.8W	149.92	-428	-11.6
Eq. (9-4)	U = 1.2D + 1.6W + 0.5L+ .5L_r	152.75	856	23.2
	U = 1.2D - 1.6W + 0.5L+ 0.5L_r	152.75	-856	-23.2
Eq. (9-5)	U = 1.2D + E + 0.5L	151.65	3560	79
	U = 1.2D - E + 0.5L	151.65	-3560	-79
Eq. (9-6)	U = 0.9D + 1.6W	109.8	856	23.2
	U = 0.9D - 1.6W	109.8	-856	-23.2
Eq (9-7)	U = 0.9D + 1.0E	109.8	3560	79
	U = 0.9D - 1.0E	109.8	-3560	-79

Design for shear

The maximum factored shear force From Table 11-9:

$$V_u = 79 \text{ kips}$$

Determine ϕV_c and maximum allowable ϕV_n

From Table 6-5 $\quad \phi V_c = 7.3 \times 8 = 58.4$ kips

maximum $\phi V_n = 36.4 \times 8 = 291.2$ kips

Wall cross section is adequate ($V_u <$ maximum ϕV_n); however, shear reinforcement must be determined

Determine required horizontal shear reinforcement

$\phi V_s = V_u - \phi V_c = 79 - 58.4 \quad = 20.6$ kips

$\phi V_s = 20.6/8 \quad = 2.6$ kips/ft length of wall

Use minimum shear reinforcement per ACI 14.3.3 similar to the 20 ft- 8 in. portion of the wall i.e. No.4 @12 in spacing (r = 0.0021). From Table 6-4 for No.4 @ 12 in.,
$\phi V_s = 7.2$ kips/ft > 2.6 OK

Determine required vertical shear reinforcement

$$\begin{aligned}\rho_v &= 0.0025 + 0.5(2.5 - h_w/\ell_w)(\rho_h - 0.0025)\\ &= 0.0025 + 0.5(2.5 - 7.88)(0.0021 - 0.0025)\\ &= 0.0036\end{aligned}$$

where $h_w/\ell_w = 63/8 = 7.88$

Required $A_{vn}/s_1 = \rho_v h = 0.0036 \times 8 = 0.029$ in.^{2}/in.

For No. 6 bars: $s_1 = 0.44/0.029 = 15.2$ in. < 18 in. $\qquad$ O.K.

Use No. 6 @ 15 in. vertical reinforcement.

Design for flexure:

Check moment strength for required vertical shear reinforcement No. 6 @ 15 in

$\rho_v = 0.44/(15X8) = 0.0037$

For the 8 ft wall segment at first floor level:

Consider $\quad P_u = 109.8$ kips
$M_u = 3560$ ft-kips
$\ell_w = 96$ in.
$h = 8$ in.

For No. 6 @ 15 in.:

$A_{st} = 0.0037 \times 96(8) = 2.82 \text{ in.}^2$

$$\omega = \left(\frac{2.82}{96 \times 8}\right)\frac{60}{4} = 0.055$$

$$\alpha = \frac{109.8}{96 \times 8 \times 4} = 0.036$$

$$\frac{c}{\ell_w} = \frac{0.055 + 0.036}{2(0.055) + 0.72} = 0.109$$

$$M_n = 0.5 \times 2.82 \times 60 \times 96\left(1 + \frac{109.8}{2.82 \times 60}\right)(1 - 0.109)/12 = 994 \text{ ft-kips}$$

$$\phi M_n = 0.9(994) = 894 \text{ ft-kips} < M_u = 3560 \text{ ft-kips.}$$

In order to compensate for the big difference between ϕM_n and M_u the thickness and reinforcement of this segment of the wall need to be increased. Increasing the thickness to 10 in. and use two layers off reinforcements.

Use two layers No. 6 @ 4 in. spacing

$\rho = 2 \times 0.44/(10 \times 4) = 0.022$

$A_{st} = 0.0022 \times 96(10) = 21.12 \text{in.}^2$

$$\omega = \left(\frac{21.12}{96 \times 10}\right)\frac{60}{4} = 0.33$$

$$\alpha = \frac{109.8}{96 \times 10 \times 4} = 0.33$$

$$\frac{c}{\ell_w} = \frac{0.33 + 0.029}{2(0.33) + 0.72} = 0.259$$

$$M_n = 0.5 \times 21.12 \times 60 \times 96\left(1 + \frac{109.8}{21.12 \times 60}\right)(1 - 0.259)/12 = 4079 \text{ ft-kips}$$

$$\phi M_n = 0.9(4079) = 3671 \text{ ft-kips} > M_u = 3560 \text{ ft-kips. O.K.}$$

11.11.2 Example 2 – Building # 1 Alternate (1) Standard Pan Joist

To illustrate the seismic design analysis for moment resisting frames Building 1 alternate (1) will be analyzed for the north-south direction seismic forces.

Seismic design data

Assuming that the building is located in north Illinois with the maximum considered earthquake spectral response accelerations for short period (0.2 second) and one second period determined from IBC maps (2003 IBC Figures 1615(1) through 1615(10)) as follows:

$S_s = 0.20$ g
$S_1 = 0.10$ g

Soil site class definition is C

Seismic use group I

Determination of the Seismic Design Category SDC

(1) Based on the values of S_s and S_1 and soil site class C, determine the site coefficients F_a and F_v from 2003 IBC Tables 1615.2(1) and 1615.2(2). Notice that linear interpolation is performed.

$F_a = 1.20$
$F_v = 1.70$

(2) Calculate the adjusted maximum considered earthquake spectral response acceleration:

$S_{MS} = F_a\, S_s = 1.20 \times 0.20 = 0.24g$
$S_{M1} = F_v\, S_1 = 1.70 \times 0.10 = 0.17g$

(3) Calculate the design earthquake spectral response accelerations:

$S_{DS} = 2/3\ S_{MS} = 2/3(0.24) = 0.16g$
$S_{D1} = 2/3\ S_{M1} = 2/3(0.17) = 0.11g$

(4) For $S_{DS} = 0.16g$ 2003 IBC Tables 1616.3(1) shows that the SDC is A

For $S_{D1} = 0.11g$ 2003 IBC Table 1616.3(2) shows that the SDC is B
Use the highest category from both tables
SDC = B

Considering the exception in IBC 1616.3 (see Section 11.2) check the building SDC:

$T_s = S_{D1}/S_{DS} = 0.11/0.16 = 0.69$
Building T (see below) = $0.43 < 0.8(T_s) = 0.8(0.69) = 0.55$ O.K.

The SDC can be determined based on Table 1616.3(1) and the building assigned SDC A. This will reduce the seismic forces drastically since the index force (IBC 1616.4.1) is permitted for SDC A. For illustrative purposes only this building example will continue using SDC B.

Table 11-1 shows that the building frame system with ordinary frame can be used for SDC B.

Fundamental Period

The approximate natural period of the structure can be calculated as follows:

$$T = C_t h_n^x$$

For moment resisting frame system $C_t = 0.016$ and $x = 0.9$

$$T = 0.016(39)^{0.9} = 0.43 \text{ second (also T can be obtained from Fig. 11-4)}$$

Seismic Response Coefficient:

$$C_s = \frac{S_{DI}}{\left(\frac{R}{I_E}\right)T}$$

For ordinary moment resisting frame system the modification factor R = 3 Table 11-1 (ASCE Table 9.5.2.2). The occupancy importance factor $I_E = 1$ (2003 IBC Table 1604.85).

$$C_s = \frac{0.11}{\left(\frac{3}{1}\right)0.43} = 0.087$$

C_s should not be taken less than:

$$C_s = 0.044 S_{DS} I_E = 0.044 \times 0.16 \times 1 = 0.007$$

C_s need not exceed:

$$C_s = \frac{S_{DS}}{\left(\frac{R}{I_E}\right)} = \frac{0.16}{\left(\frac{3}{1}\right)} = 0.053\ (5.3\%)$$

Use $\quad C_s = 0.053$

Effective Seismic Weight

The effective seismic weight for this case includes the total dead load and the partition weight. The effective seismic loads for different floors are calculated as follows:

For the first and second floors:

Slab	= (151.34 ft)(91.34 ft)(0.130 ksf)	=1797 kips
Interior columns (8 columns)	= 8 (1.50 ft)(1.50 ft)(13.0 ft)(0.15 kcf)	= 35.1 kips
Exterior columns (16 columns)	= 16(1.34 ft)(1.34 ft)(13.0 ft)(0.15 kcf)	= 56.0 kips
Effective seismic weight		= 1888.1 kips

For the roof:

Slab	= (151.34 ft)(91.34 ft)(0.105 ksf)	= 1451.5 kips
Interior columns (8 columns)	= 8 (1.50 ft)(1.50 ft)(6.5 ft)(0.15 kcf)	= 17.6 kips
Exterior columns (16 columns)	= 16(1.34 ft)(1.34 ft)(6.5 ft)(0.15kcf)	= 28.0 kips
Effective seismic weight (roof)	= 900 + 12.8 + 10.8 + 42	= 1497.1 kips

The total effective seismic weight:

$$W = 1497.1 + 2(1888.1) = 5273.3 \text{ kips}$$

Seismic Base Shear

$$V = C_S W$$

$$V = 0.053 \times 5273.3 = 279.5 \text{ kips}$$

Vertical Distribution of the Base Shear

The vertical distribution of the base shear V can be calculated using Eq. 9.5.5.4-2 (Section 11.6.2). For T = 0.43 second the distribution exponent k = 1. The calculations for the lateral forces and story shear are shown in Table 11-10:

Table 11-10 Forces and Story Shear Calculations

Level	Height h_x (ft)	Story Weight W_x (kips)	$W_x h_x^k$	Lateral Force F_x (kips)	Story Shear V_x (kips)
3	39	1,497	58,387	124	124
2	26	1,888	49,091	105	229
1	13	1,888	24,545	52	281
	Σ	5,273	132,023	281	

Story Drift and P-Δ Effect

To check that the maximum allowable limits for story drift are not exceeded (ASCE Table 9.5.2.8) the displacement at each floor level δ_{xe} need to be calculated and amplified (See Section 11.8). Then the stability index θ should be calculated and checked against 0.1 where the P-Δ effects are not required to be considered.

Distribution of the Seismic Forces to the Frames in N-S Direction

For north south direction, the lateral forces are carried by the six frames shown in figure 1-3. For the symmetrical floor layouts the center of rigidity coincides with the center of mass at the geometric center of the 150' by 90' floor area.

The seismic force transmitted to each frame must be increased due to assumed displacement of the center of application of the applied lateral force (5 percent of the building plan dimension perpendicular to the force direction). For the N-S direction the displacement considered = 0.05 × 150ft = 7.5ft. To calculate the force in each frame the equation $(F_i)_y = \frac{(k_i)_y}{\sum(k_i)_y} F_x + \frac{x_i(k_i)_y}{J_r} F_x e_x$ can be used.

Assuming equal stiffness for the six frames the equation simplifies to:

simplifies to $(F_i)_y = \frac{1}{6} F_x + \frac{1}{J_r} F_x e_x$ for this case.

where $J_r = \sum x^2 = (15)^2 + (45)^2 + (75)^2 + (-15)^2 + (-45)^2 + (-75)^2 = 15750$

The seismic forces at each floor level are shown in Table 11-11:

Table 11-11 Shear Forces and Moments at each Frames (N-S)

Level	Height h_x (ft)	Lateral Force F_x (kips)	$F_x e_x$	Frame 1	Frame 2	Frame 3	Frame 4	Frame 5	Frame 6
3	39	124	932.8	16.3	18.1	19.8	21.6	23.4	25.2
2	26	105	784.3	30.0	15.2	16.7	18.2	19.7	21.2
1	13	52	392.2	36.8	7.6	8.3	9.1	9.8	10.6

The axial, forces, shear forces, and bending moment for interior frame are calculated using the portal frame method (see Chapter 2). The results of the calculations are shown in Figure 11-9

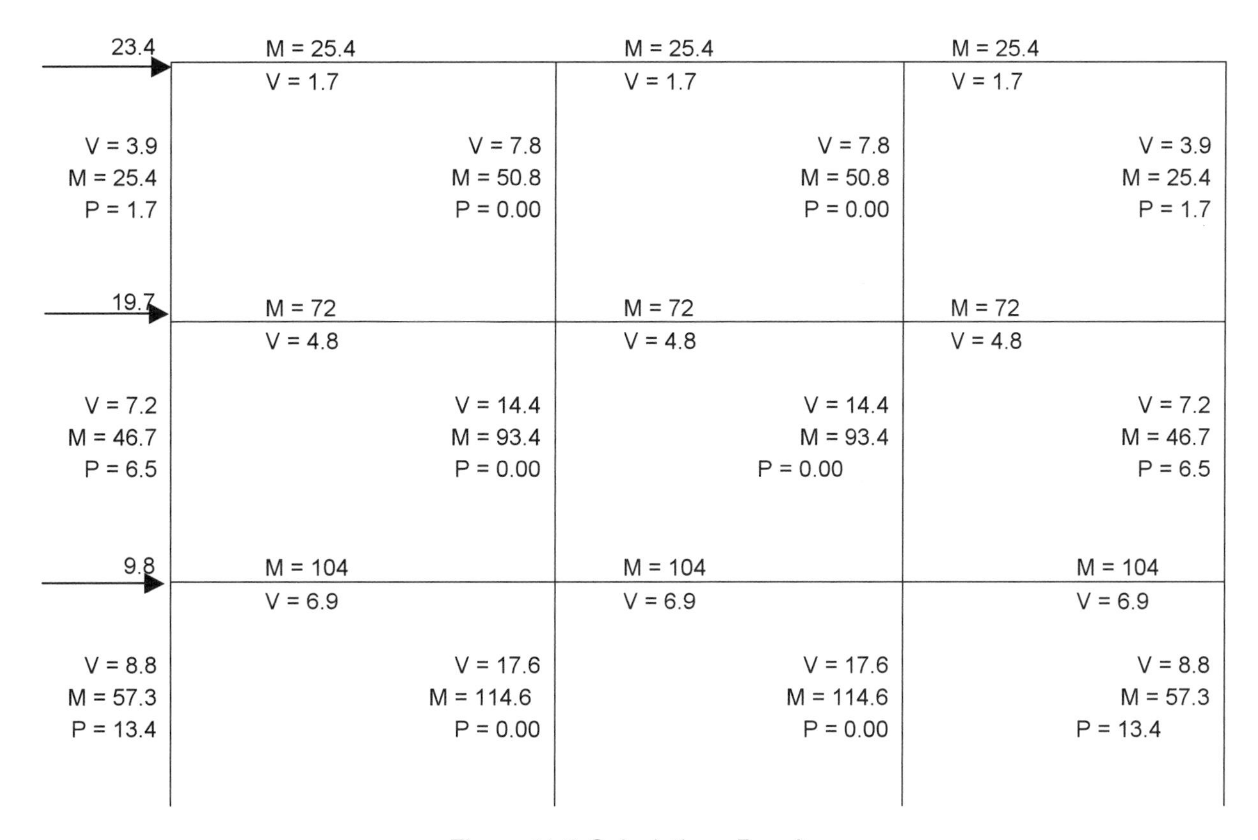

Figure 11-9 Calculations Results

Design of exterior span of the first interior beam N –S direction (first floor)

For live load and dead load calculations see Section 3.8.3

Dead load = 130(30)/1000 = 3.9 klf
Live load = 36.2(30)/1000= 1.1 klf

Moment and shear due to gravity loads may be calculated using the approximate coefficients from Figure 2-3 through Figure 2-7. The moment and shear due to wind loads were calculated in Chapters 2. The following equations are used to determine the earthquake effects used for the ACI load combinations:

$E = Q_E + 0.2\ S_{DS}\ D = Q_E + 0.2(0.16)D = Q_E + 0.032D$ For ACI Eq (9-5)
$E = Q_E - 0.2S_{\ DS}\ D = Q_E - 0.2(0.16)D = Q_E - 0.032D$ For ACI Eq (9-7)

For the first exterior span

For ACI Eq. (9-5)

$M = 104 + 0.032M_{DL}$
$V\ = 6.9 + 0.032V_{DL}$

For ACI Eq (9-7)

$$M = 104 - 0.032M_{DL}$$
$$V = 6.9 - 0.032V_{DL}$$

A summary for dead, live, wind and seismic loads moment and shear forces are presented in Table 11-12.

Table 11-12 Exterior Span Forces

		D	L	W	E (ACI Eq 9-5)	E (ACI Eq 9-7)
Exterior support	M (ft-kip)	-199	-56.2	± 99.56	± 110.4	± 97.6
	V (kips)	55.8	15.7	± 6.64	± 8.68	± 5.1
Midspan	M (ft-kip)	227.5	64.2			
Interior support	M (ft-kip)	318.6	-89.9	± 99.56	± 114.2	± 93.8
	V (kips)	64.3	18.1	± 6.64	± 9	± 4.8

Load Combinations

Table 11-13 shows the factored axial force, bending moment and shear.

Table 11-13 Factored Axial Forces, Moment and Shear

Load Combinations		Exterior Support		Midspan	Interior Support	
		M_u (ft-kips)	V_u (kips)	M_u (ft-kips)	M_u (ft-kips)	V_u (kips)
Eq. (9-1)	U = 1.4D	-278.6	78.12	318.5	-446.0	90.0
Eq. (9-2)	U = 1.2D + 1.6L + 0.5L_r	-328.7	92.08	375.7	-526.2	106.1
Eq. (9-3)	U = 1.2D + 1.6L_r + 0.5L	-266.9	74.81	305.1	-427.3	86.2
	U = 1.2D + 1.6L_r + 0.8W	-159.2	72.27	273.0	-302.7	82.5
	U = 1.2D + 1.6L_r - 0.8W	-318.4	61.64	273.0	-462.0	71.8
Eq. (9-4)	U = 1.2D + 1.6W + 0.5L+ 0.5L_r	-107.6	85.43	305.1	-268.0	96.8
	U = 1.2D - 1.6W + 0.5L+ 0.5L_r	-426.2	64.18	305.1	-586.6	75.6
Eq. (9-5)	U = 1.2D + E + 0.5L	-156.5	79.91	305.1	-313.1	95.2
	U = 1.2D - E + 0.5L	-377.3	69.71	305.1	-541.5	77.2
Eq. (9-6)	U = 0.9D + 1.6W	-19.8	60.84	204.8	-127.4	68.5
	U = 0.9D - 1.6W	-338.4	39.59	204.8	-446.0	47.2
Eq (9-7)	U = 0.9D + 1.0E	-81.5	55.32	204.8	-192.94	62.7
	U = 0.9D - 1.0E	-276.7	45.12	204.8	-380.54	53.1

Design for flexure:

Check beam size for moment strength

Preliminary beam size = 19.5 in. × 36 in.

For negative moment section:

$$b_w = \frac{20M_u}{d^2} = \frac{20(586.6)}{17^2} = 40.6 \text{ in.} > 36 \text{ in.}$$

where d = 19.5 – 2.5 = 17.0 in. = 1.42 ft

For positive moment section:

$$b_w = 20\,(375.7)/17^2 = 26.0 \text{ in.} < 36 \text{ in.}$$

Check minimum size permitted:

$$b_w = 14.6(586.6)/17^2 = 29.6 \text{ in.} < 36 \text{ in. O.K.}$$

Use 36 in. wide beam and provide slightly higher percentage of reinforcement at interior columns.

Top reinforcement at interior support:

$$A_s = \frac{M_u}{4d} = \frac{426.2}{4(17)} = 6.27 \text{ in.}^2$$

From Table 3-6: Use 9-No. 9 bars (A_s = 9.0 in.^{2})

Check $\rho = A_s/bd = 9.0/(36 \times 17) = 0.0147 > \rho_{min} = 0.0033$ O.K.

Top reinforcement at exterior support:

$$A_s = 375.7/4(17) = 5.53 \text{ in.}^2$$

From Table 3-6: Use 7-No. 9 bars (A_s = 7.0 in.^{2})
Check $\rho = A_s/bd = 7.0/(36 \times 17) = 0.0114 > \rho_{min} = 0.0033$ O.K

Bottom bars:

$$A_s = 375.7/4(17) = 5.53 \text{ in.}^2$$

Use 8-No. 8 bars (A_s = 6.32 in.^{2})

Design for Shear :

V_u at distance d from column face = 106.1 – w_u(1.42) = 97.0 kips

where

w_u = 1.2(3.9) + 1.6(1.1) = 6.4 klf

$(\phi V_c + \phi V_s)_{max}$ = 0.48 $b_w d$ = 0.48(36)17 = 293.8 kips > 97.0 kips   O.K.

ϕV_c = 0.095 $b_w d$ = 0.095(36)17 = 58.1 kips

$\phi V_c/2$ = 29.1 kips

Length over which stirrups are required = (106.1 – 29.1)/6.4 = 12.0 ft

ϕV_s (required) = 97.0 – 58.1 = 38.9 kips

Try No. 4 U-stirrups

From Fig. 3-4, use No. 4 @ 8 in. over the entire length where stirrups are required.

References

11.1 International Code Council, *International Building Code*, 2003

11.2 American Society of Civil Engineers, *ASCE Standard Minimum Design Loads for Buildings and other Structures*, ASCE 7-02, 2002

11.3 S.K. Ghosh, and David A. Fanella, *Seismic and Wind Design of Concrete Buildings* (PCA LT276), June 2003

11.4 ACI, *Building Code Requirements for Structural Concrete (ACI 318-02) and Commentary (ACI 318R-02)*, American Concrete Institute, 2002.

11.5 David A. Fanella, and Javeed A. Munshi, *Design of Low-Rise Concrete Buildings for Earthquake Forces*, EB004, Portland Cement Association, 1998

11.6 David A. Fanella, *Seismic Detailing of Concrete Buildings*, SP382, Portland Cement Association, 2000